Philosophy of Experimental Biology

Philosophy of Experimental Biology explores some central philosophical issues concerning scientific research in modern experimental biology, including genetics, biochemistry, molecular biology, developmental biology, neurobiology, and microbiology. It seeks to make sense of the explanatory strategies, concepts, ways of reasoning, approaches to discovery and problem solving, tools, models, and experimental systems deployed by modern life science researchers and also integrates recent developments in historical scholarship, in particular the New Experimentalism. It concludes that historical explanations of scientific change that are based on local laboratory practice need to be supplemented with an account of the epistemic norms and standards that are operative in science. This book should be of interest to philosophers and historians of science as well as to scientists.

Marcel Weber is Swiss National Science Foundation Professor of Philosophy of Science at the University of Basel, Switzerland.

Philosophy of
Experimental Biology

MARCEL WEBER

University of Basel

CAMBRIDGE
UNIVERSITY PRESS

CAMBRIDGE
UNIVERSITY PRESS

32 Avenue of the Americas, New York NY 10013-2473, USA

Cambridge University Press is part of the University of Cambridge.

It furthers the University's mission by disseminating knowledge in the pursuit of education, learning and research at the highest international levels of excellence.

www.cambridge.org
Information on this title: www.cambridge.org/9780521829458

© Marcel Weber 2005

First published 2005

A catalogue record for this publication is available from the British Library

ISBN 978-0-521-82945-8 Hardback
ISBN 978-0-521-14344-8 Paperback

For Andrea, Liliane, and Ardian

Contents

Contents

Preface

In the century between the rediscovery of Mendel's laws (1900) and the completion of the Human Genome Project (2001), biology has come a very long way. During this time, biologists have made spectacular advances in understanding the cellular and molecular basis of life. This knowledge has considerable potential for improving people's lives – a potential that is only beginning to be realized. Thus, in terms both of the knowledge it produces and of the technological opportunities it offers, modern experimental biology is one of the most successful scientific endeavors of all time. Yet remarkably little is known about its epistemology and the underlying metaphysics.

This book explores some central philosophical issues concerning scientific research in modern experimental biology, that is, in areas such as genetics, biochemistry, molecular biology, microbiology, neurobiology, and developmental biology. Evolutionary theory – traditionally the center of attention in the philosophy of biology – only appears marginally, and only where it is relevant to understanding experimental biology. There already exists a massive amount of philosophical literature on evolutionary theory (including a book by this author, Weber 1998a), while experimental biology has not received the philosophical attention that it deserves and needs.

In addition to filling this gap, the present work also reveals a certain independence of experimental biology from evolutionary theory. This is not necessarily to imply that Theodosius Dobzhansky was mistaken when he said, "Nothing in biology makes sense except in the light of evolution," but perhaps that there are different ways of making sense of things. This book is an attempt to make sense of the explanatory strategies, concepts, ways of reasoning, approaches to discovery and problem solving, tools, models, and experimental systems deployed by modern life science researchers. To a considerable extent, this can be done independent of evolutionary theory. However, I will also show where evolutionary thinking is indispensable.

The readers I have in mind are philosophers and historians of science as well as practicing scientists. To the latter group of potential readers, apologies are extended for the strange questions sometimes asked by philosophers. Philosophy is a long and highly elaborate discussion that has been going on for more than 2,000 years now, and some of the questions raised by philosophers today are a result of things that were said by other philosophers centuries ago. But philosophy and modern science are birds of a feather. Their common historical origins should ensure that communication is possible across the boundaries of today's academic specialties. I have tried to facilitate this by providing brief introductions to the philosophical issues at the beginnings of the chapters, and to the scientific principles involved when I discuss actual examples from experimental biology.

Teachers may find the book helpful for a graduate-level course in philosophy of biology. In particular, it could complement the standard readings in philosophy of evolutionary biology. For courses in general philosophy of science, this book covers some of the central problems in the field, such as laws and explanation, reduction, scientific inference, experimentation, discovery and problem solving, and scientific realism. Traditionally, these issues are treated in the context of physics. But classical physics is outdated, while contemporary physics is incomprehensible to anyone without a Ph.D. in physics. By contrast, the science covered in this book should be accessible to everyone with some basic knowledge of biology.

If the book should lead students, teachers, and researchers to a deeper appreciation of some of the exciting philosophical problems that lurk behind the headline-making scientific discoveries of modern biology, my aim would be fulfilled.

Acknowledgements

Special thanks to Paul Hoyningen-Huene for his continuing support. The great team spirit at the center for Philosophy and Ethics of Science at the University of Hannover has been a major inspiration during the genesis of this book. A part of the book was written in spring 2001 at the Max-Planck-Institute for the History of Science in Berlin. I am grateful to Hans-Jörg Rheinberger and his group for giving me access to their stimulating intellectual community and for their patience in listening to a philosopher's problems. Martin Carrier, Paul Hoyningen-Huene, Peter McLaughlin, and three anonymous readers for Cambridge University Press have read drafts of the whole work and have provided valuable criticism. Individual chapters were read by Roberta Millstein, Tatjana Tarkian, Hanne Andersen, Douglas Allchin, Jay Aronson, Hans-Jörg Rheinberger, Howard Sankey, Ken Waters, Daniel Sirtes, Werner Eisner, Ingo Brigandt, Helmut Heit, Renato Paro, Ulrich Stegmann, and Eric Oberheim. Their numerous objections, criticisms, comments, and suggestions (as well as some kind words and a few insults) allowed me to write a much better book than I could have accomplished all by myself. Of course, none of these colleagues should be held responsible for any of the shortcomings that remain.

My research has also benefited from discussions with a number of scientists, especially Gottfried Schatz, Walter Gehring, Renato Paro, and Bernhard Dobberstein. Thanks also to Greta Backhaus and Liliane Devaja for their help with the illustrations.

Furthermore, I wish to thank audiences at the Minnesota Center for Philosophy of Science and the University of Minnesota's Studies of Science and Technology Program in Minneapolis, the Max-Planck-Institute for the History of Science in Berlin, the Department of Medical Philosophy and Clinical Theory at the University of Copenhagen, the 1999 Spring School in Science and Technology Studies at the University of Zürich, the Zentrum für Molekulare Biologie (ZMBH) at the University of Heidelberg, and the philosophy departments at the Universities of Bielefeld, Heidelberg, Erfurt, and Münster for stimulating discussion of paper presentations related to this book. Last but not least, I thank my editor, Michael Ruse, for his encouragement, and Stephanie Achard for seeing the book through the press.

My apologies go to those colleagues whose original work on the relevant topics is not cited. The literature in the history and philosophy of biology as well as in general philosophy of science has grown vast; I could not read it all.

1

Introduction

How can scientists understand the intricate processes that occur inside the tiny cells of a living organism? They tell us about strange-shaped protein molecules that chemically transform foodstuffs to provide the cell with energy, molecules that rotate like propellers, molecules that scan the DNA double helix for structural damage, molecules that turn genes on and off, molecules that pull chromosomes apart when the cell divides, molecules that make the cell crawl around on a surface, and so on. It appears like magic that humans should be able to look inside the cell and unveil all this minuscule clockwork. Yet scientists seem to have found ways of doing exactly that. How is this possible? If scientists are not magicians but people with the ordinary range of human cognitive abilities, how do they deploy these abilities in order to understand life itself?

A traditional answer would be that scientists invent speculative theories or hypotheses, which are then tested by experiments in accordance with the rules of the scientific method. At least this is how an experimental science proceeds according to two major traditions in the philosophy of science: Critical Rationalism and Logical Empiricism. The former approach was championed by Karl Popper, while the latter grew out of the logical positivism of the Vienna Circle. Both Critical Rationalists and Logical Empiricists thought that they could find out about the scientific method on the basis of logical considerations alone. These logical considerations would show what it means to reason scientifically, that is, to reason *rationally*. However, both of these approaches have proven to be inadequate to deal with real science. First, they leave many questions unanswered, for example, questions about how scientists generate new hypotheses (formerly known as the "context of discovery") and research problems, or questions about the exact conditions under which a theory or theoretical framework is abandoned or retained. Second, a wealth of historical and sociological studies has shown that scientists rarely abide by the prescriptions that have been issued from philosophical armchairs. While this could simply mean that scientists fail to act rationally, such a conclusion would be premature. The failure of real science to conform to philosophical accounts is just as likely to be the philosophers' fault and should be taken as a stimulus to doing better epistemology.

If, today, many historians and sociologists of science do not see any rationality in science, a possible reason is that the existing accounts of scientific rationality are too simple.

The great twentieth-century debates on the rationality of science mostly ignored biology. It was assumed that the most advanced science is physics, which was probably true during the first half of the century. However, in the meantime, biology has come of age and it is today at least equal to physics and chemistry in terms of maturity and the reliability of its knowledge claims. Thus, a philosophy of science that cannot deal with experimental biology would be missing some of the best examples of sound scientific knowledge that we have.

This book is partly a result of my conviction that modern science, even though it exhibits a considerable amount of internal diversity, is a well-defined entity from an epistemological point of view. The unity of science may have been exaggerated in the past – especially by the philosophers of the Vienna Circle and their heirs (e.g., Carnap 1938; Oppenheim and Putnam 1958) – but science is not quite as disunified or local as current opinion in science studies (e.g., Galison and Stump 1996; Keller 2002) would have us believe. There are ways of reasoning that can be found in many different scientific disciplines. This, of course, does not mean that there are no differences in the exact ways in which these approaches are applied or that the relative significance of different methods is the same everywhere. What it does mean is that the philosophy of biology should not detach itself from the general debates in philosophy of science to the extent that it has in recent years.[1]

When the philosophy of biology began to establish itself as a professional field of inquiry in the 1970s (mainly in North America), its main concern was evolutionary theory. I think there are several reasons for this. First, evolutionary theory has profound philosophical and religious implications for such issues as human origins, the nature of the human mind, love, sex, culture, and morality. Second, parts of evolutionary biology – especially population genetics – are more closely related to physics than areas such as molecular biology. The reason is that there are structural similarities in mathematically formulated theories and models. Until the 1970s, most philosophy of science was philosophy of physics. As a consequence, many philosophers of biology have simply transferred some of the issues concerning the structure of scientific

2

theories from physics to biology. Third, several eminent evolutionary biologists have written extensively on historical and philosophical issues in biology, thus challenging philosophers of science to take up these issues. The most important figure in this respect was clearly Ernst Mayr, who has unceasingly defended biology as a science in its own right that is in several ways located between the physical and the social sciences (Mayr 1982, 1997).

An important contribution to the philosophy of biology that is due to Mayr (1961) is the distinction between *proximate* and *ultimate* causes (cf. Ariew 2003). On Mayr's view, evolutionary biology is concerned with ultimate causes or historical explanations of an organism's properties. Evolutionary biology explains how species of organisms came to have the properties they do. It is concerned with *phylogeny*. By contrast, areas such as genetics, neurobiology, and biochemistry are concerned with proximate causes, in other words, with the processes that occur within an individual organism. Proximate causes explain how an individual organism develops during its own lifetime, that is, by virtue of its physiological makeup, its genes, and its environment. This kind of biology is interested mainly in *ontogeny*. It is the biology of proximate causes that this book is concerned with.

It will become clear in the course of this study that, by and large, to make sense of the practice of experimental biology, it is not necessary to refer to evolutionary theory. Thus, I want to counter a certain tendency in recent philosophy of biology to see evolutionary theory as some sort of master theory of biology. This is not to question the scientific status or importance of evolutionary theory. What I want to claim is that there is a certain conceptual, explanatory, and foundational independence of experimental biology from evolutionary theory, which mirrors its institutional independence.[2] However, this independence is not complete, as I show (see Chapter 6).

Apart from evolutionary biology, a predominant concern in the philosophy of biology has been the issue of reduction and reductionism, mainly in the context of genetics. Because the relationship between classical, Mendelian genetics and molecular biology has been widely viewed as a paradigm case of reduction in biology, philosophers of science have taken it as an important test case for the theory of reduction developed by logical empiricist philosophers, in particular Ernest Nagel (1961). A student of Nagel's, Kenneth Schaffner (1969), has claimed that a slightly modified account of reduction can accommodate this case. This claim was subsequently challenged by David Hull (1972, 1974, 1976), which sparked a voluminous debate. Eventually, a consensus formed that classical genetics is not reducible to molecular biology – in either Nagel's or any other sense. Only a few authors have opposed this consensus, most forcefully Kenneth Waters (1990, 1994, 2000). Both reductionists and antireductionists have given little attention to the possibility that genetics may be a special case within experimental biology.

Since the 1980s, the field has opened up considerably. Many authors have directed their attention to other areas of experimental biology, as well as to issues other than reduction. Important monographs in this respect include Lindley Darden's *Theory Change in Science* (1991), Kenneth Schaffner's *Discovery and Explanation in Biology and Medicine* (1993), and William Bechtel and Robert Richardson's *Discovering Complexity* (1993). In addition, there is now a considerable body of individual articles dealing with diverse philosophical issues in various areas of experimental biology, including molecular biology, biochemistry, cell biology, immunology, neurobiology, and developmental biology. However, apart from Schaffner's *Discovery and Explanation in Biology and Medicine*, there is (to my knowledge) currently no monograph available that treats the central problems of philosophy of science in connection with these experimental disciplines. Hence the present book.

As I mentioned previously, the philosophy of biology has shown a strong tendency to detach itself from general philosophy of science as well as from the philosophy of the physical sciences. Thus, philosophers of biology have mostly kept themselves busy with philosophical issues that arise from within biology or from areas where biology engages social and political issues (e.g., issues related to race or gender). And indeed, this development has been fruitful and has advanced the debates beyond the initial attempts to simply apply logical empiricist philosophy of science to biology. However, the time is now ripe to reconnect some of the issues to more general philosophical problems and to other areas of the history and philosophy of science. This is one of the main goals of this book.

The historical literature on experimental biology has really exploded in recent years. Historians of biology have produced a considerable number of detailed studies, especially of twentieth-century developments in genetics, biochemistry, molecular biology, and immunology. Even though it is generally accepted today that good philosophy of science should take the history of science very seriously, there have not been many attempts by philosophers of biology to assess the philosophical implications of recent developments in historical scholarship. One of the most important such developments is clearly the increasing emphasis that historians have placed on experimental practice, especially the "material culture" of biology, such as experimental systems and model organisms. Thus, the "New Experimentalism" has found an especially strong resonance in the history of biology. This work raises a number of challenges for philosophers of science that must be addressed.

Therefore, I want to critically assess the implications of some of this work for the philosophy of science. This is another goal of this book.

In the following chapters, I address a series of central epistemological and metaphysical issues, some of which also arise in other sciences, while others are specific to experimental biology. These issues concern the ways in which scientific knowledge is structured, how it explains natural phenomena, how it is generated and evaluated, and how it connects to the world. All of these issues are dealt with in the present book, roughly in this order. The book is thus organized according to the philosophical issues, not according to biological disciplines or historical epochs.

Concerning the selection of examples and historical case studies, I chose a middle path between two different approaches. One approach would have been to present a single case study in great detail and then treat all the philosophical issues on the basis of this example. Something could be said for such an approach; however, experimental biology shows considerable internal diversity and probably no single example instantiates all the main philosophical problems equally well. The alternative approach is to treat each philosophical issue using a different example that seems especially well suited. This approach is somewhat uneconomical in terms of the amount of technical discussion that has to be presented and digested. Furthermore, this approach threatens to paint an overly disunified picture of experimental biology. For these reasons, I have chosen to focus on a few different examples, but not to introduce a new one for every issue addressed. The cases I ended up with are derived from neurobiology (Chapter 2), genetics and molecular biology (Chapters 3, 6, 7, 9), biochemistry (Chapters 3, 4, 5), developmental biology (Chapter 8), and microbiology (Chapter 9).

Chapters 2, 3, 4, 5, 7, and 9 address issues that also arise in other sciences (e.g., the physical sciences or the earth sciences). By contrast, Chapters 6 and 8 deal with philosophical issues that are specific to modern experimental biology and where (to my knowledge) no corresponding problems exist in other sciences.

Chapter 2 examines the reductionistic explanations of biological phenomena given by modern experimental biologists. As already mentioned, a consensus has developed among philosophers of biology that reductionism fails even in those areas of biology that are generally considered to be successes of reductionism, such as molecular genetics. However, this consensus is based on too restrictive an account of reduction, namely the logical empiricist account. The time is ripe to develop an alternative account that captures the main sense in which much of modern experimental biology is truly reductionistic (for better or for worse).

My approach is to first present an example from the area of neurobiology. This example is the mechanism of nerve signal propagation (action potentials), which has been known in its basic outlines since the 1950s. The mechanism is fundamental to the entire discipline of neurobiology, as it explains how nerve fibers conduct signals. This is the basis of information processing in the nervous system.

I begin by investigating what kind of theoretical structure is instantiated by this example. I suggest that the most salient question is whether we should look at the case as a self-contained biological *theory* – as most philosophers of biology do – or, instead, as an *explanation* that applies theories from physical chemistry to a certain type of biological system. To affirm the latter defines a strongly reductionistic position. It is this position that I try to defend, at least for parts of experimental biology (but not for evolutionary theory or population biology). This defense involves a discussion of the nature of the *laws* that carry the explanatory burden in such examples.

A number of philosophers of physics have argued that the traditional concept of natural law should be dropped altogether (Cartwright 1989, 1999; van Fraassen 1989; Giere 1995). The idea that there are natural laws could be a relic from the theistic worldview that was popular during the formative years of modern science in the seventeenth century. In this worldview, God acted as a lawgiver in both the moral and natural realms. However, God has been banned from the explanations of natural science, and perhaps the concept of natural law should go with Him. I examine whether experimental biology is ready for this, in other words, whether its explanations can dispense with laws of nature.

I also address the issue of whether there are genuinely *biological* laws of nature. It has been suggested that all distinctly biological generalizations are contingent and therefore not laws of nature. This claim is known as the Evolutionary Contingency Thesis (Beatty 1995). I assess the validity of this thesis in the context of experimental biology. In addition, I want to exhibit its strong affinity to the kind of reductionism that I defend. Furthermore, I discuss the status of natural kinds in experimental biology. As it turns out, this issue is also strongly connected both to the existence of laws and to reductionism.

Functional explanations are viewed as a hallmark of biology, a feature that seems to distinguish it from the physical sciences. In giving functional explanations of some organismic traits, biology – even molecular biology – seems to be closer to psychology and the social sciences than to physics and chemistry. Thus, functional explanations pose a challenge for reductionism. Functional explanation is *teleological* in that it shows that some structure or capacity is a means to some end. Aristotle thought that each organism has an intrinsic

6

end or *telos*, and his entire biology (as well as his general metaphysics) was permeated by this idea. However, modern science rejects intrinsic *tele* in the realm of nature. This raises the question of whether functional explanation should be eliminated. Most twentieth-century philosophers have been reluctant to issue such a radical recommendation and have instead tried to explicate functions in a metaphysically unproblematic way by somehow relating them to causal explanations (though some have argued that functional analysis is at best of heuristic value in science). In the literature, we find two different kinds of accounts, known as *etiological functions* and *causal role functions*. The main difference between the two is that the former attempts to explain why function bearers are present in a system, whereas the latter only states the specific contribution that function bearers make toward some system capacity.

According to one version of the etiological account, the presence of the function bearers can be explained with the help of evolution by natural selection. This view faces considerable difficulties. A more recent and more viable version of the etiological account (McLaughlin 2001) states that functions explain the presence of function bearers via the latter's contribution to self-reproduction, that is, the continuous self-regeneration of individual organisms. I argue that the etiological view does not capture the use of the term "function" in experimental biology. Instead, I adopt a causal role account. The main difficulty of the causal role account lies in the selection criteria for the significant system capacities to which the function bearers are supposed to contribute. Based on my neurobiological example, I suggest a way of solving this problem of the causal role account of biological functions.

In the final part of the chapter, I examine how the basic mechanism of action potentials is embedded in higher-level mechanisms that explain animal behavior. A strategy like this is often involved when experimental biologists explain some complex property of organisms. The specific question I discuss is whether a well-known argument against reductionism known as the "multiple realization" argument is relevant to this example. This argument is quite powerful for showing that theories like classical Mendelian genetics or Darwinian evolutionary theory are irreducible. However, these theories could be special in this respect, while multiple realization may not be an obstacle to reductionism in other areas.

Chapter 3 turns to an issue that is generally known as "scientific discovery." Clearly a misnomer, this term is widely used to refer to the reasoning processes by which scientists construct or generate new theories, hypotheses, or explanations. There is a long tradition of rejecting this as a legitimate subject for philosophical inquiry. Most twentieth-century philosophers thought that the proper domain for epistemology is only the justification or validity of scientific explanations, not their genesis. The reason behind this was the idea that only justification is subject to the constraints of rationality, such as the rules of deductive and (perhaps) inductive logic. The process of generating ideas was viewed as a psychological process not subject to normative considerations. Since the 1970s, this view has increasingly come under attack. A number of philosophers, together with some cognitive scientists, have argued that "discovery" is a rational process after all and that it can be analyzed as such. Since then, the search has been on for a "logic of discovery."

Some participants in this quest took the term "logic" quite literally and started to design computer programs that generated hypotheses after being fed experimental data (Gillies 1996). The goal of these attempts was both to design expert systems that might assist working scientists and to remove some of the mystery surrounding scientific creativity. Initial attempts at simulating discovery were hardly more than computer-aided curve-fitting; however, more advanced programs are also capable of introducing theoretical entities and suggesting experiments. Of course, all this research in artificial intelligence cannot show that human scientists *actually* reason like these computer programs. Thus, the most that computer simulations could show is that some machine-implementable rules or heuristics are *sufficient* for generating plausible theories or hypotheses from some given inputs. There is no denying that such proofs of sufficiency would be interesting, provided that the rules implemented do not already contain the solution in some way. However, it seems to me that the computer programs in existence today suffer from exactly this drawback. In spite of this, artificial intelligence research on scientific discovery may have its applications. At any rate, it is not my goal to provide a systematic appraisal of this work. I am concerned mainly with flesh-and-blood attempts to understand the genesis of scientific ideas.

I critically examine three different attempts to use the historical record in order to draw conclusions concerning generative reasoning in experimental biology. The first is Kenneth Schaffner's (1974a) early attempt to reconstruct the reasoning behind the genesis of François Jacob and Jacques Monod's repressor model of enzyme induction in bacteria. Schaffner argued that this model was basically deduced from experimental results with the help of theoretical background assumptions. Schaffner suggested that this is indicative of an identity of generative and justificatory reasoning. Even though Schaffner has since revised these conclusions, a critical examination of his thesis is quite revealing.

The second attempt I examine is Lindley Darden's (1991) account of the genesis of the "classical theory of the gene" due to Thomas Hunt Morgan and his students. Like most biological theories, the theory of classical genetics was not created by a single stroke of genius. Instead, it originated from a very simple precursor by a series of modifications triggered by experimental anomalies. Darden is concerned mainly with the way in which these anomalies necessitated revisions in genetic theory. Based on the published record, Darden inferred some heuristics or "strategies" that could have generated a number of the anomaly-driven revisions that occurred between 1900 and 1926. I critically examine in particular Darden's claim that her strategies are both sufficient *and* general.

The third attempt I examine concerns what should be one of the best understood episodes in the history of science, namely, the discovery of the urea cycle by the biochemist Hans Krebs. In comparison to the two other examples, this case offers a very rich historical record. Not only have most of Krebs's laboratory notebooks and his letters been preserved, but the historian Frederic Holmes was also able to conduct extensive interviews with Krebs in the late 1970s. A true master of his trade, Holmes has assembled one of the most detailed accounts of any episode in the history of science, tracing Krebs's activities from day to day (Holmes 1991, 1993a). In spite of this, it has been remarkably difficult to reconstruct how Krebs generated some of his more important ideas. The problem is that the historical record reveals in great detail what Krebs was *doing*, but not much of what he was *thinking*. Thus, the exact sequence of mental steps that Krebs took remains a matter of historical debate. Nevertheless, Holmes thinks that it is possible to make Krebs's discovery "intelligible to reason." I consider in particular the questions of what this means and whether the case supports the idea that there are general and domain-unspecific problem-solving heuristics operative in science.

Chapter 4 turns to the question of how hypotheses are subjected to experimental test. After reviewing some extant accounts of scientific inference, I present a detailed case study of a historical episode. This episode is known as the "oxidative phosphorylation controversy" in biochemistry (ca. 1961–1977). It involved two competing hypotheses that explain how cells generate biological energy. Several years of experimental research and theoretical debate failed to resolve this controversy. Finally, the development of novel experimental techniques allowed biochemists to stage a crucial test that provided definitive reasons for choosing one of the two theories. This account is somewhat simplified, and I point out some complicating factors.

I then examine whether this case instantiates a philosophical conundrum known as the "Duhem–Quine thesis" or underdetermination. Actually, this "thesis" needs to be differentiated into Duhem's problem, which is the problem of allocating praise and blame between a theory and the auxiliary assumptions needed for connecting the theory to observable consequences, and the problem of underdetermination, which concerns the possibility that two theories are empirically equivalent. I examine whether the two competing theories involved in the ox-phos controversy instantiate any of these problems.

Next, I critically examine whether a currently popular account of scientific inference and theory comparison does a good job of explicating this case. This account is known as "Bayesian confirmation theory" and is thought to provide some sort of an inductive logic. More precisely, Bayesian confirmation theory claims that rational cognitive agents ought to attach a probability value to empirical statements. These probability values should reflect the agent's personal probability, which measures how much the agent is willing to bet on the truth of the statement. Then the theory says that rational agents ought to update these subjective probabilities under the impact of incoming evidence according to Bayes's theorem. This theory is ridden with difficulties, and Bayesians have been quite ingenious in their attempts to fix these difficulties. I examine whether the scientists involved in the ox-phos controversy qualify as rational agents in the sense of Bayesian confirmation theory. Of course, any discrepancies will be blamed on the Bayesians, not on my biochemists. I give a justification for this reversal of the traditional order of things.

Another theory of scientific inference I critically examine on the basis of my case study is Deborah Mayo's (1996) error statistical approach. Mayo has attempted to provide an alternative to Bayesianism. She also advocates the use of probabilities in scientific inference, but not the standard Bayesian personal probabilities. Mayo's probabilities are objective and measure the relative frequency with which an experimental test procedure passes a hypothesis given that the hypothesis is false. If this *error probability* is very low, then the test earns the label "severe" from Mayo. I examine in particular to what extent this account of scientific inference can be applied to a case such as ox-phos, where the experimental test procedures were not of a statistical nature.

Having found fault with all these accounts of theory testing, I then try to develop my own account. I take on board some valuable insights from Mayo, especially the central role she accords to the control of errors. However, "error" should not be interpreted in a formal, statistical sense. I place particular emphasis on the practice of controlled experiments in biochemistry, which I think holds the key to understanding experimental reasoning.

Chapter 5 critically examines a perspective on experimentation that radically differs from the perspective of Chapter 4. The historian of biology Hans-Jörg Rheinberger (1997) has presented a novel account of the role of experimental systems in biological research. His notion of experimental system is very broad, as it includes a constellation of various material and cognitive resources needed to do research in biology. Based on his detailed historical studies of the development of research on protein synthesis after World War II, Rheinberger suggests that experimental systems play a crucial role in moving research into new directions, sometimes in very unexpected ways. On Rheinberger's account, experimentation is not guided by theories or antecedently formulated research problems, but by the scientists' attempts to explore the intrinsic capacities that experimental systems offer.

Rheinberger's approach exemplifies a recent move away from theory-dominated history and philosophy of science to the study of experimental practice and laboratory settings, which distinguishes the recent work of historians of science. The predominant view among historians today appears to be that, even if not all agree with the specifics of Rheinberger's account, this way of reconstructing scientific research is more adequate than the older methodological approach. According to these historians, it is a mistake to look for a "scientific method," because scientific practice is largely a matter of local cultures, not of general principles or methodological norms. It is thus clear that this recent work in history of science presents a serious challenge to any philosophy of science that thinks in terms of general methodological norms. For example, considerations of the conditions under which experiments test a theory (such as the ones presented in Chapter 4) seem to be utterly irrelevant under the new approach.

I adopt Rheinberger's "experimental systems perspective" (my term) in order to look at the oxidative phosphorylation controversy from Chapter 4 from a different angle. Even though it might seem at first that the latter case is more theory-driven than Rheinberger's example of protein synthesis research, there are interesting parallels. However, these findings do not mean that the theory-testing perspective from Chapter 4 is irrelevant to understanding a complex scientific episode such as ox-phos. I examine more closely whether the two approaches are really in conflict or whether they rather provide mutually complementary perspectives on experimentation in biology. For this purpose, it is crucial to appreciate that experiments can play different roles in research, and that testing a theory is only one possible role. Furthermore, the role that an experiment ends up playing may not correspond at all to the experimenter's intentions. But all this does not mean that epistemic norms are irrelevant to explaining scientific change. I discuss what kinds of epistemic norms were involved in the ox-phos case.

Continuing the focus on experimental practice, Chapter 6 turns to the role of so-called model organisms in biological research. Unlike zoologists or botanists, experimental biologists do not usually immerse themselves in the diversity of life. In fact, most of them study just one or two species of organisms during their entire career. Historians have noted this peculiarity of twentieth-century experimental biology and have used model organisms as a focus for studying what they take to be a central aspect of scientific culture, namely its so-called material culture. One of the reasons for the historians' interest in model organisms is that the major laboratory organisms, such as the fruit fly *Drosophila melanogaster* and the soil nematode *Caenorhabditis elegans*, are associated with complex social networks that organize the exchange of research materials (strains, mutants). Thus, model organisms seem to be part of a peculiar "economy" that governs social interactions between scientists.

Furthermore, Robert Kohler (1994) has argued that the fruit fly *Drosophila* was ecologically adapted to life in the laboratory because of its life cycle and because of the large number of mutants it produces. Kohler describes *Drosophila* as a "breeder reactor" or a "system of production." He has rewritten the history of *Drosophila* genetics by focusing almost entirely on such economic and ecological aspects of the major laboratory organism without paying much attention to developments in genetic theory. As he shows, theoretical innovations often occur at stages in the research process where a reorganization of laboratory practice becomes necessary. For example, he traces the origin of genetic mapping back to the need to classify the large number of mutants that *Drosophila* kept producing.

These and similar historical studies show that it would be naive to think that biologists start out with some theoretical problem or research question and then select a suitable laboratory organism to solve this problem. Specific research problems often arise only in the course of developing a model organism. Thus, it has been argued that laboratory organisms and the problems they help to solve are "co-constructed." This conclusion challenges the more common view that a model organism is some kind of biological "Rosetta stone" that is selected for deciphering general biological principles. For this reason, I examine a recent historical episode from the history of *Drosophila* genetics, namely the adaptation of *Drosophila* to molecular biology.

Even though the fruit fly was instrumental in the growth of twentieth-century genetics, the molecularization of genetics that began in the 1950s was largely a consequence of research on microorganisms, namely the bacterium *E. coli* and its bacteriophages. Genetic research on *Drosophila* continued during the molecular revolution, but the fly was clearly not as prominent as in Morgan's days. However, beginning in the 1970s, the fly has made a spectacular comeback as a major laboratory organism of molecular biology. It turned out to be invaluable for studying the molecular basis of development and other biological processes. This transition was mainly accomplished with the help of recombinant DNA technology. I examine this, as I call it, "moleculariza-tion" of the fruit fly with an eye toward the "material culture" perspective on model organisms advanced by historians.

I want to work out the merits as well as the limits of this approach. I focus especially on the use of economic notions such as "production," but also the idea that model organisms are "tools" or "instruments." Even though scientific research clearly has an economic aspect, it also has an epistemic aspect. After all, model organisms are deemed valuable because they help scientists to produce knowledge. What are the epistemic reasons for the importance of model organisms in research? Historians have largely ignored this aspect of model organisms, and I try to bring it back into focus. My own case study of the molecularization of *Drosophila* allows me to identify a number of previously unnoticed epistemological aspects of experimentation in biology.

In Chapter 7, I return to more familiar ground in the philosophy of science – the classic issue of reference and conceptual change. According to a vener-able tradition in philosophy of language, concepts come with *meaning* and *reference* (or intension and extension). The former is the concept's content, whereas the latter is the class of objects to which the concept applies. Thomas Kuhn (1970) and Paul Feyerabend (1962), in particular, have drawn attention to the fact that the meaning of scientific terms is subject to historical change. For example, the concept of mass changed its meaning in the transition from Newtonian to relativistic mechanics. Kuhn and Feyerabend thought that such meaning changes can be subtle and at the same time radical, so that it is not possible to translate one concept into another. This phenomenon is known as "incommensurability" and has generated a vast body of philosophical lit-erature, because it was thought by some to lead to relativism. The reason is that the alleged translation failure between incommensurable theories might preclude an empirical comparison of these theories.

In response to this incommensurability thesis, several philosophers have argued that the reference of scientific terms could be more stable than their meaning, which could rescue the comparability of theories. One way of de-fending such a stability of reference is by adopting a so-called causal theory of reference that was developed by Saul Kripke (1971, 1980) and Hilary Putnam (1975c). The main characteristic of this theory is that it holds reference to be determined independent of a term's meaning. One consequence of this is that a theoretical term can have reference even though all the theories about the term's referents are false. For example, the term "electron" could have had a determinate reference even back when physicists produced false descriptions of these entities. To put it differently, they could have referred to electrons without being able to give an adequate *definition*.

One problem with the original version of the causal theory is that it did not allow any referential change. This is an undesirable consequence, because there are clear cases of reference failure in the history of science. The most famous example is the term "phlogiston." Closer to home, we could mention Darwin's "gemmules" (freely circulating particles that Darwin postulated to explain the origin of heritable variations), but also early versions of the concept of Mendelian "factors." Thus, the causal theory of reference had to be modified to allow for reference change. For this purpose, Philip Kitcher (1978, 1982) has introduced the concept of *reference potential*, according to which there are always a multiplicity of ways in which a scientific community fixes the reference of theoretical terms.

My goal in Chapter 7 is to examine the reference stability of the gene concept between 1900 and the introduction of the contemporary, molecular gene concept. Here, I do not get involved with issues of incommensurability and theory comparison; rather I focus on reference stability. It is widely agreed that the meaning of the gene concept has changed considerably since the beginning of the twentieth century. Biologists are of course aware of this; however, they seem to assume that the gene concept has always referred to the same class of entities (more or less). By contrast, historians of genetics have tended to see little or no continuity in the historical development of the gene concept. The truth may well lie between these positions.

What I want to examine is how different generations of geneticists fixed the concept's reference and how the reference was affected by shifting experimental practices, theoretical innovations, and even new model organisms. Chapter 7 examines some of the major transitions that occurred in the reference potential of the gene concept. In addition, my case study from Chapter 6 on the molecularization of *Drosophila* allows me to point out some previously unnoticed aspects of the relationship between classical and molecular genetics. Finally, I show that the case of the gene differs from those of theoretical entities in the physical sciences because of some special metaphysical features of biological entities. The extent to which the nature of biological entities *as* biological entities has consequences for concept formation has not been sufficiently appreciated in recent philosophy (except in the philosophy of taxonomy).

Chapter 8 turns to a set of issues that are specific to a certain biological subdiscipline, namely developmental biology. The starting point of this chapter is provided by the popular notion of a developmental program that is thought to be written into an organism's DNA sequence. I examine two major difficulties with this idea. One difficulty is that this notion appears to be *intentional*. But it is open to question what legitimate place intentionality can have in a purely causal mechanism such as a developing organism. This general difficulty thus arises against the background of the reductionism defended in Chapter 2.

Another difficulty was raised by the defenders of a currently popular philosophical account of development known as "developmental systems theory" (DST). DST is a set of interrelated claims about the respective roles of genes and DNA and other components of a developing organism. DS theorists draw attention to the fact that no cellular constituent alone is capable of initiating and maintaining a developmental process leading from, for example, a fertilized egg to an adult individual. Development is an immensely complex, tightly orchestrated process that generates a new organism out of a whole developmental system. Although the latter claim is indisputable, the question arises of whether there is indeed anything that programs or instructs this process, perhaps in the manner in which a computer program instructs a computer. The standard view in biology, of course, is that DNA and genes take such a programming role. In contrast, DS theorists argue that all attempts to identify a special role for genes and DNA in this process fail. According to them, DNA is just one developmental resource among many.

DST offers an interesting challenge to current biological thinking. For this reason, I begin this chapter by subjecting it to critical scrutiny. As in the other chapters, I proceed by paying close attention to concrete examples. Philosophical defenders of DST have mainly worked with caricatures of biological theory, which obviously harbors the danger of investing DST with a plausibility that it may not have. The main case study I examine is the early development of *Drosophila* – the model organism that already plays an eminent role in this book (in Chapters 3, 6, and 7). Particular attention is given to the concept of information in developmental biology, because some DS theorists have claimed that the nucleotide sequence of DNA does not constitute the only kind of genetic information.

Chapter 9 turns to the question of whether experimental biology has access to a mind-independent reality – the issue of scientific realism. This issue is controversial today, with most historians and sociologists of science taking an antirealist position, while most contemporary philosophers of science prefer realism (unlike some of their predecessors, the positivists).

I begin by examining the relationship between realism in experimental biology and other forms of realism, in particular realism about space and time, realism about theoretical entities and unobservable structures, and realism about laws, natural kinds, biological functions, and biological information. This establishes that there is a philosophically interesting realism issue in experimental biology, even though the issue of realism concerning space and time is more fundamental. Space and time are omnipresent in biology, as organisms are complex spatiotemporal processes.

Next, I examine the standard argumentative strategies for scientific realism. Recent debate in philosophy has centered on the so-called miracle argument. This argument purports to show that we are justified in accepting well-confirmed scientific theories as (approximately) true and their theoretical terms as referring because this is the best explanation for the theories' empirical success. If theories were not true and their terms were nonreferring, it would be a miracle that such theories make successful novel predictions. This "miracle argument" is riddled with various difficulties, the most serious being the fact that there are false historical theories with nonreferring terms that were nevertheless empirically successful. This fact may also be viewed as a base for a so-called pessimistic meta-induction to the effect that we have no reason to think that our best current theories are true and their terms referring. I examine ways in which scientific realism about experimental biology could be defended in the face of such epistemological difficulties.

One possibility would be to adapt Ian Hacking's (1983) "experimentalist" argument, according to which a theoretical entity can be accepted as real once it can be used to do things in the laboratory, for example, building a new apparatus. Surely, in the age of genetic engineering and molecular biology kits, such an argument could provide grounds for the reality of entities such as genes? I examine this argument critically.

Another possible approach consists in arguing for the reality of unobservable structures or entities in biological systems (e.g., molecules or subcellular structures such as organelles) from the agreement of experimental results obtained with independent methods. Physicists successfully used this kind of argument in the early twentieth century to argue for the reality of atoms. The late Wesley Salmon (1984) reconstructed this argument in terms of Reichenbach's principle of common cause. I show that this is problematic. The argument from independent determinations (as I call it) is better rendered as an *inference to the best explanation*. Thus, I reconstruct the argument from independent determinations as a variant of the "miracle argument." This variant avoids some of the difficulties that beset the classic form of the "miracle argument," which relies on a scientific theory's ability to issue novel predictions.

I also examine some pitfalls of this argument, for example, the possibility that the agreement of experimental results has a trivial explanation. Furthermore, I apply the argument to an example from the history of genetic mapping.

Experimental biologists always have to look out for *experimental artifacts*. In other words, almost any phenomenon they observe could have been created in the laboratory by the experimental techniques used. A famous historical example of such an artifact was the bacterial "mesosome." This structure looked like a membrane-bounded organelle and appeared consistently and reproducibly in electron micrographs of bacterial cells. However, after several years of research, microbiologists suddenly judged that this structure was not real; it is thought to be an artifact today. Historians and philosophers disagree over why biologists declared the mesosome an artifact, for example, whether they used a criterion known as "robustness" (which is basically the criterion of independent determinations mentioned earlier). I reexamine the case and present what I think is the correct interpretation. My analysis shows that ruling out artifacts is a procedure that resembles the testing of theories and theoretical hypotheses.

2

Reductionism and the Nature of Explanations

Reductionism is one of the oldest and most controversially debated issues in the philosophy of biology. Many arguments have been proposed for and against it, and many attempts have been made to spell out what exactly reductionism entails.[1] The traditional opponent to reductionism is *holism*, a general approach to the study of complex systems that grants the whole a special ontological or epistemological significance that the parts of the system allegedly lack. Reductionists, by contrast, believe that once the parts of a system and their interactions are understood, there is nothing left for science to explain. The debate between holists on the one side and reductionists on the other side has accompanied biology's maturation as a scientific specialty since the nineteenth century (Weber and Esfeld 2003).

In modern philosophy of biology, the debate over reductionism has taken a somewhat narrow direction, in spite of the broad range of epistemological and ontological issues that are connected to reductionism. In the tradition that emanated from the logical positivism of the Vienna circle, the main issue has been the problem of *theory reduction*. Since the influential book by Ernest Nagel (1961), this term has designated the derivation of the laws of some higher-level theory from the laws of some more fundamental theory with the help of so-called bridge principles. Such principles relate the terms of the theory to be reduced to the terms of the reducing theory, which – presumably – is necessary in order to effect the derivation. The standard example, which has been discussed at length, is the reduction of classical thermodynamics to statistical mechanics. Classical thermodynamics contains theoretical terms such as "temperature" or "entropy." These terms, initially, were alien to mechanics, which stood in the way of a reduction. However, thanks to the work of Maxwell, Boltzmann, and others, it became possible to connect these terms to purely statistical mechanical terms, such as the mean kinetic energy of the atoms of an ideal gas. Hence, a derivation of some thermodynamic laws from first mechanical principles became possible.

There has been extensive debate over whether Nagel's account of reduction is adequate for this case (Sklar 1993; see also the introduction to Chapter 7). What is relevant here is only the tremendous effect that this account has had on the philosophy of biology: Most of the debates on reduction and reductionism in this area revolved around the question of whether Nagel's account of theory reduction could be applied to biology. The main example that has been studied in this question has been the reduction of classical, Mendelian genetics to molecular biology.[2]

In these debates, something like an antireductionist consensus has emerged (Waters 1990). Roughly, this consensus runs as follows. Bridge principles of the kind that are required by Nagel's account of reduction are not available in genetics. The reason is that classical genetic terms like "dominance" or "gene" cannot be defined in purely molecular terms. The genetic phenomenon of dominance can be realized, in theory, by a vast range of different molecular mechanisms. These molecular mechanisms have nothing in common at the molecular level. Thus, no molecular property corresponds to the classical property of dominance.[3] The same is true for other theoretical terms from classical genetics, for example, the term "gene" itself (see Chapter 7). Thus, reduction fails because of the lack of bridge principles. As a consequence, according to the antireductionist consensus, the theories of classical genetics should be viewed as autonomous, which also means that they have explanatory virtues that molecular biology lacks.

This antireductionist consensus, which I have presented in a nutshell here, has some merits because it has led to some interesting insights concerning the logical relations of theories that apply to different levels of organization. However, as an attempt to settle the age-old issue of reductionism in biology, it leaves much to be desired. For it frustrates the main intuition behind reductionism, namely that it has something to do with explaining the properties of complex systems (such as organisms or ecosystems) with the help of the properties of the *parts* of the system. But surely, some of the most exciting biology today accomplishes exactly this; namely, it explains certain biological phenomena (e.g., the production of genetically identical sister chromosomes before cell divisions) with the help of the properties of the molecular constituents of the cell (in this case, the DNA double helix). Why would there be so many discontents with modern experimental biology if it were not reductionistic in this sense?

19

My goal in this chapter is to work out a sense in which large parts of modern experimental biology are reductionistic. In other words, I am interested here in a sense of reductionism that has been unduly neglected in the modern debates in philosophy of biology. Specifically, this chapter concerns what Sahotra Sarkar (1998, 10) has termed "physical reduction," which exemplifies his general notion of "strong reduction" (1998, 45). He means by this the claim that the properties of complex systems can be explained by the properties of the parts and their interactions, where the parts are *spatial* parts (see also Wimsatt 1976b). In the natural sciences, the interactions will be physical and chemical interactions. Sarkar shows for five examples that this is an important sense of reduction in molecular biology (1998, 142–146). I want to go even further than Sarkar and claim that this is true for most of modern experimental biology. For reasons that I address in Section 2.5, classical genetics is an exception.

I claim that where biologists have not yet accomplished physical reduction, they are trying to move closer to it. Of course, in order to really establish these general claims we would have to systematically examine a variety of examples in considerable detail. As this would result in a book-length account of its own, I have to limit my discussion to a single example that I take to exemplify this kind of reduction almost ideally. But the reader may note that, for example, the case study discussed in Chapters 4 and 5 (oxidative phosphorylation) is quite similar to the example discussed in this chapter in terms of theoretical and explanatory structure.

Special difficulties for the reductionist are presented by current developmental biology. These are addressed in Chapter 8.

The example I choose here comes from neurobiology: It concerns how biologists explain the transmission of nerve signals by neurons. The basic mechanism is presented in Section 2.1. In Section 2.2, I discuss a phenomenon that I call *explanatory heteronomy*, which characterizes physical reductionism in a way that has (to my knowledge) been missed in previous philosophical accounts. After this, I address some metaphysical issues that are closely related to physical reductionism. This provides a rationale for the kind of reductionism enunciated here. In Section 2.3, I turn to the issue of biological laws and natural kinds. The relevance of laws in biology has been increasingly questioned in recent years, while natural kinds are an age-old problem of biology (since Aristotle held biological species to be the paradigm for natural kinds). It turns out that these issues are tightly connected to the sense of reductionism that I am advocating here. In Section 2.4, I examine a mode of explanation that is thought to be specific to biology and that seems to exhibit some tensions with reductionism: functional explanation. Finally, in Section 2.5 I discuss the problem of multiple realization, which many philosophers view as a major problem for reductionism.

Mechanisms are thought to play a central role in scientific explanations.[4] Peter Machamer, Lindley Darden, and Carl Craver (2000, 3) define mechanisms as "entities and activities organized such that they are productive of regular changes from start or set-up conditions to finish or termination conditions." I use this definition for the following discussion of my main example for this chapter. The example is the mechanism that explains how neurons can conduct action potentials ("nerve impulses") along their axons (nerve fibers).

The basic mechanism of action potentials was worked out by the neuro-physiologists A. L. Hodgkin and A. F. Huxley in the 1950s, working with giant axons from squid. This mechanism is crucial for the operation of the nervous system in all animals including humans (Alberts et al. 1983, 1018–1031). This mechanism includes the following entities (see Table 2.1): (1) the *cell membrane* that encloses both the cell body and the axons (nerve fibers) of neurons, (2) *Na^+/K^+-ATPase*, a protein complex located in the cell membrane that generates a *membrane potential* by actively pumping Na^+ and K^+ ions across the membrane, (3) *voltage-gated Na^+ channels* (also proteins) that open and close under certain conditions, (4) *K^+-leak channels* (another protein complex) that allow the limited passage of K^+ but not of Na^+ ions, (5) *Cl^- leak channels* that allow the limited passage of chloride ions across the membrane, and (6) *ligand-gated Ca^{++} channels* that open upon binding of a neurotransmitter (the ligand). The Na^+/K^+-ATPase expends a lot of biological energy in the form of ATP (adenosine triphosphate; see Section 4.1) to pump Na^+ ions out and K^+ ions in. As a result, the concentration of Na^+ is about ten times higher outside than inside the cell, while the concentration of K^+ is about ten times higher on the inside. Because of the selectivity of the leak channels, which allow partial equilibration of K^+ and Cl^- but not Na^+, an electric potential of about -70 mV is created across the cell membrane (the sign is purely conventional). This is called the neuron's *resting potential*.[5] This state corresponds to the start or set-up conditions.

An action potential is initiated when a part of the membrane enclosing the cell body is *depolarized* by incoming stimuli (e.g., by the action of neurotransmitters at the synapses that are connected to the neuron). This can happen by a neurotransmitter-induced opening of Ca^{++}-channels. The resulting influx

Table 2.1. *Entities and Activities in the Mechanism of Action Potentials in Neurotransmission*

Entity	Composition	Activities
Na^+/K^+-ATPase	Protein complex	Active cation transport
Voltage-gated Na^+-channels	Protein complex	Passive, voltage-dependent cation transport
K^+-leak channels	Protein complex	Passive cation transport
Cl^- channels	Protein complex	Passive anion transport
Membrane	Phospholipid bilayer	Phase barrier, insulator
Membrane potential	Chemiosmotic gradient	Control of voltage-gated channels, driving force for ion transport
Na^+, K^+, Ca^{++}	Hydrated cations	Freely diffusible positive charge carriers
Cl^-	Hydrated anion	Freely diffusible negative charge carrier

of calcium ions locally depolarizes the membrane. This depolarization causes nearby voltage-gated Na^+-channels to open. As their designation suggests, these channels respond to the voltage of the electric potential across the membrane. A sudden drop in the voltage difference causes a change of the protein's *conformation* (i.e., its three-dimensional structure) via electrostatic interactions with the protein molecule, which forms an electric dipole due to an asymmetric distribution of positively and negatively charged amino acid residues. The conformational change opens a pore in the protein molecule, allowing Na^+ ions to pass. The resulting influx of positive charge depolarizes the membrane further, thus allowing other voltage-gated Na^+-channels in the vicinity to open, causing more depolarization, which causes still more channels to open, and so on. In this way, local depolarization rapidly spreads across the membrane's surface area. The membrane potential can change from -70 mV to up to $+50$ mV, which corresponds to the equilibrium value of Na^+ set by the Na^+/K^+-ATPase and the K^+ and CL^- leak channels.

So far, the mechanism explains how local depolarization can spread across the surface of the membrane. Now, the question arises of how this process can be *directional*, for example, spreading away from the neuron's body along the axon. This can be explained by the fact that the voltage-gated Na^+ channel has three different states: open, closed, and inactivated. In the resting state, the channel is closed, thus not allowing the passage of Na^+. In this state, the channel can be opened by a depolarization of the membrane. However, with a certain probability, the channels fall from the open state into an inactive state.

In this state, they cannot be opened by a depolarizing event; they stay closed for a while, allowing the Na^+/K^+-ATPase to restore the resting potential. As a consequence, a depolarizing event does not simply spread across the entire membrane surface until the Na^+-equilibrium potential is reached everywhere. The automatic inactivation of the channels after a certain time makes sure that the depolarization spreads like a wave away from already depolarized areas into areas not yet depolarized, while the initially depolarized area becomes polarized again due to the activity of the Na^+/K^+-ATPase. In this way, waves of depolarization travel along the axon away from the cell's main body. Such a traveling depolarization wave is called an *action potential*. The emission of an action potential is also referred to as the "firing" of the neuron. Action potentials can cause the release of neurotransmitters at synapses located at the far end of the axon. These neurotransmitters can cause depolarization in the postsynaptic cell, thus transmitting a signal from one neuron to the next. This is the basic process by which the nervous system transmits signals. For the purposes of my discussion, I ignore the other complex processes that constitute neural transmission, like synaptic transmission. Instead, I have focused only on the basic mechanism of the propagation of action potential. This mechanism only explains the passive, cablelike properties that nerve fibers possess, namely, the property of transmitting signals along axons. The "smart" information-processing properties of neurons, specifically their capability of adding or subtracting signals, are ignored for the sake of simplicity.

I have summarized the entities that are part of this mechanism in Table 2.1. According to Machamer et al. (2000), mechanisms also comprise activities, start or setup conditions, finish conditions, and sets of regular changes in between. In the mechanism under discussion, the setup and finish conditions are identical[6]; they both correspond to the resting potential. The set of regular changes in between are represented by the opening of Na^+-channels, their subsequent inactivation, and their eventual transition to the closed state.[7] The obvious *activities* that these entities exhibit include the active ion pumping of the Na^+/K^+-ATPase, the passive ion movements though the various channels, and the various state transitions of the voltage-gated channels (see Table 2.1). In addition there are less obvious activities involved. For example, does the membrane also exhibit an activity? Intuitively, we are inclined to say that the membrane is not "doing" anything; it simply holds the various pumps and channels and passively prevents the ions from equilibrating their concentrations. But it is an important entity in the mechanism and should, perhaps, also be attributed with an activity. If we go further down, the question arises of whether the *membrane potential* is an entity, and whether it has an activity. I defer these questions to Section 2.3.

The mechanism of action potentials provides an *explanation* for the transmission of signals along nerve fibers. There have been many attempts to characterize scientific explanations philosophically (see Salmon 1989 for a review). Of all the accounts that have been given to date, Wesley Salmon's (1984) causal-mechanical approach is probably best suited to examples such as the one discussed in this section. According to this approach, scientific explanations must show how some events or processes fit into the causal structure of the world. The mechanism of action-potential propagation provides such a causal structure: it states all the relevant causal relations that contribute to action potentials. These causal relations include, for example, the relations between the states of the various ion transporters and the concentrations of the different anions and cations on both sides of the membrane. Other sets of causal relations exist among different parts of the ion channel molecules, between the electric field component of the membrane potential and the voltage-gated channels, and so on. Once these causal relations are understood, the propagation of action potentials is explained.

There are many interesting philosophical issues concerning scientific explanations.[8] For the issues to be discussed in this chapter, it does not matter what philosophical account of explanation is preferred. I focus on a particular aspect of such explanations, namely, their heavy reliance on physical and chemical principles. Now, I show that this is the heart of the reductionism of modern experimental biology.

2.2 EXPLANATORY HETERONOMY

What is the structure of theories and explanations in experimental biology? Based on his detailed study of various examples from molecular biology, immunology, and neurobiology, Kenneth Schaffner has concluded that in most areas of experimental biology we find "theories of the middle range which are best characterized as overlapping series of interlevel temporal models" (1993, 125). By "middle range," Schaffner means that, in terms of generality, these theories "fall between biochemistry at one extreme and evolutionary theory at the other extreme on the continuum of levels of aggregation, from molecules to populations" (98). Thus, according to Schaffner, universality or near-universality is found only at very low and very high levels of aggregation (119). By "temporal models," Schaffner means "collections of entities that undergo a process" (119). Biological entities, for example, protein or DNA molecules, or larger structures such as membranes, organelles, or whole cells, are represented in the models in an idealized way, that is, not in their full complexity. By "interlevel," Schaffner means that the models incorporate entities from different levels of organization, for example, the molecular, macromolecular, cellular, or even tissue and organ levels (87). Finally, the "overlapping" aspect of these models has to do with the fact that the biological systems represented by these models are typically *variable* and that most models represent *prototypes* of such systems. Thus, by varying aspects of a model to accommodate, for example, different mutants, a series of overlapping models is obtained.

Our example also exhibits some of the features that Schaffner found in biological theories. First, the mechanism of action-potential propagation seems to be described by a middle-range theory. It lacks the kind of generality that either evolutionary theory or physicochemical theories possess, but it does have considerable scope in applying to all animals with nervous systems. Second, the mechanism has the interlevel aspect; in other words, it includes entities from different levels of aggregation. For example, the model makes reference to whole membranes and axons, that is, to the macromolecular and cellular levels, as well as to ions and proteins, that is, to the molecular level. Third, the mechanism has an important interfield character (see also Darden and Maull 1977) in that it incorporates knowledge from biochemistry, physical chemistry (electrochemistry and thermodynamics), cell biology, and neuroanatomy.

Schaffner thus views theoretical knowledge in experimental biology as fairly self-contained: Biological middle-range theories and interlevel models serve as the *explanans* (that which does the explaining) in biological explanations. His account certainly brings out some important features of theoretical explanations in experimental biology. However, there is one important aspect that it does not capture: the dominant role played by physical and chemical principles. For this reason, I am going to suggest an altogether different way of thinking about theories and explanations in experimental biology.

My alternative proposal is to view the model of the mechanism of action potentials not as a self-contained *theory* like, for example, Maxwell's theory of electrodynamics, but as a physicochemical *explanation* of a special type of phenomenon. On this view, it is merely an application of a physicochemical theory (or set of theories) to a specific kind of system rather than a self-contained theory.

According to the classical deductive–nomological (D–N) account of explanation, scientific explanations involve sentences expressing laws of nature and statements of the initial and boundary conditions that apply to a specific situation (Hempel and Oppenheim 1948). From these premises, the *explanandum* is deduced. In our example, the description of the mechanism (especially the entities and their spatial arrangement) could be viewed as playing the role of the initial and boundary conditions. The "covering laws" (as they are also known) would be the physicochemical laws that govern the behavior of the molecules involved. An example of such a law that is involved in explaining action potentials is provided by the Nernst equation (Zubay 1983, 1148–1149), which derives from electrochemistry:

$$V = \frac{RT}{F} \ln \frac{c_o}{c_i}.$$

V is the electric potential across the membrane, R the gas constant, T the temperature, F the Faraday constant, and c_o and c_i the ion concentrations on the outside and inside of the membrane, respectively. This equation relates the electric potential to the concentration ratio of an ionic substance (of charge $+1$ in the simplified form that I have given) at equilibrium, for example, the concentration ratio of Na^+ ions across a neural cell membrane. Thus, there is no net flow of, for example, Na^+ ions across the membrane if the concentration ratio and the electric potential satisfy the Nernst equation. The electric potential that creates an equilibrium for Na^+ ions is known as the equilibrium potential, V_{Na}. If the Na^+ concentrations are not at equilibrium with the membrane potential V, then there is a force driving ions out of the cell that is proportional to the potential difference $V - V_{Na}$.

Thus, it is a physicochemical law that ultimately explains why ions move across membranes when action potentials spread. Most of the residual parts of the mechanistic model discussed above basically describe the *conditions* under which this ion transport occurs. Thus, the entire model could be viewed as an explanation that uses covering laws from physical chemistry in combination with a description of the initial and boundary conditions that obtain in neural cell membranes. It is these covering laws that have explanatory force; the rest of the mechanistic model basically states how the physicochemical theory should be applied.

This account raises several difficulties that need to be addressed. First, my account as stated seems to presuppose the D–N model of scientific explanation. This model is known to have its difficulties (see Salmon 1989). However, the same point about mechanistic models in experimental biology being physicochemical explanations could be made on the basis of a different account of explanation, for example, a unification (Kitcher 1981, 1989) or the causal–mechanical (Salmon 1984) account mentioned above. Even if laws of nature are rejected altogether as parts of explanations, as some philosophers do (van Fraassen 1989; Cartwright 1989; Giere 1995), my point could be argued. For example, we could use the so-called semantic view of theories (e.g., Giere 1988) to say that the model discussed in Section 2.1 is really a model *of* a physicochemical theory that incorporates some biological information. These nonphysical parts of the model merely provide the information necessary for the deployment of the physicochemical principles in model building.

Thus, my thesis is not tied to a particular account of scientific explanation. No matter how explanations are analyzed logically, the model of action potentials is basically an application of principles from physical chemistry to biological systems, not a theory in its own right. If it is a model, then it is a model of certain theories from physical chemistry.

Second, it might be objected that the mechanistic model discussed is not fully cast in a physicochemical vocabulary. The terms "neuron" and "axon" stand for bona fide biological concepts, and their descriptive content plays a role in the mechanistic explanation. However, this role is not much different from that of wiring diagrams or technical drawings of the coils and capacitors used when physicists explain the behavior of electric circuits. These concepts are *descriptive* rather than explanatory. They serve to identify the kind of system that is to be explained. The terms that do real explanatory work are all physical and chemical terms (e.g., membrane potential, ion channel, conductance); they refer either to molecular species (e.g., Na^+/K^+-ATPase), to species of macromolecular aggregates (phospholipid membranes), or to purely physical entities (membrane potential).

Third and perhaps most importantly, the example I have chosen may be much closer to some *explanatory ideal* than other examples from cellular and molecular biology. Other cases such as the ones examined by Schaffner in his 1993 book or by Machamer, Darden, and Craver (Craver and Darden 2001; Machamer et al. 2000) in their recent works contain more gaps and involve more interlevel and interfield components. Another example is provided by the mechanism of synaptic transmission, which explains how an action potential can be transmitted from one neuron to the next across a chemical synapse. There, we have entities such as secretory vesicles that empty their content (neurotransmitters) into the synaptic cleft and excite the postsynaptic membrane with the help of ligand-gated channels (see Alberts et al. 1983, 1042). Such components of biological mechanisms seem to be further removed from physical chemistry than the components of the more basic mechanism of action potentials that I have used as an example. However, this difference could be seen as a matter of degree instead of a qualitative difference. Nevertheless, it could be maintained that, even in these cases, it is the interactions described by physicochemical theory that do the real explanatory work, while all the biological details give the initial and boundary conditions under which these interactions unfold.

What I am proposing here is a thesis of *explanatory heteronomy* for some parts of experimental biology. While experimental biology may still be autonomous with respect to its disciplinary organization, methodological standards, investigative procedures, and so forth, it does not usually produce theories of its own (at least not in the usual sense of this term), only applications of physicochemical theories.[9]

Of course, I do not mean to question the ingenuity needed by experimental biologists to go about their business (which is the subject of Chapter 3). To figure out what kind of system they are dealing with, what entities these systems contain, and what physical and chemical principles govern their behavior can take years of painstaking research (for another example, see Chapter 4). In this respect, biology may strongly differ from physics, where the correct specification of the systems in question and the corresponding initial and boundary conditions can be comparatively easy, sometimes even trivial.[10] But the thesis of explanatory heteronomy does not concern the way in which biological explanations are generated or arrived at (this is the subject of Chapter 3). It concerns the structure of the *finished* explanations. In the final analysis, I claim, all the explanatory principles in a case like the action-potential mechanism will be physical or chemical.

It must be stressed that the account given here does not entail that *all* biological theories are really physicochemical explanations. What I have argued here is clearly not true for evolutionary theory, classical genetics, or population ecology, which have self-contained theories, lawlike generalizations, and explanatory force of their own (see Weber 1998a, 1999a). The intended domain of my present thesis of explanatory heteronomy includes those parts of biological knowledge that have been characterized as "interlevel" and "interfield." These areas rely strongly on the explanatory force of principles from physics and chemistry.

The thesis of explanatory heteronomy captures an important sense in which this kind of science is reductionistic: Experimental biologists must apply theories from physics and chemistry in order to provide explanations of biological phenomena. The explanatory force is provided solely by the physicochemical theories – theories that describe how molecules interact and how they behave in bulk. There is no specifically *biological* explanatory import; all the explanatory concepts are physicochemical ones (osmosis, electrochemical gradient, ion transport, conductance, etc.). If there are genuinely biological terms (like axon or synapse), they are purely *descriptive* terms that are used to identify the type of system to which the explanatory, physicochemical concepts are applied.[11]

It may very well be that other examples are further removed from the explanatory ideal that the mechanism of action potentials exemplifies. But, I claim, this kind of explanation – explanation that really takes biological systems down to physicochemical laws – is the *goal* of much twentieth- and twenty-first-century biological research. Where biologists have not yet reached it, they are trying to move closer to this ideal. This is the sense in which much of modern experimental biology has a strong reductionistic tendency.

In the following section, I present some considerations that provide a metaphysical rationale for the reductionistic explanatory approach that I have just outlined.

2.3 LAWS AND NATURAL KINDS

Machamer, Darden, and Craver have argued that biological knowledge is typically organized around *mechanisms*. As we have already seen, they define mechanisms as "entities and activities organized such that they are productive of regular changes from start or set-up conditions to finish or termination conditions" (2000, 3). Typical mechanistic explanations in experimental biology then show how such entities and activities interact to produce regular changes from some setup to some termination conditions in a biological system. These regular changes, according to Machamer et al. (2000, 7–8), are a consequence of the operation of *activities*. They take these activities to be ontologically basic; that is, they are not to be explained in terms of anything else. Specifically, Machamer et al. think that no laws of nature must be assumed in order "to make things happen" in biology.[12]

Kenneth Schaffner thinks that the view that biological explanations are accomplished by descriptions of mechanisms is correct "as a first approximation" (1993, 287). However, he has argued that something else is needed to make such explanations complete, namely "laws of working." In this, Schaffner follows the analysis of causation given by John Mackie (1980). For both Schaffner and Mackie, laws are essential for causation. In particular, it is laws that provide the nomic necessity that distinguishes causal from noncausal regularities. Thus, for Schaffner and Mackie, laws are constitutive for causation, and causation is constitutive for mechanisms. In other words, laws are *ontologically prior* to both causation and mechanisms. By contrast, Machamer et al. seem to want to take *activities* as ontologically prior.[13]

In this section, I want to show that Schaffner was right in surmising that laws of nature are necessary parts of explanations in experimental biology. Furthermore, I show that these laws are necessarily physicochemical laws. Let me explain.

It is crucial for the notion of biological mechanisms that these mechanisms exhibit some *regularity* in their behavior. Biological mechanisms are such that they take a biological system or part of a system from some setup to some termination conditions *whenever* these setup conditions obtain (assuming that the mechanism is deterministic). For example, according to the model discussed in the previous section, a neuron will fire an action potential whenever a sufficient number of stimuli cause a local membrane depolarization. Furthermore, this model works under the assumption that a voltage-gated Na^+ channel will open whenever the membrane potential drops under a certain threshold value. Thus, both the constituent entities and the whole mechanism exhibit a regular behavior. Furthermore, the kind of regularities exhibited by mechanisms and their constituent entities are not contingent regularities; they display *nomic necessity*.[14] The best indication of nomic necessity is thought to be the fact that these regularities support counterfactuals. We can say that, for example, a resting neuron *would* fire an action potential if it were to receive a sufficient number of stimuli. In fact, this kind of counterfactual claim is not merely supplementary with respect to the mechanism of action potential; it seems to be an essential part of the explanation of how a neuron works.

These considerations suggest that the behavior of biological mechanisms and their constituent entities is *lawful*. This raises the question of what the source is for this kind of lawful behavior. Are the regularities exhibited by biological mechanisms due to some laws of nature? Or can the same regularities be explicated without laws? In order to examine this, I now go back to my main example from Section 2.1.

I will focus on a specific process that is part of the mechanism of action potentials, namely passive ion transport across the membrane (see Table 2.1). Passive ion transport is not driven by an ion pump (which would be active transport), but instead by two different forces. The first force is *electromagnetic*: positive and negative charges are distributed unevenly across the membrane. This creates a net Coulomb force for charges such as Na^+ ions to move across the membrane. The second force is *osmotic*. It arises whenever Na^+ ions are present in a higher concentration on one side of the membrane. Because ions are always in thermic random motion, they constantly bounce against the membrane. Because more ions will bounce against the membrane from the side where the concentration is higher, more will cross the membrane from that side. There exists an *equilibrium state* at which the two forces cancel each other out. As we have seen (Section 2.2), this equilibrium is given by the Nernst equation. Thus, there is no net flow of, for example, Na^+ ions across the membrane if the concentration ratio and the electric potential satisfy the Nernst equation. The electric potential that creates an equilibrium for Na^+ ions is known as the *equilibrium potential* V_{Na}. If the Na^+ concentrations are not at equilibrium with the membrane potential V, then there is a force driving the concentrations toward equilibrium that is proportional to the potential difference $V - V_{Na}$.

What we have done now is to explain the activity of passive membrane transport in terms of two laws of nature, namely Coulomb's law and the Nernst equation. The former gives the force by which charged bodies attract or repel each other, while the latter gives the equilibrium state for passive ion transport. These laws explain the regular behavior of ions – their "activity" – in passive transport. Thus, laws of nature seem to be required after all in order to understand how certain activities arise.

At this point, opponents of laws could argue that purported laws such as Coulomb's or the one expressed by the Nernst equation are really descriptions of *capacities* or *causal dispositions*[15] of a certain kind of entity. For example, Nancy Cartwright (1999, 59) argues that Coulomb's law is merely a way of describing the capacity of charged bodies to be moved by other charged bodies under certain conditions. Thus, opponents of laws could say that what is needed to explicate activities is not laws but Cartwrightian capacities.[16] On this view, the activities that are constitutive of biological mechanisms are really the exercise of the physical and chemical capacities of the component entities.

Although such a defense could work for Coulomb's law, it is not open for the Nernst equation. This equation states the relation between the electric potential difference and the concentration difference across the membrane in a state where osmotic and electrostatic forces cancel each other out. But the osmotic forces involved here cannot be viewed as capacities of any particular entity (nor is the term "force" applied in its proper sense here). Ions may have a capacity to move or be moved by other charged bodies and a capacity to be pushed around by solvent molecules. However, they have *no* capacity to try to equilibrate across a membrane. Yet this is exactly what they appear to be doing. Perhaps we could say that the electrochemical *gradient* across the membrane has the capacity to equilibrate ions. However, a gradient is not a material thing.

31

The reason behind these peculiarities is that the Nernst equation expresses a thermodynamical law. Laws of this kind have the strange property that they seem to determine the behavior of bulk matter without paying any attention to the causal details of specific processes (see Chalmers 1999, 12–14). Thus, such laws cannot be explicated by intrinsic capacities possessed by individual objects. It thus seems that we need laws of nature in order to explicate regularities due to processes like osmosis, which play an extremely important role in biology.[17]

Thus, the idea that the necessity behind certain natural regularities can be explained by citing a hidden mechanism that connects the events is attractive, but it is known to run into difficulties as soon as this approach is applied to fundamental physical laws (Glennan 1996). It seems that the same problem occurs when we inquire further what makes biological mechanisms behave regularly.

The next question to be addressed in relation to laws of nature is whether the laws that feature in typical explanations in experimental biology are just physicochemical laws – as my account given in Section 2.2 suggests – or if distinctly biological kinds of laws are involved.

John Beatty (1995) has argued that all *distinctly* biological generalizations describe contingent outcomes of evolution and thus fail to be necessary. According to Beatty, if there are nomologically necessary generalizations in biology, then they are really disguised physicochemical generalizations. Beatty's "evolutionary contingency thesis" has been criticized by several authors (Carrier 1995; Waters 1998; Weber 1999a). However, there are actually strong points of contact between Beatty's thesis and my thesis of the explanatory heteronomy of parts of experimental biology (see Section 2.2). This requires some elaboration.

Beatty argues that whatever generalizations appear in the context of biology either are those of physics and chemistry, or describe contingent outcomes of evolution, in which case they lack the necessity or counterfactual-supporting force traditionally required of lawlike generalizations. The main reason for this, according to Beatty, is the fact that the evolutionary process displays "high-level contingency." This means that if evolution were a tape that could be replayed, it would play a different tune each time. Even under the same environmental conditions, selection pressures, and so forth, evolution would take a species of organisms to a different adaptive peak (or to no adaptive peak at all, or the adaptive peaks might not even be the same) if it could be replayed. This means that any biological generalization that holds in our world cannot be expected to hold in a possible world sufficiently different from ours to bestow lawlikeness on it. Therefore, it is accidental. Frequently cited examples of contingent biological generalizations are the genetic code, which is widely seen as a "frozen" historical accident, or the most famous "laws" of biology: those associated with the name of Mendel, which are supposedly contingent on some rather strange historical accidents that led to the evolution of sex.

Kenneth Waters (1998) counters the Evolutionary Contingency Thesis by drawing a distinction between biological generalizations that describe the *distribution* of some trait among groups of organisms and generalizations that describe *dispositions* or *causal regularities*. Using Waters's distinction, Mendel's laws (note the absence of scare quotes) could be interpreted as describing a disposition of organisms with a certain internal genetic/physiological makeup to pass on their genes in a certain way. Any organism with this internal genetic/physiological makeup has this disposition and will *necessarily* behave in this peculiar way, which renders Mendel's laws lawlike. But Mendel's laws say nothing about the distribution of this disposition within the phylogenetic tree. This distribution is indeed historically contingent. In other words, the domain of application of Mendel's laws is a certain genetic/physiological type, and organisms from any region in the phylogenetic tree may contingently evolve into or out of this domain. Hence, the Evolutionary Contingency Thesis only applies to some biological generalizations, namely those that describe distributions, but not to others, namely those that posit dispositions and causal regularities.

To this move, Beatty (1995, 60–62) responds as follows. He first accepts Waters's point that biological regularities such as Mendel's laws could, at least in principle, be rendered lawlike (i.e., noncontingent) by specifying a *natural kind* (see below) in terms of the underlying physiological mechanisms, although he considers this to be difficult. But he then goes on to claim that such a reformulation will always destroy the laws' *distinctively biological* character; in other words, the laws transmogrify into physicochemical ones. Thus, according to Beatty, there is some kind of necessary tradeoff between a regularity's lawlikeness and its distinctively biological character: "[T]he closer one's generalizations [...] come to describing sequences of chemical reactions, the more certain one can be that they are laws, because one can be more certain that no evolutionary outcomes can contradict them. But at the same time, the generalizations will become less and less distinctively biological" (Beatty 1995, 62).

Indeed, Waters's (1998, 19) example of a lawlike causal regularity, "Blood vessels with a high content of elastin expand as internal fluid pressure increases [...]," is vulnerable to exactly this objection. Waters views this as a *biological* lawlike statement, where "blood vessels" refers to a somewhat sloppy natural kind defined by the shared internal makeup of these anatomical structures. But Beatty would view it as a disguised *physical* law about a certain kind of macromolecular aggregate, where the disguise is provided by the biological term "blood vessel." Furthermore, the natural kind in question is a *physical* natural kind for Beatty. According to Beatty, as soon as one tries to specify the "shared internal makeup" of these structures in order to pick out their causal dispositions, one ends up with molecules, and thus leaves the distinctively biological realm.

This thesis concerning natural kinds in biology deserves to be examined more closely. Natural kinds are thought to be classes of things (or processes, or events) that are clearly distinct from other things by sharing an internal makeup and, therefore, a set of causal dispositions. Now, it seems to be an implicit assumption of Beatty's that all natural kinds of the sort that support genuine laws of nature must be physical or chemical natural kinds. None of Beatty's commentators (nor John Beatty himself) seem to have noticed just how reductionistic this assumption is. I am not aware that Beatty has given any justification for it. However, other philosophers have done so. For example, Brian Ellis has argued that only physical and chemical kinds, for example, the kinds denoted by "electron" or "ethanol," satisfy the criteria that an adequate metaphysics must require for natural kinds (Ellis 2002, 26–32). In particular, the members of natural kinds must share a set of unchanging causal powers. Presumably, only physical and chemical kinds satisfy this requirement. Even though Ellis allows that something similar to physical/chemical natural kinds exist in biology, these latter kinds are fundamentally different, in that their causal powers are not fixed. But this means that if a law is genuinely biological, then it must have a different status from physical and chemical laws, which are about natural kinds that are individuated by immutable causal dispositions. This, I take it, is also the essence of Beatty's thesis. Furthermore, I claim that it is the essence of the kind of reductionism that characterizes much of modern experimental biology.[18]

In an earlier work, I argue that there are counterexamples to Beatty's thesis in population biology (Weber 1999a). What I want to claim here is that in large parts of modern experimental biology, Beatty's thesis is correct. That is, all the *genuine* laws of nature that feature in experimental biology are physicochemical laws. If they describe the behavior of some natural kind of system, then the natural kind in question is a physicochemical kind, for example, a molecular species. If they do not range over such natural kinds, then biological generalizations describe contingent outcomes of evolution and are, therefore, not genuine laws of nature.

I come back to the issue of natural kinds in Sections 7.4 and 9.1. Here, I am concerned mainly with the strong affinity between Beatty's Evolutionary Contingency Thesis and my thesis of explanatory heteronomy from Section 2.2. Physics and chemistry are in the business of describing laws of nature. Reductionistic experimental biology, by contrast, is in the business of *applying* such laws to explaining biological phenomena; it does not look for any generic biological laws. This reductionistic practice makes sense if viewed from a metaphysical standpoint such as Ellis's, according to which genuine laws and natural kinds, that is, laws about the fixed causal powers of natural kinds of things, are always physical or chemical (e.g., molecular structures). Thus, reductionism as I have characterized it in Section 2.2 and some of the underlying metaphysical assumptions of modern experimental biology go hand in hand.

In the following section, we need to attend to a special mode of explanation that is specific to biology, namely *functional* explanation. For this explanatory mode seems to be at loggerheads with my thesis that the explanatory approach of some parts of experimental biology consists in applying physicochemical laws to biological systems, since functional explanation is alien to the physical sciences. It would be senseless to ask for the function of gravity or the Higgs boson, while the function of respiration or mitochondria is a perfectly fine subject for scientific inquiry. Why is this so?

2.4   FUNCTIONS AND FUNCTIONAL EXPLANATION

Biology has always been thought to deploy a special kind of explanation, namely functional explanation.[19] To a first approximation, a functional explanation shows what something is *for*, for example, that the heart is for pumping blood. Modern philosophers have been somewhat uneasy with admitting this kind of explanation as a legitimate mode of scientific explanation. The reason is that asking what something is for is uncomfortably similar to asking what the purpose or final cause of something is. But purposes and final causes have been banned from modern science, because such causes require intentions. The purpose of something is whatever the thing was intended for, for example, by a divine Creator. Since invoking divine intentions is not considered to be a legitimate strategy in scientific explanation anymore, intentions and purposes are rejected by modern science, at least in the realm of nature. But what about functions and functional explanation? Is it epistemologically legitimate to explain some biological trait or structure by showing what it is for?

In the more recent philosophical literature, there have been two major approaches to naturalize functions, that is, to show that functions are natural properties of certain kinds of systems. One such approach is known as the *etiological* account of function (Wright 1973). On this account, exhibiting the function of a trait or a structure is equivalent to showing that this trait or structure is present in a given system because of what it does within this system. To be more precise, the *tokens* of a certain type of function bearers are present in a system because of what other tokens of the same type did in systems of the same type in the past. In the biological domain, of course, one way in which this is possible is by evolution by natural selection.

Traits that have a positive effect on fitness afford their carriers with an advantage in natural selection, thus securing their own fixation and maintenance in a population of organisms. The defining feature of this approach is that functions owe their presence in an organism to their effect on the organism's reproduction and survival. For example, most animals have hearts because they facilitate the supply of oxygen and nutrients to the body cells. This afforded animals with hearts (or primitive heart-like structures) a selective advantage over animals without hearts, which explains why most animals have hearts. To say that the function of the heart is to pump blood means that hearts are there because they have this effect on circulation. Thus, ascriptions of etiological functions are hypotheses about the evolutionary history of the structure in question.

There are several difficulties with the etiological account. One difficulty is best brought out with the help of a fictional but not unrealistic example. Imagine that a mutation occurs in a bacterial gene that allows the bacterium to utilize a previously indigestible sugar. The newly created gene is advantageous and spreads in the bacterial population. After it has become fixed by natural selection, we can say with the etiological account that the corresponding enzyme's function is to metabolize the sugar, because its capacity to metabolize the sugar is the reason why the enzyme is present in the bacterium. But what about the ancestral bacterial cell that enjoyed the benefits of this enzyme for the very first time, that is, the one in which the new mutation occurred? Does the enzyme perform the same function there? The etiological account says no, because the enzyme does not owe its presence in that bacterial cell to natural selection. Yet intuition prompts us to say that the enzyme nevertheless already performed the function that would later lead to its establishment in the population. This suggests that an appropriate evolutionary history is not necessary for something to have a particular function.

36

Another question is whether biologists are always making a claim about a trait's evolutionary history when they ascribe a function to it. The etiological account of functions is committed to *adaptationism*, because all biological functions are viewed as evolutionary adaptations. Adaptationism has been criticized, for example, by Stephen Jay Gould and Richard Lewontin (1979). But even if their strong antiadaptationism is rejected, it could (and should) be maintained that a functional ascription does not always imply a claim about evolutionary history. In fact, adaptive claims are notoriously difficult to test empirically, and molecular and cell biologists, in particular, rarely provide such tests. Furthermore, if "function" is synonymous with "adaptation," biologists should use these terms interchangeably, which seems not to be the case.

A different kind of etiological account of functions has recently been developed by Peter McLaughlin (2001). On McLaughlin's view, it is neither sufficient nor necessary for something to be a beneficiary of natural selection to qualify as a biological function. Prebiotic replicators or growing crystals are subject to natural selection or differential reproduction, too, but they have no functions. A system can only have functions if it is capable of *self-reproduction*. By this, McLaughlin means the maintenance of an organism's form over time, which is not to be confused with the production of tokens of the same type (i.e., procreation or propagation). Organisms maintain their form not by preserving their material constituents, but by constantly regenerating their parts. My body does not contain the same carbon atoms as it did 20 years ago, but the individual organism that I am sustains metabolic activities continually replacing all the atoms and molecules that make up my body.

McLaughlin argues that there is a sense in which tokens of function bearers, by virtue of what they do, indirectly cause their own continuing presence in an individual organism, namely by contributing to the maintenance of the whole system (token). Thus, McLaughlin's account can be classified as etiological. But he has replaced the function bearer's contribution to evolutionary fitness by its contribution to self-reproduction as the relevant feedback mechanism that accounts for the presence of tokens of function bearers in the system. Importantly, this account avoids some of the problems of natural-selection-based accounts of function mentioned above, such as the commitment to adaptationism or the problem of the first appearance of a function.[20] However, McLaughlin's account might incur metaphysical costs such as a certain form of holism (McLaughlin 2001, 210–212).

In addition to etiological accounts, there is a whole different tradition of thinking about functions. This tradition sees the goal of functional ascriptions not as the explanation of the presence of the function bearers, but as the identification of the *causal role* played by a system's or organism's components.

The most influential account of this kind is due to Robert Cummins (1975). He has attempted to give an analysis of expressions like "X functions as a ϕ in System S." An instance of such a functional claim would be "the heart functions as a blood pump in the circulatory system." Thus, X is the thing that has a function, ϕ is an activity of X, and S is a system. On Cummins's analysis, X functions as a ϕ in System S exactly if X's capacity to ϕ is part of an adequate analytic account of a capacity Ψ of the whole system S.

For example, the heart (X) functions as a blood pump in the circulatory system (S) exactly if the heart's capacity for pumping blood (ϕ-ing) is part of an adequate analytic account of the circulatory system's capacity for delivering oxygen and nutrients to body cells (Ψ-ing). By an "adequate analytic account," Cummins means an explanation of a system's capacities that appeals to the capacities of the system's constitutive parts such as X's capacity to ϕ. In the example of the heart, the heart's capacity to pump blood is part of an explanation of the circulatory system's capacity to deliver oxygen and nutrients to body cells. Thus, on Cummins's account, functions can be viewed as causal roles that parts of a system play within the working of the system, which is why Cummins-functions are also known as (causal) *role functions*.

A central difficulty with Cummins's account has to do with his requirement of an overall system capacity Ψ to which the function bearers contribute in some way. Some commentators have pointed out that this requirement makes role functions relative to the investigator's *interests* (McLaughlin 2001, 120–124; see also Schaffner 1993, 400). For the overall system, capacity Ψ may be any capacity that "arouses our analytical interest" (McLaughlin). Since functions are individuated according to the contribution they make to some overall system capacity, which may be chosen arbitrarily, a subjective element is introduced into functions; their status as objective natural properties is not guaranteed by Cummins's account.

In spite of this difficulty, I think that Cummins-functions provide the best analysis of the function concept in the areas of biology currently under investigation (see also Craver 2001). However, to make a Cummins-style account of function work, we need some way of picking out *significant system capacities* to which role functions contribute. In order to examine this problem, I now return to my example from Section 2.1.

A Cummins-style functional analysis of the mechanism of action potentials looks like this. The system S we are considering is the neuron with its nerve fiber. The overall system capacity Ψ to be explained is the neuron's ability to propagate an action potential from the cell body down the axon upon receiving sufficiently strong stimuli. The model mechanism presented in Section 2.1 explains this capacity by decomposing the neuron into a set of entities $X_1, X_2, \ldots, X_n$ with capacities or activities $\phi_1, \phi_2, \ldots, \phi_n$. These entities include the cell membrane and various selective ion channels and pumps located in the membrane. Their activities include active and passive selective ion transport, voltage-gated or not. Furthermore, we may count the Na^+ and K^+-ions and their capacity to diffuse as part of the mechanism. Now, according to Cummins, to show that entity X has function ϕ means to show that X's capacity to ϕ is part of an analytic account of the neuron's capacity to propagate an action potential. For example, Na^+/K^+-ATPase has the function of actively transporting Na^+ and K^+ exactly if its capacity to do so is part of an analytic account of the neuron's capacity to fire action potentials – which is actually the case. The question now is whether the overall system capacity Ψ – in this case the capacity of neurons to fire action potentials – was chosen arbitrarily for functional analysis.

I think the example clearly shows that this is not the case. Far from being just any old capacity that has aroused biologists' interest for contingent reasons, the overall system capacity Ψ, propagation of action potentials, was chosen as a salient capacity for functional analysis for *physiological* reasons. What this suggests is that role functions are ascribed to individual capacities ϕ because the ϕ-ing of these function bearers contributes to an overall system capacity Ψ that *is itself a role function*.[21] In our example, the propagation of action potentials is a capacity that contributes to some higher-level system capacity, namely information processing in the nervous system. On this view, some ϕ is a role function only if it contributes to an overall system capacity Ψ that is itself a role function in some higher-level mechanism.

The additional constraint on role functions that I have introduced here is basically a *coherence condition*.[22] It implies that nothing is a role function all by itself; role functions are individuated with respect to other role functions.

As should be obvious, my construal of role functions creates a regress. If the attribution of role functions is always done with an eye to some higher-level system capacity that is itself a function, then the question arises of what makes this higher-level function a function. If we answer that it is a function if it contributes to a capacity that is itself a function at yet a higher level, then we have a regress. However, I suggest that it is a terminating regress, since the most inclusive mechanism, as it were, is the whole organism. Thus, the question arises if there is some ultimate system capacity or system goal in relation to which all role functions at different levels of the biological hierarchy acquire their functional status. One candidate for such an ultimate system goal is fitness (compare Bigelow and Pargetter 1987). Another is McLaughlin's capacity for self-reproduction. I think the latter is preferable, because we can attribute functions even to organisms whose reproductive fitness is zero (e.g., mules). Thus, I conclude that a biological (role) function is a capacity that either contributes to a higher-level system capacity that is itself a role function or contributes directly to an organism's self-reproduction.[23]

Applying this hierarchical picture to our example, the constituent parts of the mechanism of action potentials are role functions with respect to the capacity of neurons to transmit signals. This capacity is a role function with respect to the nervous system's capacity to process information. This information-processing capacity directly contributes to an organism's self-reproduction, for example, by allowing the organism to locate and ingest food.

It is important to appreciate that, on this account of functional explanation, the ascription of a function to a biological entity implies nothing about why tokens of this entity are *present* in the systems that are given a functional analysis. Functional explanation is not about phylogeny, nor about ontogeny. When we say that the Na^+/K^+-ATPase has the function of generating a membrane potential for neurotransmission we do not thereby imply that this capacity is the *reason* that the neuronal membrane contains such an enzyme. In fact, the latter statement is likely to be false, as all cell types (not just neurons) contain Na^+/K^+-ATPases (Alberts et al. 1983, 291); neurons just make special use of them.

Thus, I think that that Cummins's account of function correctly represents the use of the term "function" in experimental biology, provided that it is augmented, as I have done, by a coherence condition and an account of the ultimate system goal of self-reproduction. By contrast, the etiological function concept mentioned at the beginning of this section is not relevant here. Experimental biologists do not automatically imply a hypothesis as to the selection history of a biological entity when they buttress functional claims. Of course, hypotheses about the evolution of some biological structure are sometimes entertained or even evaluated by experimental biologists. But elucidating the evolutionary history of some system or subsystem is supplementary to analyzing its function; it is not part of it.

As far as McLaughlin's version of the etiological function concept is concerned, I think it is very worthwhile in directing us to the ultimate system goal of self-reproduction. However, I think this account does not correctly capture the use of the term "function" in experimental biology either, because, as I have argued, functional analysis does not at all attempt to explain why tokens of function bearers are present in a biological system. To give a functional analysis of some biological system differs both from studying its phylogeny and from studying its ontogeny (the latter being the domain of developmental biology; see Chapter 8).

The final point that needs to be addressed is how functional explanation is to be reconciled with my reductionistic thesis from Section 2.2, to wit, that some explanations in experimental biology are really physicochemical explanations of a special kind of system. The problem is that such explanations do not usually support functional attributions. Why is it that we attribute (role) functions to some processes occurring in a living organism, but not to typical physicochemical processes? To put it differently, why should there be an ultimate system goal in living organisms, in relation to which some causal roles acquire functional status?[24] After all, we would not attribute such a goal or ultimate capacity to a nonliving system such as a lake that self-regulates its level. I think this difference has to do with the pragmatics of explanation, in other words, with the kinds of "why" questions that we want to have answered (see van Fraassen 1980, 134–143). The capacity for self-reproduction is the most salient capacity that we want to understand in biological organisms – it makes the difference between living things and dead matter. Thus, we decompose this ultimate system capacity into role functions, in some cases with the help of physicochemical theory (see Section 2.2). By contrast, in the case of the solar system, the most salient capacity we want to understand is the regular recurrence of the planets, which is not an instance of self-reproduction but simply of the numerical persistence of the objects and states of motion involved. Therefore, in contrast to living systems, such a system does not call for a functional analysis. Even if we take a physical system with self-regulating properties such as a lake that regulates its level, such systems hardly qualify as self-reproducing. This is the reason that they are not said to have functions, at least not in the same sense as in biology.[25]

Thus, my main reductionistic thesis of this chapter is compatible with the fact that biologists also give functional explanations. While it remains true that some areas of experimental biology apply physical and chemical laws to biological systems, when they give functional explanations, biologists make special use of such laws. This is by no means out of the ordinary, as engineers (for example) also use physical and chemical principles in a different way than physical scientists do.

2.5 MULTIPLE REALIZATION AND THE NEW REDUCTIONISM

The antireductionist consensus in the philosophy of biology mentioned in the beginning of this chapter states that reduction in a case like genetics fails because there are no bridge principles to connect the theoretical terms of classical and molecular genetics, respectively. To this argument, it can be objected that it is based on too restrictive an account of reduction, namely Nagel's account. Many authors have argued that a more permissive account might accommodate the case (e.g., Kimbrough 1979; Waters 1990; Schaffner 1993). Thus, the problem for reductionism may be with a particular philosophical *theory* of reduction, not with reductionism itself. Therefore, I have tried to elaborate a different sense in which large parts of experimental biology may be said to be reductionistic, namely my thesis of explanatory heteronomy (Section 2.2), in combination with the claim that there are no genuinely biological laws of nature or biological natural kinds of things with fixed causal dispositions (Section 2.3). In this section, we need to examine an informal version of the most popular antireductionistic argument that can be stated without any reference to Nagel's model of reduction.

This argument is an application of an argument pattern that was originally due to Hilary Putnam (1975d) and that also continues to be one of the most successful arguments against reductive physicalism in the philosophy of mind (Kim 1996). It is known as the "multiple realization argument." In general, this argument proceeds by first showing that some so-called special science (the usual suspects are biology and psychology) deploys functional concepts, that is, concepts that individuate their referents by the causal role they play within a system.[26] Examples of functional concepts are the concept of dominance in genetics or the concept of emotion in psychology. Such functional concepts may play important explanatory roles in their respective domains and may enter into generalizations, even lawlike ones.

The epistemic role that functional concepts play is entirely independent of the material realization of the relevant functions. In other words, it does not matter *what* plays the causal role specified by a functional concept. These causal roles could be played by different mechanisms. For example, a specific computer algorithm can be implemented either on a silicon-based digital machine or on a computer that works with vacuum tubes. The algorithm is the same no matter what kind of physical system executes it. Most importantly, the material realizers of functions can be heterogeneous from the perspective of the lower-level theory. In other words, the natural kinds of a special science such as biology can cross-classify those of a more fundamental science such as chemistry (Fodor 1974; Carrier 2000).

A nice example of such a cross-classification is the concept of pheromone (Hoyningen-Huene 1997). This concept denotes a class of chemicals that can play a certain causal role, namely the initiation of sexual behavior in certain animal species. Thus, from the perspective of biology, pheromones form a natural kind. By contrast, pheromones are clearly not a natural kind from the perspective of chemistry. Not only is the class of pheromones chemically quite heterogeneous, but some pheromones are even mixtures of several compounds.

The heterogeneity of the realizers of a functional equivalence class, some argue, creates an "explanatory hiatus" (Carrier 2000, 182) between biology and chemistry; in other words, the generality of certain explanations given by biology cannot be reproduced at the chemical level. The reason is that the functional concepts range over a large class of otherwise disparate natural phenomena. But this unity is invisible from the chemical level; therefore, the standard line of argument goes, important explanatory accomplishments are lost when we focus on the chemical level alone.

However, in the case of pheromones, I do not see what exactly we lose if we focus on the chemistry and the neurobiology of the pheromone-triggered behavior alone. Where is the general higher-level theory that has autonomous explanatory value? In evolutionary theory and in population biology, as well as classical genetics, there are such theories, but these sciences are not reductionistic in the sense that I am examining here. At any rate, there appears not to be a general theory of pheromone action at a level that makes no reference to some specific realization. There is only the phenomenon itself, that some animals attract individuals from the opposite sex with the help of chemicals. This is hardly explanatory. By contrast, there are general, explanatory theories of natural selection (for example) that make no reference to any particular physical realizer of fitness differences (Sober 1984, 51; cf. Weber 1996). There, multiple realization is significant. But here, the multiple realization of the functional kind of pheromones appears not to be relevant to reductionism in the sense that I advocate. The alleged explanatory hiatus does not occur, even if for somewhat trivial reasons.

What about the typical mechanistic explanations given in modern experimental biology, such as the one discussed in Section 2.1? How does the multiple realization argument fare in cases, for example, from neurobiology?

The example that I introduced in Section 2.1 describes at the molecular level how neurons conduct signals (action potentials) along their nerve fibers (axons). As I have argued, most of the theoretical concepts used in treating this mechanism are straightforward physicochemical concepts such as molecule, diffusion, osmosis, electric potential, concentration gradient, selective ion transport, and membrane permeability. The only biological concepts that appear in the description of the mechanism are those of neuron, axon, synapse, and dendrite. These concepts are all structural; in other words, they describe certain kinds of microanatomical structures. However, these structural concepts are not different in kind from the concepts that physicists and engineers use in order to describe the structure of some electrical mechanism, for example, wire, bridge, coil, or switch (see Section 2.2). The only difference is that the former structures are of biological origin, while the latter are human artifacts. Thus, I have concluded that the mechanism of nerve signal transmission is really a physicochemical system that is being given a functional analysis.

I fail to see how the explanatory value of this analysis is in any way affected by the fact that the same type of mechanism could have been realized by an altogether different physical-chemical structure. (I accept this premise of the multiple realization argument for the sake of discussion.) For most purposes, the behavior of neurons can be mimicked on silicon-based computers. On the basis of this fact, some antireductionists would argue that this multiple realization prevents a reduction of certain neurological (or computational) principles to physics and chemistry. For neuronlike behavior allegedly constitutes a higher-level natural kind to which no physical or chemical kind corresponds.

An important point to note about this objection is that neuronlike behavior does not constitute a natural kind of the same type as, say, a molecular structure. The latter type of natural kinds are characterized by a set of common causal dispositions that are invariant across the natural kind and that all and only the members of this natural kind possess. These dispositions are fixed. By contrast, the class of things that behave like neurons may differ wildly in their causal dispositions (compare, for example, a silicon chip and a neuron). If this is a natural kind, it is a lot fuzzier than a chemical species that is defined by a molecular structure. To use Ellis's (2002) terminology, we have here a *variable* natural kind (see Section 2.3). I do not know whether there are any significant generalizations over this kind *other* than the ones that define it (i.e., other than the statements that describe what it means to behave like a neuron). But, as in the case of the pheromones, we seem to have no rich, explanatory theory (such as, to use the standard counterexample, natural selection theory) that could generate an "explanatory hiatus" that would make multiple realization relevant.

Perhaps we could even go further and claim that even if there *were* significant generalizations over the class of things that behave like neurons, these generalizations could, as a matter of principle, not have the kind of explanatory force that physical and chemical principles do. For the latter capture the essential causal dispositions that are shared by the members of natural kinds, which are manifested in counterfactual-supporting regularities. These causal dispositions (e.g., those of an electron or photon) are unchanging and related to fundamental laws of nature.[27] By contrast, the natural kinds whose causal dispositions become manifest only at higher levels of organization (e.g., the class of neurons or the class of pheromones) do not share such unchanging essences. They could be different even if the fundamental laws of nature remained the same. Thus, explanations that remain at the level of such variable natural kinds and the corresponding generalizations are unable to explain something by showing that it occurs *necessarily*.

I do not want to claim that all scientific explanations should show that things are the way they are necessarily. There are different ways of explaining things, and what counts as a good explanation partly depends on what questions are being asked. Furthermore, I do not want to deny that some of the higher-level classifications provided by sciences such as biology (or psychology, the social sciences, and so on) have epistemic value. My point is only that some parts of experimental biology are interested in showing what happens necessarily and that for these parts the variable, higher-level natural kinds and the corresponding generalizations cannot have the same status as physical and chemical kinds and laws.

Thus, the standard multiple realization argument can be charged with an ill-conceived egalitarianism with respect to laws and natural kinds. Higher-level kinds and the corresponding regularities do not have the same kind of explanatory force as the fundamental kinds and laws of physics and chemistry. But without such explanatory force, the standard multiple realization argument against reductionism fails to sting.

It will be objected that the explanatory value of higher-level kinds lies in their generality and the fact that they unify a large number of otherwise disparate phenomena (see Sober 1984, 47–59). This objection is indeed a strong one where such unifying generalizations actually exist, that is, in areas such as evolutionary theory. But these are explicitly excluded from my considerations. In those parts of biology that are under investigation in this book (with the exception of classical genetics), I am not aware of any higher-level generalizations with this kind of scope that have an autonomous explanatory value.[28]

We now need to examine one final possibility of how multiple realization could be relevant for reductionism. Namely, we need to examine cases where some basic mechanism such as the mechanism of action-potential propagation becomes *embedded* in higher-level mechanisms.

The mechanism of nerve signal propagation that I have discussed in Section 2.1 plays a crucial role in a variety of important biological mechanisms in animals, namely all mechanisms that involve the animal's nervous system. The primary and evolutionarily ancient functions of the nervous system are probably the control of body movements and sensory perception. Since all activities of the nervous system involve the propagation of action potentials according to the mechanism discussed in Section 2.1, this mechanism will be part of any complete explanation of animal behavior and sensory information processing. Of course, the way in which the central nervous systems of higher animals such as mammals control the extremely complex behavior and the high-resolution sensory systems of such animals is not yet fully understood. However, fortunately there exist a number of very simple animals with simple nervous systems that can serve as model systems (Schaffner 1998, 2000, 2001; Weber 2001c; see also Chapter 6, this volume).

One such organism is the nematode worm *Caenorhabditis elegans*. This worm comes in two sexes, male and hermaphrodite. The adult hermaphrodite consists of exactly 959 somatic cells, of which 302 are neurons and 95 are muscle cells. Thanks to the comparative simplicity of this organism, it was possible to construct wiring diagrams representing all the synaptic connections between the animal's individual neurons, as well as between the neurons and the muscle cells (Ankeny 2000). With this information and the full genomic DNA sequence available, *C. elegans* is the most completely mapped multicellular organism. In theory, the available information on the worm's nervous system should hold the key to understanding the animal's complete behavioral repertoire, a promise that turned out to be more difficult to realize than biologists had hoped initially (Schaffner 2000). Nevertheless, the present state of the art in explaining the worm's behavior provides an interesting case for reexamining the issues of reduction and reductionism in light of what we have learnt about the nature of theories and explanations in experimental biology.

C. elegans engages in a series of simple behavioral patterns that mostly consist of undulatory movements, such as chemotaxis (moving away from noxious substances or towards nutritious ones), thermotaxis, a tap withdrawal reflex, touch sensitivity, mating, feeding, and a very simple form of social behavior (social foraging). All of these behaviors are controlled by the nervous system and a few sensory cells. The nervous system is the control center for the worm's behavior. For some of these behaviors, it has been possible to identify the neurons that are involved and to construct computational models of how these neurons control certain behaviors such as chemotaxis (Bargmann 1993; Schaffner 1998, 2000, 2001). In addition, *C. elegans* researchers have identified a number of genes that can affect behavior. For example, the gene *npr-1* seems to be partly responsible for the worm's social behavior, as mutations in this gene can turn the worms from social into solitary foragers and vice versa (de Bono and Bargmann 1998; Schaffner 2001; Weber 2001c). Thus, there is no other multicellular organism where behavior is better understood (*Drosophila* being much more complex in terms of behavior). Indeed, *C. elegans* is the "reductionist's delight" (Cook-Deegan 1994, 53) – the first organism where experimental biologists were able to describe in considerable detail some of the mechanisms controlling behavior at the molecular and cellular levels.

Explanations of behavior in *C. elegans* exhibit a distinct multilevel character. If we take the example of chemotaxis and similar behaviors in *C. elegans*, there are four different levels involved. First, there is the level of the whole animal. The worms act as if they were following a number of innate rules such as "move away from phenol," "move toward normal temperature," and "if starving, move toward conspecifics." Second, one level down from the whole organism is the nervous system – by far the animal's largest organ. Third, in this system, groups of neurons were identified as playing special roles, for example, chemosensory neurons, interneurons (i.e., neurons that connect to other neurons), and motor neurons (neurons that connect to muscle cells). Fourth, there is the level of molecules, especially the various ion channels and receptors that control the voltage across the excitable cell and axon membranes. At this level, we will find the molecules involved in synaptic transmission (e.g., neurotransmitters and their receptors), as well as the molecules involved in the propagation of action potentials (see Section 2.1).

At each of these levels, biologists have identified mechanisms (see Craver and Darden 2001, 118). These mechanisms consist of entities and capacities, some of which are themselves mechanisms residing at the next lower level. For example, the mechanism of action potential propagation is a self-contained mechanism in the sense discussed in Section 2.1, but it is also part of neurological mechanisms at the next higher level – the level of neurons – where it takes part in the exchange of signals between neurons. A group of neurons may constitute a mechanism that is part of the whole nervous system, which, in turn, is embedded in the whole organism. Thus, a nested hierarchy of mechanisms stretches from the whole organism all the way down to the level of molecules (Wimsatt 1974, 1976a, 1976b).

According to the old view of reduction (i.e., that of Nagel), the reducing theory is connected to the reduced theory via "bridge principles." The reducing theory, then, does all the explanatory work, while the bridge principles are only needed to connect the theoretical vocabularies, which would otherwise be logically disjoint (Causey 1972). But this picture does not fit our example well. Here, there is no theory to be reduced, and there is no reducing theory. What we have, instead, are some basic mechanisms of neurotransmission that are shown to be part of more inclusive higher-level mechanisms. The lowest level mechanisms are basically molecular systems and their behavior has been explained by applying laws from physics and chemistry (see Sections 2.1 and 2.2). The higher-level mechanisms have these basic mechanisms as their (mereological) parts, in addition to some extra components (e.g., synapses or neuromuscular junctions). Once it is understood how these parts play out their causal dispositions in order to create the complex behavior of the whole system (e.g., a nematode worm), the behavior of the system has been explained. No reference to any higher-level theories, natural kinds, explanatory concepts, or laws is necessary. At the same time, there is no need to postulate any autonomous role for organic wholes or top-down causation (Sperry 1986; cf. Hoyningen-Huene 1994) or anything of this sort. In this sense, modern experimental biology is thoroughly reductionistic. I have argued that this is the relevant sense of reductionism here.

What are the implications of this new picture of reduction for the old multiple realization argument? I claim that this argument against reductionism is irrelevant. In order to see this, consider the concept of chemotaxis again. Chemotaxis is a functional property that can be realized in different ways. For example, bacteria are also capable of chemotactic behavior. However, the mechanisms controlling chemotaxis in bacteria are clearly different from those of *C. elegans*. For example, chemotaxis in bacteria does not involve neurons, or sensory cells. The chemical sensing and signal transduction is all done by a single bacterial cell. Furthermore, we could imagine chemotactic robots that used a completely different kind of mechanism. However, none of this seems to affect the reductionistic explanation of the behavioral biology of *C. elegans* at all. It does not even matter whether the mechanisms responsible for chemotaxis in *C. elegans* are widely conserved in evolution or whether they are unique to soil nematodes; the fact that chemotactic behavior is multiply realizable does not affect the reductionistic explanation of *this* organism's behavioral biology.

Of course, nobody would want to claim that chemotaxis as a general phenomenon has been reduced to the molecular level on the basis of understanding the mechanism of chemotaxis in just one species. But once it is understood in a few species, that is, once a number of different types of mechanisms that can give rise to this kind of behavior have been understood, we would probably want to call it a success for reductionism. For this, it is simply not necessary to find an equivalence class at some lower level that exactly corresponds to the behavioral equivalence class of chemotaxis (as some antireductionists require).

Why is multiple realization a problem for the reduction of classical genetics, but not for fields such as behavioral biology? I suspect that this has to do with the fact that classical genetic concepts such as dominance, like evolutionary concepts such as fitness (Rosenberg 1978; Sober 1984, 48–50; cf. Weber 1996), enter into generalizations of very broad scope such as Mendel's laws, that is, generalizations that range over a far larger number of cases than any generalization about chemotaxis. If these generalizations range over equivalence classes or natural kinds that are not matched by any lower-level equivalence classes (e.g., molecular natural kinds), then those generalizations remain irreducible. By contrast, if a functional kind does not enter into wide-ranging and significant generalizations, like chemotaxis, then multiple realization is inconsequential for reductionism.

SUMMARY

I have defined a strong sense in which large areas of modern experimental biology are reductionistic. According to this sense, a reductionistic strategy does not proceed by deriving the laws of some higher-level theory from the laws of a more fundamental theory with the help of bridge principles, as the older view of reduction required. Instead, experimental biologists directly explain biological phenomena by applying laws and theories from physics and chemistry to the specific kinds of systems that they study. The biological or interlevel parts of such explanations serve to identify the type of system to which the physical and chemical principles are to be applied. I have thus proposed a thesis of explanatory heteronomy of experimental biology. This thesis does not apply to the whole of biology, just to the biological fields that are under discussion here (with the exception of classical genetics).

This kind of reductionism makes sense in light of the metaphysical thesis that there are no genuinely biological laws of nature (evolutionary contingency thesis). Laws of nature, as traditionally conceived, range over natural kinds of things (events, processes) with immutable causal dispositions (e.g., electrons or chemical compounds). But biology's kinds are not fixed; they are subject to all kinds of variation. While the discovery of such kinds is epistemically significant, explanations that show why something occurs *necessarily* must treat the systems to be explained with physical and chemical principles.

With respect to the question of biological functions and functional analysis, I have argued that experimental biology does not generally employ an etiological conception of functions (that is, a concept according to which functional explanation must explain the presence of the function bearers). A modified concept of role function does the job, if it is supplemented with a coherence condition and an account of the ultimate system capacity of organisms, which is self-reproduction (i.e., the maintenance of the individual's form over time, not procreation). Such an account of functions is fully compatible with the kind of reductionism defended here.

Finally, I have shown in this chapter that the well-known multiple realization argument against biological reductionism is not relevant to the typical reductionistic explanatory strategies deployed in experimental biology. The relation between classical and molecular genetics, where the multiple realization argument has some force, is an odd case.

3

Discovery: Solving Biological Problems

The topic of this chapter is how biologists generate new theories, hypotheses, and explanations. This activity appears to be a form of reasoning, and reasoning is a traditional topic for philosophical inquiry. Some philosophers have referred to the process of generating theories as "discovery." This is a misnomer because theories are not discovered. Even though we hear sometimes that Darwin "discovered" the theory of evolution, or that Einstein "discovered" relativity theory, this notion is highly problematic. The reason is that this way of talking presupposes that these theories somehow *preexisted* the work of Darwin or Einstein. But surely, this is nonsense. Although we could say, perhaps, that the *law* or *mechanism* of natural selection preexisted Darwin's work, or that the equivalence relation of mass and energy expressed by Einstein's famous equation $E = mc^2$ preexisted Einstein's formulation of it, it makes no sense to say that Darwin's or Einstein's *theories* preexisted the scientific work of these men. To claim that they did would amount to some form of Platonism, according to which ideas exist independently of human thought. If we reject Platonism, we cannot meaningfully say that theories are discovered. Theories are indeed *constructed* by the creative activity of the human mind.[1] The kinds of reasoning possibly involved in this constructive, creative activity are what the present chapter is about.

The question of how biologists generate new theories[2] has recently been established as a subject for philosophical inquiry (Darden 1991; Schaffner 1993, Chapter 2). The philosophy of biology has thus followed a more general trend that was initiated with Norwood Russell Hanson's book *Patterns of Discovery* (Hanson 1958) and that, since then, has generated a considerable literature in philosophy and cognitive science.[3] However, until well into the 1970s, the predominant view still was that the generation of new theories, while it may be a subject for historical, psychological, or sociological investigation, is not amenable to philosophical analysis. The historical roots of this

51

attitude are not hard to find. In *The Logic of Scientific Discovery* – a woefully misleading translation of the original German title *Logik der Forschung* ("logic of research") – Karl Popper famously wrote: "The initial stage, the act of conceiving or inventing a theory, seems to me neither to call for logical analysis, nor to be susceptible of it" (Popper 1959, 31). The reason for this was that Popper viewed the proper task for the philosophy of science as "questions of *justification* or *validity*" (ibid.). Popper and several other philosophers thought that for the question of whether a scientific theory is justified or *validated*, it is irrelevant how this theory was generated. Even a theory that is the result of a nonconscious experience, for example, Kekulé's famous dream that gave him the idea for the ring structure of the benzene molecule, can be experimentally tested and shown to be valid. This is undoubtedly correct. However, whether the image of a nocturnal illumination that can only be explained psychologically (if at all) is an adequate model for the way in which scientists conceive their theoretical ideas is open to question. Albert Einstein, during the difficult period of several years during which he developed the general theory of relativity, was hardly fantasizing or dreaming; surely, he was *reasoning*. Why should this form of reasoning not be susceptible to philosophical analysis, even if it is accepted that the genesis of a theory is irrelevant to its validity?

The traditional way of pushing the genesis of theories outside the proper purview of the philosophy of science is the distinction between the *context of discovery* and *context of justification*, usually attributed to the logical empiricist Hans Reichenbach (see Giere 1999, 227–230 for some historical background). In the logical empiricist (as well as Popper's critical rationalist) program, the philosophy of science was to be concerned only with the way in which theories are justified or corroborated (for example, by experimental tests) and not with the way in which they are "discovered" (i.e., generated or constructed, to use the more adequate terminology). However, the so-called two-context distinction was a conceptual muddle from the outset. Paul Hoyningen-Huene has shown that the two-context distinction conflates at least three distinctions that had better be kept separate (Hoyningen-Huene 1987). These are (1) the distinction between generation and justification as *distinct processes*, (2) the distinction between *factual* and *normative*, and (3) the distinction between *logical* and *empirical*. For subsequent discussions (also in Chapters 4 and 5), it is helpful to briefly clarify these distinctions.

(1) We can distinguish between two different reasoning processes, namely the process by which theories are *generated* and the process by which they are *evaluated*. The former process includes the generation of new scientific

concepts and theories, whereas the latter is the process by which the validity of the theories so generated is established. This distinction seems to require that the two processes occur in a certain temporal order, namely discovery before justification. I take it that Kenneth Schaffner (Schaffner 1993, 20) has this distinction in mind when he argues that what he calls "discovery" also needs to be differentiated into a *generative* and an *evaluative* phase. In Schaffner's scheme, discovery involves, first, the generation of *potential solutions* to a given scientific problem. Second, the solutions generated are then subject to *weak evaluation*; that is, scientists then determine whether a given solution qualifies at all, for example, as an acceptable theoretical explanation of some phenomenon. These two processes, according to Schaffner, constitute the "context of discovery." The context of justification, then, involves the *strong evaluation* of proposed problem solutions, for example, by critical testing of theories or hypotheses. This way of talking about different "contexts," it seems to me, is really about different processes that are somehow separated. These processes can then be further differentiated in the way that Schaffner has outlined.

(2) We can distinguish between two *different perspectives* that can be taken on the practice of science, namely a *factual* and a *normative* perspective. The former perspective focuses on how scientists *actually* reason, or how historical actors have actually reasoned. The latter perspective is interested in how scientists *ought* to reason according to some rational principles. Note that this distinction is orthogonal to distinction (1), since it is prima facie meaningful to ask what constitutes a *good* way for scientists to generate new hypotheses or theories, just as we want to know how they ought to subject their theories to critical tests (Langley et al. 1987, 45). However, various philosophers have argued that the generation of new hypotheses is not subject to normative considerations, since discovery is a creative process that is not constrained by rationality (e.g., Hempel 1965, 12).

(3) We can distinguish between logical or a priori and empirical or a posteriori analyses of scientific practice. According to this distinction, there is a fundamental difference between analyzing the logical relations between different scientific statements (e.g., deductive entailment or inductive support) and analyzing the causal relations between events that occurred in the history of science, or that occur in individual minds. The former analysis is based on a priori grounds, namely logical analysis, whereas the latter is based on empirical research (e.g., historical, psychological, or sociological). Furthermore, logical principles are ahistorical, but the causal relations between mental or social events always have a temporal dimension. The classical view was that the study of scientific method must rely on a priori reasoning only, a view

that has also fallen into disregard in the philosophy of science (see Section 5.4).

The traditional context distinction has tended to identify the process of justification according to (1) with the normative perspective according to (2) and the logical analysis according to (3). The process of generation according to (1), as well as the factual perspective according to (2) and the empirical analysis of scientific reasoning according to (3), were all relegated to the context of discovery and claimed to be subject to empirical research, but not to philosophical analysis. However, both critical rationalists and logical empiricists had a rather narrow conception of philosophical analysis, equating it essentially with logical analysis. This view has long been abandoned in the field. The more liberal view prevailing today is that, while philosophical analysis continues to be concerned with rationality, the concept of rationality includes more than just formal logic. This opened the possibility of analyzing the genesis of theories as a process that may be rational in a wider sense. In other words, it may be analyzed as a form of reasoning after all (Nickles 1980a, 41).

The upshot of this discussion is that we need not declare invalid the three distinctions discussed above when we begin to analyze the way in which biologists generate theories. Quite the contrary, to distinguish between generation and justification, between factual and normative, and between logical and empirical remains meaningful and important. What is open to question is not these distinctions themselves, but some of the *substantive claims* that philosophers of science have made with their help, especially the claim that the generation of theories cannot be explained rationally and the claim that philosophy must be restricted to logical analysis. This recognition frees us to seek answers to the following questions.

First, *is* the reasoning involved in generation different from that typically used in justification (evaluation) of theories? (Note that to acknowledge a conceptual distinction between generation and evaluation does not imply that real research processes actually exhibit this dichotomy.) Second, can we exhibit the rationality of the generative reasoning involved in the construction of biological theories? To put it differently, is the generation of theories actually a form of *reasoning*, or is it some mysterious mental process that cannot be made public? Third, what exactly does it mean, in the present context, to reveal the rationality of a mental process?

The third question is somewhat thorny. The objective of considerations of rationality in philosophy of science has traditionally been the attainment of truth. But the attainment of truth, by contrast, is not the objective in the quest for an underlying rationality in the genesis of theories. Generative reasoning should produce plausible or fruitful solutions of scientific problems, not true theories. Thus, it is far from clear what it means to reason rationally in the construction of new theories. I suggest that the minimal assumption we can make concerning generative rationality is that a prospective account should show that novel solutions to scientific problems do not just hit scientists at random. Instead, they are the result of the deployment of procedures that are in some way *efficient* with respect to the goal of producing plausible hypotheses or fruitful problem solutions. If this can be shown, I suggest, then we can say that the rationality of generative reasoning has been exhibited.

Without further ado, I now begin to critically examine three interesting attempts to reconstruct the reasoning that governed the genesis of three historically important biological theories. The first attempt is Kenneth Schaffner's account of the genesis of François Jacob and Jacques Monod's 1959 repressor model of enzyme induction in bacterial genetics (Section 3.1). The second is Lindley Darden's reconstruction of the generative steps leading to the theory of classical genetics, as formulated by Thomas Hunt Morgan and his associates in 1926 (Section 3.2). The third attempt concerns the discovery of the urea cycle by the biochemist Hans Krebs in 1932 (Section 3.3), which has been studied in great detail by several historians. In the concluding section (Section 3.4), I raise some critical points in regard to the claim that the reasoning underlying the generation of these biological theories can be explained rationally.

3.1   MOLECULAR GENETICS: DEDUCED FROM EXPERIMENTS?

Some years ago, Kenneth Schaffner analyzed a case from the history of molecular genetics and came to the conclusion that "the reasoning displayed in the genesis of a theory is in a large measure identical to that utilized in evaluating a theory" (Schaffner 1974a, 384). Thus, according to Schaffner, there is a "unitary logic" that covers the generation as well as the evaluation of scientific theories. Even though Schaffner has since revised this claim in light of his more recent work on discovery in the biomedical sciences (see Schaffner 1993, Chapter 2), I think that a critical examination of this thesis will prove to be illuminating for the present discussion. Specifically, this examination will allow us to see – against Schaffner's initial intentions – that the reasoning strategies underlying the generation and evaluation of theories are, in fact, quite different.

The theory in question is the so-called repressor model, which was part of the operon theory developed in large part by François Jacob and Jacques Monod in the late 1950s and early 1960s. This model provided a first glimpse at the molecular mechanisms that regulate gene expression and is now viewed as one of the landmark discoveries in the history of molecular biology. The phenomenon that gave rise to this discovery is called "enzyme induction" (initially called "enzyme adaptation," but this term was dropped for its teleological connotations) and was first observed around 1900. The basic observation is that microorganisms such as yeasts and bacteria only display an ability to ferment specific carbohydrates if they have been kept in a medium that contains these compounds. Alternatively, microorganisms can develop the ability to ferment specific carbohydrates within a few hours after the compounds are added to the medium.

In the 1950s, Jacob and Monod studied the genes that allow *E. coli* cells to metabolize the carbohydrate lactose. They found two genes that are involved in this process, termed z and y. The latter gene was shown to be required for the uptake of lactose into the cell; it makes an enzyme called galactoside permease. The z gene, by contrast, was shown to directly code for the enzyme β-galactosidase, which breaks down lactose into a glucose and a galactose molecule. The two genes are activated simultaneously either in the presence of lactose itself or in the presence of certain synthetic galactosides (derivatives of galactose). Compounds that stimulated the production of β-galactosidase and galactoside permease were termed "inducers."

What Jacob and Monod showed was that the induction of enzymes for lactose utilization in *E. coli* works as follows: When no lactose and no inducers are present, the expression of the z and y genes is *repressed* by the product of a third gene termed i (for "inducible"). Thus, the i gene makes a *repressor*.[4] The repressor binds to a specific site on the DNA termed o (for "operator"), which is located in the close vicinity of z and y. As long as the repressor is bound to o, the z and y genes are inactivated. In other words, no enzyme is made. When lactose or an inducer enters the bacterial cell, the repressor is displaced from o and the genes begin to make enzyme. Later, it was shown that what the repressor blocks is transcription, that is, the copying of DNA into messenger RNA (mRNA), but the role of RNA in protein synthesis was not fully understood when Jacob and Monod first presented the repressor model.[5]

An important element in the development of the repressor model was the so-called PaJaMo experiment, named after the three investigators who were involved (A. B. Pardee, F. Jacob, and J. Monod). This experiment made use of the primitive form of sexual reproduction that bacteria are capable of, namely bacterial conjugation. In this process, a "male" donor and a "female" recipient cell pair up and a bridge is formed between their cytoplasms. Then the "male" injects a DNA molecule termed the "F-factor" into the "female," which thus turns into a "male" itself. Now, there exist bacterial strains termed *Hfr* where the F-factor is integrated into the cell's main chromosome. If such a strain mates, it transfers its entire chromosome into the recipient cell. This process takes about 100 minutes, and the time it takes for a gene to appear in the recipient cell was used for gene mapping. But in addition, mating experiments using *Hfr* strains allowed the geneticists to perform genetic analysis that strongly resembles Mendelian crossing experiments. The reason is that, for a short time, a bacterial cell is created that contains two copies of some genes, much like the diploid cells of higher organisms. With this trick, it can be determined, for example, whether certain mutations are dominant or recessive.

The PaJaMo experiment, which was first done in December 1957 at the Pasteur Institute, involved two kinds of *E. coli* mutants. One class of mutants, known as z^-, had a defective gene for β-galactosidase; it was unable to synthesize an active enzyme. The other class of mutants was called i^- and showed an interesting phenotype: It synthesized β-galactosidase as well as galactoside permease at the full rate even when neither lactose nor inducers were present. This class of mutants was also termed "constitutive." In the PaJaMo experiment, a *Hfr* strain that contained the wild-type genes z^+ and i^+ was mated with a "female" that was both z^- and i^-. Thus, the "female" was unable to make functional β-galactosidase, and it was constitutive for galactoside permease. When these cells conjugated with the *Hfr* z^+i^+ "males," they received normal copies of the genes. And indeed, only a few minutes after these genes entered the cell, the recipient cells started to synthesize β-galactosidase, even in the absence of inducer. However, this synthesis stopped after about 2 hours. If inducer was added, by contrast, enzyme synthesis continued. Thus, the recipient cells were turned from constitutive into inducible cells by the transfer of the i^+ genes. If the experiment was done the other way around, that is, if male z^-i^- cells were mated with female z^+i^+ cells, no enzyme synthesis was observed even after several hours.

The PaJaMo results were at variance with the then prevailing theory concerning the mechanism of enzyme induction, the so-called generalized induction theory. According to this theory, enzyme induction is a universal ("generalized") mechanism that is also involved in the synthesis of "constitutive" enzymes (i.e., enzymes that are made at all times, no matter whether an inducer is present). On the generalized induction theory, all enzyme synthesis requires an inducer. An especially popular idea at that time was that enzymes could only assume their correct three-dimensional structures in the presence of their substrates.[6] On this hypothesis, an enzyme would be constitutive if there was an endogenous substrate present which could assist in the folding of the protein, whereas inducible enzymes required that such a substrate be provided externally (e.g., in the case of *lac* genes). The generalized induction theory predicts an outcome of the PaJaMo experiment different from the one that was actually observed. On this theory, it was expected that, in the presence of both i^+ and i^- alleles in a cell, the cell would be constitutive rather than inducible. This is because, on the generalized induction theory, i^- must have a stimulating effect on enzyme production. This effect should prevail even in the presence of an i^+ copy of the gene, thus making i^+i^- heterozygotes constitutive.

By contrast to the generalized induction theory, the repressor model can readily explain the observed outcome of the PaJaMo experiment. The recipient z^-i^- cells lack the repressor; therefore, the z gene starts to make enzyme as soon as it enters the recipient cell. However, once the i^+ gene is transferred and expressed as well, its gene product – the repressor – is made and blocks enzyme synthesis from the z gene. Synthesis resumes when an inducer is added, which inactivates the repressor. The asymmetry of the PaJaMo experiment (i.e., the different results obtained by mating a z^+i^+ donor with a z^-i^- recipient and a z^-i^- donor with a z^+i^+ recipient) is explained by the fact that, in bacterial conjugation, only DNA is transferred, no cytoplasmic components. This showed that the repressor is cytoplasmic. When the recipient cell is i^+, the repressor is thus already in place in the cytoplasm; therefore, the cell cannot make enzyme in the absence of inducer.

The establishment of the repressor model and the associated developments in molecular biology (especially the roles of RNA in protein synthesis) were complex historical episodes[7] that I have simplified considerably. I focus specifically on those aspects that are relevant to an appraisal of Schaffner's main thesis. How did Schaffner purport to show the identity of the reasoning involved in generation and justification of the repressor model?

The center of Schaffner's (1974a, 370–371) case is that something he called the *repressor hypothesis* follows *deductively* from the results of the PaJaMo experiment. "Deduction" is meant in the strict sense of logical consequence. Schaffner's claim is not that a hypothesis was deduced from raw experimental data, formulated in a theory-neutral observation language. The deduction of the repressor model, according to Schaffner, required that the experimental results be cast in a specific theoretical terminology and augmented by additional theoretical premises, which were part of the background knowledge of molecular genetics. The specific terminology that had to be used to make this deduction work is a part of the language of classical genetics. Using this language, the PaJaMo results can be put like this: The inducible phenotype is *dominant* over the constitutive phenotype.[8] In other words, when both a wild type and a defective copy of the *i* gene are present in a cell, this cell exhibits the normal, inducible phenotype. The main theoretical premise that has to be added to deduce the repressor hypothesis from the PaJaMo result was that dominant alleles of a gene produce a substance that the recessive allele lacks. Schaffner's deductive argument, then, can be put like this[9]:

Premise 1: For any pair of alleles $\{A, a\}$: If A is dominant over a, then A produces a substance not produced by a

Premise 2: Allele i^+ is dominant over i^-

Conclusion 1: Allele i^+ produces a substance not produced by i^-

Definition: Let a *repressor for X* be a substance that blocks expression of gene X

Premise 3: Allele i^+ produces a substance that blocks expression of z and y

Conclusion 2: Allele i^+ produces a repressor for z and y

Conclusion 2 is what Schaffner calls the "repressor hypothesis." It is to be differentiated from the repressor *model* (see below), because it makes no assumptions about *how* the repressor acts, for example, whether it acts directly or indirectly on the regulated genes. Premise 1, according to Schaffner, is "a generally accepted background hypothesis of this time" (1974a, 370). Premise 2 is the main result of PaJaMo cast in the language of classical genetics. The definition just introduces a new term; it has no empirical content. Premise 3 appears to be a theoretical interpretation of the PaJaMo experiment that relies on premise 2, but adds some content by identifying the substance produced by i^+ according to premise 2 with whatever blocks expression of z and y in the PaJaMo experiment.

On closer inspection, it is clear that even more theoretical interpretation of the experiment is required to make Schaffner's deductive argument go through. For example, Pardee, Jacob, and Monod did not directly observe that something blocks expression of z. What they measured was β-galactosidase activity by an enzymatic assay. Thus, low β-galactosidase activity could, in theory, be due to inactivation of the enzyme rather than blocking of the synthesis of the enzyme. Therefore, Schaffner's argument needs to be augmented at least with the additional premise that the β-galactosidase activity measured was strictly correlated with new protein synthesis from the z gene (which, in fact, had been shown by Monod, Pappenheimer, and Cohen-Bazire in 1952). Furthermore, the identification step contained in premise 3 does not directly follow from the PaJaMo experiment either; perhaps it does follow with the help of additional premises. Schaffner (1974a, 371) does acknowledge that additional premises might be needed to render the argument in strict deductive form. This need not be a problem for his account, since deductive reasoning might frequently suppress premises that could be made explicit if necessary. Right now, what I would like to examine is the relevance of all this to the generation of new theories.

An obvious objection arises if the historical evidence is considered. Namely, it seems that several scientists had ideas very similar to the repressor model *before* they knew about the PaJaMo experiment. One of them was the physicist Leo Szilard, who happened to be passing through Paris in December 1957, just when Pardee, Jacob, and Monod were pondering the implications of the PaJaMo experiment. Monod recalls in his Nobel Lecture:

> I had always hoped that the regulation of "constitutive" and inducible systems would be explained one day by a similar mechanism. Why not suppose, then, since the existence of repressible systems and their extreme generality were now proven, that induction could be effected by an anti-repressor rather than by repression by an anti-inducer? This is precisely the thesis that Leo Szilard, while passing through Paris, happened to propose to us during a seminar. We had only recently obtained the first results of the injection [PaJaMo, M. W.] experiment, and we were still not sure about its interpretation. I saw that our preliminary observations confirmed Szilard's penetrating intuition, and when he had finished his presentation, my doubts about the "theory of double bluff" had been removed and my faith established – once again a long time before I would achieve certainty (Monod 1966, 479).

The "theory of double bluff" was an old insider joke between Monod and his colleague Melvin Cohn that expressed the idea that induction could be due to antirepression. Thus, the idea was not new when the PaJaMo experiment was done; it was just not taken very seriously initially because of the biochemical charms of generalized induction theory. Furthermore, as Monod also hints in the above quotation, repression was a phenomenon that had been demonstrated previously in other enzyme systems, where the end product of a biosynthetic pathway repressed the enzymes required to make this product. It seems that repression was already on the minds of several scientists including Pardee,[10] Jacob, and Monod when the PaJaMo experiment was done (see Judson 1979, 409–410; Jacob 1988, 293).

Schaffner, of course, was aware of these historical circumstances. What, then, are the respective roles of Szilard's and other people's previous speculations such as the "double bluff" theory and the alleged deductive reasoning to the repressor hypothesis? Schaffner's answer to this question is subtle (1974a, 367–369). He suggested that what Jacob and Monod[11] took from Szilard was a "formal derepression hypothesis," stating that any induction effect can be due to the inactivation of a repressor. This hypothesis was somehow contained in the specific model that Szilard presented in Paris (which was considerably different from the repressor model that Jacob and Monod would later present; see Szilard (1960)). Jacob and Monod were able to "abstract the formal [derepression, M. W.] hypothesis from the Szilard model and to formulate another model in which the double bluff hypothesis is embodied involving the constituents of the *lac* region of *E. coli*" (Schaffner 1974a, 367). Thus, on Schaffner's view, the repressor hypothesis preexisted the PaJaMo experiment only as an abstract idea that needed to be instantiated by Jacob and Monod to produce the repressor model.[12] That this idea might have guided the interpretation of the PaJaMo experiment is not in conflict with Schaffner's thesis that the repressor model followed from the complete results of the experiment plus some background assumptions by a deductive argument.[13] After all, as every mathematician knows, the production of a deductive argument often requires ingenuity and might be guided by all sorts of heuristic considerations, for example, analogies.

Another objection, which was also addressed by Schaffner, is that his argument that purports to deduce the repressor model from the PaJaMo experiment belongs to the context of *justification* rather than the generation of theories. Schaffner's answer to this objection was the following. He distinguished between the repressor *model* and the repressor *hypothesis*. The former claims that the repressor acts directly on the genes, while the latter only states that there is a repressor. According to Schaffner, the deduction generated the repressor *hypothesis*, which then served as a basis for further "controlled speculation." For example, the repressor hypothesis was used to propose an alternative model, the so-called inducer model (Pardee, Jacob, and Monod 1959, 175). According to this model, the synthesis of β-galactosidase requires the presence of an endogenous inducer in both inducible and constitutive cells. However, in inducible cells, this endogenous inducer is destroyed by an enzyme produced by the i^+ gene. In i^-, presumably, this enzyme is defective, resulting in constitutive expression of z. This enzyme plays the role of the repressor. Thus, the inducer model can be seen as a modification of the repressor model that was also consistent with the PaJaMo experiment. Note that the inducer model differs from the generalized induction theory, because – as in the repressor model – the effect of the product of the i gene is to *block* expression of β-galactosidase. The difference is that, in the repressor model, the repressor acts directly on the z and y genes, whereas it acts via destruction of an endogenous inducer according to the inducer model.

Schaffner's point now is that the generation of an alternative (the inducer model) with the help of the repressor hypothesis shows that the deduction of the latter must be situated in the context of generation rather than the context of justification. The context of justification only became relevant once Jacob and Monod had to *choose* between the inducer and repressor models on the basis of additional evidence.

In my view, the crucial question on which the validity of Schaffner's conclusion turns is the following: Granted that the repressor model follows deductively from the complete results of the PaJaMo experiment plus some background assumptions, does this deductive argument reflect how Jacob and Monod *actually* generated this hypothesis? Monod's own recollections quoted above suggest that this is not the case. Rather, what Monod implies is that the repressor hypothesis was generated by an *argument from analogy*. This argument starts from the assumption that all enzyme systems share a common expression mechanism. A second assumption in this argument is that repression (i.e., inhibition of gene expression) exists in *other* enyzme systems, such as the enzymes required for tryptophan or arginine synthesis. From this assumption, it can be suggested by analogy that the expression of β-galactosidase is also subject to repression rather than generalized induction. Since generalized induction was in conflict with the PaJaMo experiment, whereas repression was not, Monod and his colleagues started to take repression seriously.

Schaffner recognized the role of this analogy and suggested that there were "two routes to the repressor hypothesis for the *lac* system" (1974a, 368): one analogical and one deductive. Schaffner's account is not quite clear as to which kind of reasoning *actually* generated the repressor hypothesis. He suggested that "once the final results of the Pajama experiment were available, [...] it became possible to reason from these results [deductively, M. W.] to a repressor hypothesis." However, granted that it is *possible* to reason deductively from the PaJaMo experiment to a repressor hypothesis, what is the historical evidence that Monod and colleagues *actually* reasoned in this way? Their publications and the correspondence cited by Judson do not contain the formal argument presented above. Of course, it is *possible* that Jacob and Monod carried out such a logical deduction. But to claim that they *actually* did is in need of historical justification. I find it more plausible to assume that the repressor model was actually generated by analogy with the known cases of repression. This is sufficient to explain the genesis of the repressor model; the deductive route is not needed.

However, the deductive argument might be viewed as a reconstruction of why the PaJaMo experiment *supports* the repressor model.[14] Although Monod indicates (see quotation above) that certainty that this mechanism is correct was reached only later, it is clear that the PaJaMo experiment was *evidence* for some kind of repressor mechanism. Thus the deductive argument belongs to the context of justification after all. As we have seen, Schaffner considered this possibility, but argued that the deductive argument was part of the genesis of theories because it was used to generate further hypotheses. But this is not a good argument. As Schaffner's own analysis amply demonstrates, scientists make liberal use of theoretical background beliefs when they generate new hypotheses. Thus, the repressor hypothesis was used by Jacob and Monod to generate an alternative to the repressor model, namely the inducer model. But for this it is irrelevant how the repressor hypothesis itself was generated. Therefore, Schaffner has not established that the deductive argument played a role in the generation of the repressor hypothesis.

To conclude, Schaffner's provocative thesis that the reasoning involved in the generation and evaluation of scientific theories in experimental biology is identical must be rejected.[15] What a critical evaluation of his account shows is quite the contrary, namely that the reasoning employed in generation and justification is *not* the same. If it is true that the generation of the repressor model was, by and large, a case of analogical reasoning, we have a clear-cut case of a reasoning strategy that is *not permissible in the context of justification*. Clearly, a scientific theory cannot be justified by analogy to a similar theory that successfully explains a different phenomenon. We will encounter an excellent example of this in Chapter 4 (Slater's chemical hypothesis, Section 4.1). Although analogical reasoning is not permissible in justification, it is perfectly fine in the generation of new theories. Thus, if there is a logic underlying the generation of scientific theories, it is different from the logic that governs the evaluation of theories.

In the following section, I discuss a more detailed attempt to explain how new theories are generated.

3.2 MENDELIAN GENETICS: RESOLVING ANOMALIES

Rome was not built in a day, nor are theories. Scientific theories that were conceived in basically finished form by a single stroke of genius are probably rare. Most theories are more like Rome; they were built piece by piece, and sometimes by tearing down edifices that were not needed any more. Theories can be continuously revised as scientists learn more about them as well as the phenomena that they cover. The theory of Mendelian genetics is such a case. In this section, I critically examine Lindley Darden's attempt to reconstruct the reasoning underlying the genesis of the classical theory of genetics (Darden 1991).

In order to fully understand this historical development, it is necessary to recall Mendel's two laws. These laws state the expected outcomes of crossing different varieties of an organism, typically ones that exhibit alternative states of one or several characters. If two organisms showing characters A and a, respectively, are hybridized, the first generation (F_1) of offspring will exhibit only one of the two characters. If no a forms appear in F_1, then a is called *recessive*, and A is called the *dominant* character. If, now, the individuals of F_1 are crossed again, the next generation (F_2) will show a character distribution of $A:a = 3:1$. Since AA and Aa exhibit the A phenotype, while only aa shows the a phenotype (A is dominant over a), A and a occur in a numerical ratio of 3:1. This regularity is known as the *law of segregation* or, sometimes, "Mendel's first law."

Mendel's other law applies to cases where two characters A and B are considered. If A is dominant over a and, furthermore, B is dominant over b, the character types are distributed in F_2 in ratios of $AB:Ab:aB:ab = 9:3:3:1$. This is the *law of independent assortment*, or "Mendel's second law." Note that, assuming that the characters A and B are inherited independent of each other, the law of independent assortment is a consequence of the law of segregation applied to cases where two different characters are involved. In fact, Darden argues that, initially, the two laws were not considered to be conceptually distinct. The focus after the rediscovery of Mendel's work was on segregation; independent assortment was viewed as a simple extension of segregation to two-character cases or "dihybrid crosses" (Darden 1991, 60). However, this view had to be abandoned when exceptions to independent assortment began to appear.

The first exceptions to independent assortment were found by W. Bateson, E. R. Saunders, and R. C. Punnett in 1905. Working with sweet peas (a genus different from Mendel's garden peas), they were first able to confirm the law of segregation (i.e., 3:1 ratios in F_2) for several characters. However, when they performed two-character crosses they encountered deviations from the 9:3:3:1 ratios. The characters they examined were purple versus red flowers and long versus round pollen. Purple was dominant over red, and long was dominant over round. Crosses of purple/long with red/round plants produced, in F_2, 1,528 purple/long, 106 purple/round, 117 red/long, and 381 red/round (Punnett 1928, Vol. II, 140–141). Compared to Mendelian expectations, there was a lack of purple/rounds, as well as of red/longs, whereas there was an excess of red/rounds. Thus, it seemed that purple and long showed a tendency to be inherited together, as did red and round. Bateson and his associates described these anomalous results as "partial coupling" between the purple and long, as well as between the red and round characters. The first theoretical explanation they offered was that there is some kind of physical interaction between the corresponding factors in the germ cells. If two characters tended to be inherited together, they postulated some attractive force ("coupling") between the factors during gametogenesis. If two factors appeared less frequently in combination than was expected from Mendel's second law, Bateson and co-workers assumed that there was a repulsive force between the factors ("repulsion"). Later, Bateson and Punnett abandoned this hypothesis and replaced it by the "reduplication hypothesis," which stated that deviations from independent assortment are due to differential duplication of germ cells (Darden 1991, 124).

In 1911, T. H. Morgan proposed an alternative explanation for deviant ratios observed in certain dihybrid crosses. Morgan's laboratory had also found interesting deviations from Mendel's laws. The first one they found involved *white*, one of the first *Drosophila* mutants to be identified. This mutant had white eyes, instead of the normal red eye color of fruit flies. Morgan found that the *white* character did segregate in Mendelian 3:1 ratios; however, all the white-eyed individuals in the F_2 generation were males (Morgan 1910). This result was anomalous because, in all previous cases of Mendelian inheritance, the F_2 generation contained an even mixture of males and females showing the recessive character. Morgan and his students thus proposed that the *white* gene was "sex-linked." Furthermore, they found other sex-linked characters in *Drosophila*. These findings played an important role in the eventual acceptance of Boveri's and Sutton's chromosome hypothesis by Morgan (Darden 1991, 135). This hypothesis stated that the Mendelian factors are located on the chromosomes – threadlike structures in the cell nucleus that are visible under a light microscope. Since chromosomes occurred in pairs and were observed to segregate during the formation of gametes, the chromosome hypothesis furnished a simple cytological explanation for Mendelian segregation. Initially, Morgan was skeptical of the chromosome hypothesis, because he thought it implied that there could be only as many Mendelian factors as there are chromosomes (*Drosophila* has only four chromosomes). However, the finding of several sex-linked characters suggested that the corresponding factors are all associated with the sex-determining factor of *Drosophila*. Since that sex-determining factor had been identified as the X-chromosome on the basis of cytological studies, this implied that the factors corresponding to the sex-linked characters are all associated with the X-chromosome.

In a brief communication in *Science*, Morgan then suggested "a comparatively simple explanation" in lieu of "attractions, repulsions and orders of precedence, and the elaborate system of coupling" (Morgan 1911, 384). Morgan's explanation of partial coupling contained four main elements. (1) He accepted Boveri and Sutton's claim that Mendelian factors are located on chromosomes. (2) Morgan added to this the idea that the factors are arranged in a *linear order* on the chromosomes.[16] Elements (1) and (2) explained why genes come in linkage groups, that is, groups of factors that tend to be inherited together in Mendelian crosses. If nothing more is said, we would expect genes that are located on the same chromosome to show *complete* linkage. However, linkage (coupling) was always partial in the cases known in 1911. Therefore, (3) Morgan suggested the occasional occurrence of crossing-over. (4) Morgan reformulated Mendel's second law, to the effect that independent assortment obtains only between factors in different linkage groups. (The notion of "linkage group," to my knowledge, is original with Morgan.)

Morgan's reference to Jannsens is important. In 1909, F. A. Jannsens had published his cytological findings showing that homologous chromosomes intertwine visibly during the "strepsinema" stage of meiosis (better known as "metaphase" today). Writing in French, Jannsens referred to this phenomenon as *chiasmatypie*. Morgan's insight was that this intertwining of chromosomes makes it possible for chromosomes to break and rejoin with their homologous partners, thus explaining why the linkage of factors located on the same chromosome was not complete. Thus, Morgan concluded that Bateson's phenomenon of coupling in dihybrid crosses was "a simple mechanical result of the location of the materials in the chromosomes, and of the method of union of homologous chromosomes" (ibid.).

Morgan's suggestion represents one of the single most important theoretical ideas in the history of genetics. It proved to be extremely fruitful and laid the theoretical foundations for several decades of genetic research on *Drosophila* and other organisms. The reason is that Morgan's idea is the basis of a genetic mapping technique that came to be known as *linkage mapping*. This technique could even be adapted to microorganisms and played an important role in the molecularization of genetics (Weber 1998b). Even today, in the age of genomic sequencing, geneticists still use the method of linkage mapping to identify genes (see Section 6.1). Linkage mapping is based on Morgan's postulate that crossing-over is more likely to occur the greater the physical separation of two genes on the chromosomes. On the basis of this postulate, the distance between two factors on the chromosome can be determined by counting the frequency of crossing-over between these two factors. This principle was used by Sturtevant (1913) to produce the first genetic map ever, where he localized six different genes on the *Drosophila* X-chromosome. This map was constructed on the assumption that the frequency of crossing-over between any two genes reflected the distance between them.

I now turn to the question of how Morgan's theory of linkage and crossing-over could have been generated. The first interesting point to note is that this theory was introduced in response to an *anomaly*, that is, an experimental result that did not meet the scientists' expectations. The findings from the experiments with sweet peas by Bateson and his associates, as well as the first results from crossing *Drosophila* mutants, disagreed with the ratios that were expected on the basis of Mendel's laws. However, this disagreement only arose if more than one character was considered; each individual pair of characters exhibited the expected 3:1 ratio. By contrast, several characters did not obey the 9:3:3:1 ratios that Mendel's laws predicted for two-character crosses. Morgan's theory of linkage explained these deviations. What is more, this theory led to a reformulation of Mendel's second law, the law of independent assortment. Although geneticists initially assumed that this law was a simple extension of the law of segregation (and, therefore, not a distinct law), Morgan introduced a distinction between segregation and independent assortment. The latter was shown to be universally valid only for certain factors, namely factors that are located on different chromosomes.

Thus, Morgan successfully resolved an anomaly to genetic theory by introducing new theoretical elements. How did he accomplish this? Darden's attempt to explain the reasoning possibly employed by Morgan involves various so-called *strategies*, which I discuss now.

A "strategy," according to Darden (1991, 20), is any procedure by which scientists generate new ideas, design experiments, evaluate theories, diagnose problems, and so forth. A strategy differs from an *algorithm* in that algorithms, in the strict sense of the term, are procedures that guarantee correct results. Clearly, procedures for generating new scientific theories need not produce any true theories, only a range of plausible candidates. Strategies for producing new ideas are more like *heuristics*, a term introduced by William Whewell. A heuristic does not guarantee correct results. Darden is also interested in artificial intelligence (AI), where problem-solving heuristics have been programmed on computers and thus acquired some of the properties of algorithms. However, I want to focus on Darden's nonalgorithmic characterization of her strategies. To what extent do they illuminate the generation of Morgan's theory of linkage?

Darden (1991, 136–138, 161–167) has introduced the following strategies for the case of linkage and crossing over: (1) *delineate and alter*, (2) *specialize and add*, (3) *propose the opposite*, (4) *use interrelations to another body of knowledge*, (5) *move to a new level of organization*, (6) *use an analog model*. I briefly discuss them in turn.

(1) *Delineate and alter*: This strategy was involved when Morgan differentiated between the laws of segregation and independent assortment. The two laws were delineated into two separate claims, the first of which obtained in all cases, whereas the second only obtained for dihybrid crosses of factors located in different linkage groups.

(2) *Specialize and add*: After it had been delineated, the law of independent assortment was specialized to a certain class of cases.

(3) *Propose the opposite*: Mendelians had initially assumed that, in hybrids, gametes of the two types involved were produced in equal numbers. Like Bateson before, Morgan reasoned that this could not be the case for the anomalous cases. When two factors are linked, they distribute unevenly in the gametes. Thus, the opposite of equal numbers of gametes was proposed, namely unequal numbers.

(4) *Use interrelations to another body of knowledge*: Morgan drew an important connection between the cytological work of Sutton and Boveri and also of Jannsens and the coupling anomalies. Looking at Jannsens's visualizations of intertwined chromosomes could have given him the idea that chromosomes occasionally break and rejoin, which was the crucial insight for reconciling the Sutton–Boveri chromosome hypothesis with the observed partial coupling of factors. Thus, Morgan used an interrelation between cytology and Mendelian experiments to generate his idea.

(5) *Move to a new level of organization*: This strategy was used by Morgan when he postulated the linkage group as a group of genes that have a tendency to be inherited together. The linkage group constituted a new level of genetic organization above the individual gene and below the set of all genes of an organism.

(6) *Use an analog model*: Morgan and his associates devised a visual model of the genes on the chromosomes, which resembled a series of beads (=genes) on a thread (=chromosome). They used this analog model to devise hypotheses about interference effects between different crossing-over events. Even though the beads-on-a-thread model of the chromosome was later severely criticized and eventually abandoned, it seems to have played an important heuristic role initially.

In addition to these strategies, Darden (1991, 201–204) proposed a general procedure for diagnosing problems and fixing faults in a theory. Basically, her suggestion is that scientists diagnose problems and fix faulty theories in a way similar to that in which a repairperson diagnoses and fixes a faulty electronic device. Repairpersons can use diagrams representing the circuits of the device to locate possible defects according to the *nature of* the malfunction. In addition, they can make use of the modular structure of such circuits to localize the defect to a specific module. Once they have located the defect, they can focus on this module or replace it if necessary. Darden suggests that there is a strong analogy between the diagrams that are used to diagnose malfunctions in an engineering device and scientific theories. Biologists also use diagrammatic representations of the processes they study, for example, diagrams tracing the transmission of genetic factors across generations (i.e., the typical diagrams representing crossing experiments in genetics textbooks). Darden then likens the occurrence of anomalous results in experimental research to the malfunction of an engineering device. Such occurrences will induce scientists to ponder diagrammatic representations of the processes under study. Like repairpersons, they will try to locate the problem in specific modules of the diagrammatic structures that represent the theory in question. Darden suggests that geneticists were able to localize the source of certain anomalies within genetic theory in this way, for example, when exceptions to Mendel's laws were found. By drawing up the steps in the transmission of genetic factors from parents to offspring, they formed hypotheses as to where the source of the anomaly might lie, for example, in Bateson's case, in the assumption that an even number of germ cells of each kind are formed. Of course, localizing the problem is only the first step in fixing a faulty theory (or engineering device). The next step is to replace the malfunctioning parts. In Bateson's case, the faulty assumption of the formation of equal numbers of gametes of different types was replaced by the "reduplication hypothesis" mentioned above. Morgan replaced this faulty assumption by his theory of linkage and crossing over. The latter theory was then shown to be superior to the reduplication hypothesis.

In sum, Darden analyzes the steps that lead to new theories as an ongoing interplay between preexisting theories and their exemplars, emerging anomalies, diagnosis of the source of these anomalies, and resolving them by adjusting the theories. Once an anomaly has been localized, adjustments are made with the help of a set of heuristics or strategies. Now, I turn to a critical assessment of Darden's account of the generation of new theories in classical genetics.

While Darden has provided a rich and illuminating account of the development of genetic theory in response to a series of anomalies, the entire approach of using strategies in order to explain how theories are constructed and revised raises a number of philosophical and historiographic questions.

The most obvious question concerns the ability of such strategies to illuminate scientific reasoning in discovery. For it is not at all guaranteed that Bateson, Morgan, and the other geneticists featuring in Darden's historical account *actually* generated their theories and problem solutions with the help of Darden's strategies.[17] The claim that they did would be a historical claim, and therefore, in need of historical justification. Darden does not provide such historical justification. However, she argues that no such justification is needed, given the aims that she is following when developing strategies for theory change. Namely, her account is intended as "(metascientific) hypotheses about strategies that could have produced the historical changes that did occur" (1991, 15). The account she gives is not intended to be "descriptive of strategies that I claim a given scientist *consciously* followed." Darden's goal is not "to find descriptively adequate strategies," even though she admits that "those are good to have when possible." However, in the case of classical genetics, descriptive adequacy is beyond reach because the historical record available is insufficient for this purpose. For example, there are no surviving laboratory notebooks from the early days of Morgan's laboratory at Columbia. Morgan cleaned out his files every five years (Darden 1998, 62). To make things worse, the main protagonists are all dead, and only a few survive who knew them personally. Thus, Darden's account is based entirely on the published record. But Darden argues that this is no handicap, since her goal is "not to study the private mental life of geneticists, but to devise strategies that characterize the conceptual relations between publicly debated hypotheses at different stages" (1991, 35).[18]

Thus, although Darden's reconstruction of the theoretical ideas developed by geneticists between 1900 and 1926 is intended to be historically adequate, she does not claim such historical adequacy for her strategies. Her claims that a certain development in genetics *exemplifies* one of her strategies is not intended as a historical claim. Rather, she wants to remove some of the mystery surrounding creative work in science by showing what kinds of strategies *could* have produced the changes in genetic theory found in the published historical record.

Thus, the criteria for adequacy for Darden's strategies for theory change are different from those that historians would apply to their reconstructions of historical episodes. This raises the question of what the criteria of adequacy are for an account such as Darden's. Clearly, we cannot ask for a procedure that would be *necessary* for producing a certain change in a theory. There is no reason to believe that any of the generative steps of the theory of classical genetics could not have been produced by a different procedure. Therefore,

Darden only requires, as a criterion of adequacy, that her strategies be *sufficient* for their products, in other words, that they can generate, without additional help, the steps actually observed on the basis of the information available at some given time. A second criterion of adequacy is that the strategies be *general*; in other words, they should be able to generate other plausible hypotheses by using different inputs. In other words, a strategy must not be tailored to produce one desired result only (Darden 1998, 64). Sufficiency can in some cases be demonstrated with the help of AI implementations (computer simulations), as Lindley Darden and Roy Rada have done for a special case of the strategy *use interrelations to another body of knowledge* mentioned above (Darden and Rada 1988a, 1988b). This work showed that a computer program based on a method for knowledge representation called "frames"[19] and a set of suitable transformation rules and constraints was able to generate Sutton's chromosome hypothesis (the claim that genes are parts of chromosomes). Furthermore, Darden has collaborated with AI researchers to implement strategies for diagnosing faults and fixing theories, using a more sophisticated way of representing knowledge than the frame-based approach (Darden 1998).

Assuming that Darden's strategies can be shown to be sufficient for generating the steps that led to Morgan's theory of the gene of 1926, what this approach shows is how this theory could have *possibly* been generated. While this may engender confidence that the creative work of constructing new scientific theories is in principle explainable, the extent to which this approach illuminates actual scientific reasoning is open to question. First, it is not obvious that the historical actors employed abstract procedures that were sufficient in Darden's sense. After all, there could be such a thing as a lucky guess. Furthermore, this approach fails to address the question of whether there are any *general* procedures (rules, strategies, heuristics) involved when real scientists generate new theories. Darden's approach only shows that historic scientific theories can *in principle* be generated by general procedures and suitable inputs; it cannot show that real scientists actually *used* general procedures. But the extent to which the reasoning employed by real scientists exemplifies *general* procedures seems to be the most salient question in the ongoing debate over the "logic of discovery." I come back to this issue in the final section (Section 3.4).

A second issue that needs to be addressed is the role of hindsight in Darden's approach. She developed most of her strategies by comparing the theories or theoretical elements in question with their historical predecessors.[20] Then she formulated various hypothetical strategies that could generate the change from one theory or theoretical element to the next in the historical succession. Thus, when developing her strategies, Darden knew the directions in which the theories should change. In other words, she knew the *result* of each generative step. This is true not only for the theories that turned out to be correct, for example, Morgan's theory of linkage, but also for theories that were abandoned, such as Bateson's reduplication hypothesis. The scientists that Darden studies, of course, did not know how their science would develop. This does not imply that they did not follow heuristic rules of the kind that Darden proposed. However, it does raise the question of whether Darden's general strategies are at all relevant to the historical reconstruction of scientific change. I think that Darden wants to claim at least that her strategies, even if they were not exactly those that were used by the historical actors, are nevertheless *of the same kind* as the heuristics actually used by these actors. But where did these heuristics come from? Were they part of a local scientific practice? Or were they embedded in some broader scientific tradition? Or are they even part of the very fabric of human reason, sort of like the categories of pure reason according to Kant? As long as these questions cannot be answered, the status of Darden's strategies remains unclear.

A third critical point has to do with the considerable *abundance* of strategies that Darden found. If so many different strategies are required to generate the various changes to genetic theory that occurred between 1900 and 1926, the historical actors, presumably, must have had a similar arsenal available to them. But how did they know which ones to choose? Did they try them all and chose the one that looked the most promising? This would appear to generate an explosion of hypotheses. Or is there some kind of meta-strategy, that is, a heuristic for choosing strategies according to the problem situation? This point seems highly relevant for AI implementations. Darden's own attempts to implement some of her strategies on a computer, to my knowledge, always required that the user choose the appropriate strategy that the computer should follow.

All this is not meant to imply that the approach examined in this section has no merits. It might be relevant for developing problem-solving toolkits, including computer-aided expert systems that could provide assistance to the working scientist. Perhaps, one day, scientists will use such tools extensively even when doing creative work. However, to date, most of the more important scientific ideas have been generated without such aides. Thus, for those of us who are more interested in how science actually works, the lack of historical evidence for the relevance of Darden's strategies remains a central weakness (compare Vicedo 1995). I now turn to the examination of a case that has been studied in much more historical detail.

3.3   THE UREA CYCLE: INTELLIGIBLE TO REASON

As we have seen in the last section, the scanty historical record of the genesis of the classical theory of the gene limits the extent to which the philosopher can reconstruct the reasoning by which Morgan and other geneticists arrived at some of their more important ideas. However, there are historical episodes where a much richer historical record is available. One such episode is the elucidation of the urea cycle in 1932 by the great biochemist Hans Krebs (who, five years later, also discovered the citric acid cycle, which was named "Krebs cycle" after him). Krebs (1900–1981) preserved all his laboratory notebooks[21] and he made them available to the historian of science Frederic L. Holmes. Furthermore, during the late 1970s, Holmes had the opportunity of conducting extensive interviews with Krebs, actually going through the notebooks with him. Based on this evidence, as well as other documents such as letters, Holmes assembled one of the most detailed accounts of a research pathway from the history of science ever (Holmes 1991, 1993a). For certain periods of Krebs's research, Holmes traced the biochemist's activities from day to day. If this account is not rich enough to reconstruct the actual scientific reasoning in historic scientific discovery, then nothing is. Therefore, it is worth examining whether it was possible to reconstruct the reasoning actually involved in the generation of new theories in this example. I begin by giving a synopsis of Holmes's account (Holmes 1991, Chapters 8–10).

The case of the urea cycle is technically quite simple compared to the intricate examples from genetics discussed in the previous two sections. Urea or NH_2CONH_2 is the main metabolic waste product from the degradation of nitrogen-containing compounds (e.g., amino acids) in land vertebrates. (Birds produce uric acid instead of urea.) It is almost exclusively formed by the liver. A young clinician and medical researcher at the University of Freiburg (Germany), Krebs was experimenting with tissue slices isolated from fresh rat liver. He measured various metabolic activities in these slices using a Warburg apparatus, a manometric device turned into a very reliable instrument for biochemical experiments by the highly influential German biochemist Otto Warburg (with whom Krebs had studied in Berlin).

Given its importance for studies of metabolism in the first half of the twentieth century, it is a remarkable fact that all a Warburg apparatus can measure is the gas pressure in a closed reaction vessel kept at constant volume and temperature. Even though a Warburg apparatus cannot differentiate between different gases, experiments can be designed in such a way that the device measures the release (or consumption) of a specific gas by a biochemical reaction occurring in the vessel. Warburg himself, for example, developed great skills in measuring respiration (oxygen consumption) in various biological materials. But a Warburg apparatus can also be used to measure other biochemical reactions, so long as these reactions consume or release a gas, for example, carbon dioxide. For example, Krebs measured the production of urea in liver slices with the help of soybean extracts containing the enzyme urease. Urease decomposes urea into carbon dioxide (CO_2) and ammonia (NH_3). If this reaction is carried out in acidic solution, gaseous CO_2 is formed, whereas ammonia is retained in the solution in its protonated form (NH_4^+). Therefore, under such conditions, the amount of urea degraded by urease is proportional to the amount of CO_2 released. This CO_2 is detected by the Warburg apparatus as an increase in pressure. Thus, Krebs had a very quick and convenient way of measuring the amount of urea formed by his liver slices.

Krebs was able to confirm some earlier findings concerning the synthesis of urea, for example, that the process is stimulated by various amino acids and by ammonium salts such as ammonium chloride. Krebs was looking for precursors of urea, that is, intermediates in the metabolic pathway from amino acids to urea. All the evidence available indicated that ammonia was such a precursor. According to a generally accepted principle at that time, a bona fide intermediate in a metabolic pathway should be transformed at least at the rate of the overall process. Ammonium satisfied this requirement, as Krebs was able to show. In October 1931, Krebs and his new assistant, the medical student Kurt Henseleit, began to test the effects of a variety of different substances on the rate of urea formation. Apart from ammonia, none of the substances tested, alone or in combination, had enough of an effect on urea formation to really qualify as an intermediate. An exception was the amino acid arginine; however, this was to be expected, as the liver was known to contain an enzyme called arginase, which cleaves arginine to give urea and ornithine. At that time, this reaction was not thought to be part of the main pathway of urea formation.

The first substance found to have a strong effect on the rate of urea formation was ornithine (an amino acid that is not part of naturally occurring proteins) in combination with ammonium chloride. Given that all the other amino acids only showed a weak effect, this result was totally unexpected. Henseleit's notebook suggests that the experiment of testing the effect of ornithine was sort of tacked onto a larger experiment systematically testing various sugars. Holmes went through considerable trouble in order to find out why Krebs and Henseleit added ornithine to this experiment. Krebs told Holmes that he had no specific reason; he just wanted to test all the amino acids, and ornithine happened to be available. However, Holmes suspected that Krebs had forgotten the exact reason that induced him to test ornithine more than 40 years before this interview. For one, Holmes's investigations (worthy of his namesake, the famous detective from London) showed that commercially available ornithine was quite expensive at that time. Furthermore, the records show little evidence that Krebs and Henseleit systematically tried to test all the naturally occurring amino acids.

Another ambiguity noted by Holmes has to do with the question of whether Krebs tested ornithine as a potential nitrogen donor or as a substance that affected the rate of urea synthesis for other reasons (e.g., like lactate, which probably stimulates the reaction by providing metabolic energy). The high concentration of ornithine used suggests that it was tested as a potential source of urea nitrogen, which is actually what Krebs remembered. However, Krebs and Henseleit also tested other substances at high concentration, even those substances that were not candidates for the nitrogen source. The entire experimental setup rather suggests that the effect of ornithine was discovered in the course of a search for substances that indirectly influence the rate of urea synthesis. These ambiguities show that, even in a case that is as densely documented as this one, there remain open questions about the reasons that motivated certain experimental actions.

Krebs and Henseleit next examined whether there are similar compounds that exert such a strong effect on urea formation. After concluding that the effect was specific to ornithine, they began to study the "ornithine effect" more closely. One of the main questions, now addressed directly, was whether ornithine was a nitrogen donor or whether it indirectly stimulated urea synthesis. In February 1932, Henseleit reproduced the effect using much lower concentrations of ornithine. Krebs and Henseleit also determined the amount of ammonia consumed, finding that two molecules of ammonia were consumed for each molecule of urea produced. This provided further evidence that both nitrogen atoms in urea came from ammonia, rather than from ornithine. Whatever ornithine was doing, it was not a source of nitrogen.

One possibility was that ornithine has some *catalytic* action. (A catalyst is defined as a substance that influences the rate of a reaction, but that emerges from that reaction unchanged.) It is a matter of historical debate exactly when Krebs formed this "catalyst hypothesis," specifically, whether he thought of it *before* or *after* testing ornithine at low concentrations (Grasshoff, Casties, and Nickelsen 2000, 319–321). As Krebs remembered it, the experiments varying the concentration of ornithine were done *in order* to test the hypothesis that it acted as a catalyst. However, examining the concentration-dependence of an effect was a routine procedure in biochemistry, and it was the only thing that Krebs and Henseleit had not tried yet in order to learn more about this mysterious ornithine effect. Thus, the catalyst hypothesis could have been a *consequence* of the concentration experiments, rather than the *reason* that this experiment was done (Grasshoff and May 1995, 56). I come back to this problem below.

After these results were in, there followed a period of several weeks during which, while still following up other lines of experimental work, Krebs tried to conceive of a theoretical explanation for the ornithine effect. According to Holmes, Krebs had a crucial insight in late March or early April 1932. Krebs does not remember a single "eureka" experience. Instead, Krebs reported to Holmes that the solution to the problem occurred to him "gradually." Krebs's solution was the following:

$$\text{ornithine} + CO_2 + 2NH_3 \rightarrow \text{arginine} + 2H_2O \qquad (1)$$

$$\text{arginine} + H_2O \rightarrow \text{ornithine} + \text{urea} \qquad (2)$$

This scheme explained all of the major experimental findings, namely, that the nitrogen source for urea formation is ammonia, that the ornithine effect is specific, that ornithine is regenerated from the reaction, and that it forms a temporary union with the substrate. Furthermore, reaction (2) was already well known to biochemists; it was the reaction catalyzed by the liver enzyme arginase. Thus, Krebs suggested that this reaction was part of the main pathway of urea formation after all. While writing a brief publication on this scheme (Krebs and Henseleit 1932), Krebs searched for an additional intermediate between ornithine and arginine. Such an intermediate was theoretically necessary because reaction (1) was unlikely to occur in a single step, as reactions where four molecules come together are extremely unlikely. By the end of April 1932, Krebs had eliminated all but citrulline as the missing intermediate; he consequently tested citrulline for its effect on the rate of urea formation. Indeed, citrulline showed an effect similar to that of ornithine. Figure 3.1 shows the scheme postulated by Krebs in 1932, which came to be known as the urea cycle. It was the first of several cyclic metabolic pathways to be found and is therefore considered to be a landmark discovery in intermediary metabolism. Indeed, the urea cycle brought Krebs instant fame, which helped him when he had to leave Nazi Germany and emigrate to Cambridge, England, in 1933.

Let us now focus on the first reaction scheme shown in Figure 3.1, which was the crucial insight leading to the urea cycle. In his memoirs, Krebs gave the following account of the reasoning that led him to this reaction scheme:

> When considering the mechanism of this catalytic action I was guided by the concept that a catalyst must take part in the reaction and form intermediates. The reactions of the intermediate must eventually regenerate ornithine and

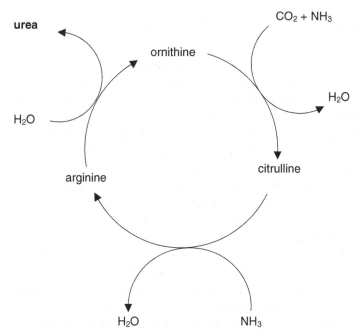

Figure 3.1. The urea cycle as formulated by Hans Krebs in 1932.

form urea. Once these postulates had been formulated it became obvious that arginine fulfilled the requirements of the expected intermediate. This meant that a formation of arginine from ornithine had to be postulated by the addition of one molecule of carbon dioxide and two molecules of ammonia, and the elimination of water from the ornithine molecule (Krebs 1981, 56–57).

On this written account given by Krebs, he reasoned *from* the experimentally supported hypothesis that ornithine has a catalytic effect. He then used what he knew about catalysts in order to derive the conclusions that ornithine had to form a temporary intermediate with the substrate (ammonia) and had to be regenerated from the reactions. Finally, he recognized that arginine meets the requirements for an intermediate formed from ornithine. The rest was balancing the equations according to the rules of organic chemistry.

But Krebs gave a somewhat different account in conversation with Holmes:

I know that after having discovered the ornithine effect and being satisfied with it, it took something like a month before it occurred to me that ornithine may give rise to arginine. And then we could do the critical experiments, namely measuring whether it disappeared. Because if that were the case, then it would be catalytic. If a small quantity would be effective and it would disappear. So

before that I was just puzzled, and carried on doing experiments (Krebs to
Holmes on Aug. 4, 1978, quoted in Holmes 1991, 325).

On this account, the order of the reasoning steps is reversed: Krebs *first*
thought of arginine as an intermediate, and *then* postulated a catalytic role for
ornithine (Krebs also slipped; he must have meant that when ornithine does
not disappear, then it must be catalytic).

Holmes (1991, 325–328) discusses at some length which one of the two
accounts we should believe. The first account was written by Krebs, probably
after some reflection. The second account was presented orally in response to
Holmes's queries. Thus, it is probably more spontaneous and unrehearsed. For
both accounts, reasons could be given for preferring them. Perhaps a carefully
written account is more reliable than a casual remark made in conversation.
On the other hand, the first account invites a certain suspicion. Scientists are
accustomed to write up their findings in a certain way. They first state some
experimental findings, and then show that these findings support or disconfirm
certain hypotheses. The whole structure of a scientific paper is designed for
presenting *justificatory reasons* for the truth of theoretical beliefs. It is thus
possible that Krebs, when he wrote down the first account, unconsciously followed this practice. Thus, he first stated the experimental fact that ornithine
acts like a catalyst and then presented the theoretical mechanism that is supported by this fact. To put it a bit loosely, the logic that governed the first
account might be the *logic of justification*. By contrast, the second account
might correspond to the actual sequence of mental events that led Krebs to
the solution; it might, therefore, be more indicative of the underlying logic
of generation (if there is such a thing). Confounding the reasons that *justify* a
certain theoretical belief with those that were involved in the *genesis* of a new
theory has been termed the "historical fallacy" (Grasshoff et al. 2000, 325).

The historical record is insufficient to definitively decide whether Krebs
reasoned from the catalytic action to arginine as an intermediate (Krebs's
written account), or whether he only saw the analogy to catalysis *after* having
thought of arginine as an intermediate (as Krebs's oral report implies). Holmes
finds the first possibility more plausible, because Krebs probably considered
arginine as an intermediate several times, but without being clear what role it
played exactly. Possibly he was only able to see that role once he started to
think in terms of catalysis. By contrast, Grasshoff and May (1995) argue for
the second possibility, according to which the concept of catalysis was not
involved when Holmes worked out his cyclic reaction scheme. At any rate,
the fact that Holmes and other historians had to resort to considerations of
plausibility instead of just using the historical record in reconstructing this

crucial step in Krebs's discovery is certainly noteworthy for our present topic. The reason is that scientists usually do not record what they think when they perform an experiment; they only write down the details of the experiments done and the results obtained. Sometimes, they may jot down a theoretical remark, usually in cryptic form. As Holmes shows, it is sometimes possible to infer why an experiment was performed from the specifics of the experiment. But there are always gaps, and sometimes all the historian can do is to choose the most plausible account among a few different possibilities. As this case also shows, the memory of scientists is not always accurate with respect to the exact sequence of certain events. This is because certain historical events may become reconfigured in a scientist's memory. As we have seen, one way in which this could happen is by presenting a story in a way that reflects the justificatory reasoning behind certain theoretical beliefs, rather than the exact sequence of a series of historical events.

Grasshoff and May (1995) have given an account of Krebs's reasoning according to which Krebs did not think in terms of catalysis from the very beginning. In my view, this account is the most plausible so far.[22] On their account, the generation of the cyclic scheme involved the following steps.

The starting assumption was the source hypothesis that Krebs took from the literature, namely that urea is formed from amino acids via ammonia:

$$\text{amino acids} \rightarrow NH_3 \rightarrow \text{urea} \qquad (3)$$

But the amino acids tested did not stimulate urea formation sufficiently, except for ornithine. According to the source hypothesis, the pathway from ornithine to urea would be

$$\text{ornithine} \rightarrow NH_3 \rightarrow \text{urea} \qquad (4)$$

However, this scheme was not consistent with the data. First, ornithine only stimulated urea formation in combination with ammonia. Second, ornithine acted at concentrations that were too low for it to be a source of urea nitrogen. And third, the measurements of the amount of urea consumed showed that two molecules of ammonia were consumed for each molecule of urea formed. Thus, the only possibility was

$$\text{ornithine} + 2NH_3 \rightarrow \text{urea} \qquad (5)$$

The low concentrations at which ornithine acts require that ornithine is *regenerated* from the reaction. The only known reaction that formed ornithine was the arginase reaction:

$$\text{arginine} + H_2O \rightarrow \text{ornithine} + \text{urea} \qquad (2)$$

Now, this reaction had to be combined with (5). However, (5) is not a chemical reaction in the strict sense, since it requires that three molecules come together, which is unlikely. Reaction (5) probably requires additional steps. By contrast, (2) could occur in a single step and should thus be retained. The only meaningful place to put (5) is *before* the arginase reaction. The crucial step now is to introduce arginine as an *intermediate*. In order to do this, (5) can be expanded as follows:

$$\text{ornithine} + 2NH_3 + X \rightarrow \text{arginine} + Y \tag{6}$$

$$\text{arginine} + H_2O \rightarrow \text{ornithine} + \text{urea} \tag{7}$$

The rest can be filled in by balancing (6) stoichometrically:

$$\text{ornithine} + 2NH_3 + CO_2 \rightarrow \text{arginine} + H_2O \tag{1}$$
$$\text{arginine} + H_2O \rightarrow \text{ornithine} + \text{urea} \tag{2}$$

This is the solution. Thus, on Grasshoff and May's account, there is no need to assume that Krebs used an analogy to the action of a catalyst. The reaction scheme that forms the core of the urea cycle can be generated by taking the source hypothesis (3), modifying it in accordance with Krebs's and Henseleit's ornithine results, postulating an intermediate (arginine) that was known to give rise to urea, and then using stoichometric rules to build a possible reaction pathway.

What this reconstruction shows is that the reasoning from the ornithine effect to the urea cycle can be rendered "intelligible to reason" (Holmes 2000a, 186). If the reasoning is reconstructed in this manner, there seem to be no mental leaps, no "illuminationist" experiences (Schaffner 1993, 19) involved. Anyone who understands some basic principles of organic chemistry can follow these steps. Holmes (2000a, 186) even suggested that Krebs's solution resembles T. S. Kuhn's descriptions of *puzzle solving* within a tradition of normal science.[23] For the solution "conformed to strict existing criteria for the formulation of a sequence of metabolic reactions: that is, in addition to the immediately relevant experimental data, they incorporated known structural formulas for organic compounds involved, and long-standing rules for balancing chemical equations" (Holmes 2000a, 186).

On this view, the paradigm within which Holmes was working contained all the resources necessary to solve the problem, and the solution did not violate any fundamental principles.

It could be objected that the urea cycle does not fit Kuhn's account of successful normal science well. According to Kuhn (1970, 52), normal science "does not aim at novelties of fact or theory and, when successful, finds none." However, the urea cycle was considered to be an important theoretical novelty in biochemistry. The urea cycle may not be at variance with *fundamental* principles of biochemistry, but it nevertheless added an important theoretical concept (the concept of a cyclic metabolic pathway). The case is thus not exactly like an instance of Kuhnian puzzle solving, but it does seem to exhibit some aspects of Kuhnian normal science.

What conclusions with regard to generative reasoning can we draw from this case? The first important conclusion, I suggest, is that the case confirms the view that most students of scientific "discovery" have adopted, namely that the generation of new theories is best described as a problem-solving activity (e.g., Simon, Langley, and Bradshaw 1981, Schaffner 1993, 18f.; Nickles 1980a, 33–38). The search for general principles of generative reasoning is thus transformed into the search for general problem-solving heuristics. However, and this is my second conclusion from the present case, the *generality* of such heuristics appears questionable.

On the account of Krebs's reasoning given above, I can find nothing that would be transferable to scientific problems outside the field of intermediary metabolism. Grasshoff and May (1995) argue that Krebs used general principles of *causal inference* in the research leading to the urea cycle. An important such principle was John Stuart Mill's "method of difference," according to which the causal relevance of a factor can be established by observing a difference between an instance where this factor is present and an instance where it is absent, but where all other factors are kept constant. This principle is indeed very important in biochemical research (see Section 4.7) and can be applied to a great variety of unrelated problems in other sciences and even in everyday life.

But in the urea cycle example, the only role that the method of difference played is in establishing the *causal relevance* of various factors in urea synthesis. When Krebs and Henseleit measured the rate of urea formation, for example, in the presence of ornithine alone, ammonia alone, ornithine and ammonia together, and without any addition, they showed that both ammonia and ornithine are causally relevant in urea synthesis. This is basically an application of Mill's method of difference. However, the only role that this general principle of causal inference plays is in demonstrating that both ornithine and ammonia are necessary for obtaining the maximal rates of urea synthesis observed. But what we want to understand here is how Krebs reasoned to his theoretical scheme *from* the finding that ornithine and ammonia are causally relevant. According to the account given above, only some principles from organic chemistry and intermediary metabolism were used in this step. But these principles are rather specific; they cannot be transferred to other scientific fields (e.g., genetics or economics). This suggests that, although scientific traditions (or Kuhnian paradigms) *do* provide some problem-solving heuristics that make the generation of novel problem solutions an efficient and thus rational process, these heuristics are far less *general* than certain other epistemological principles such as Mill's methods of causal inference (which can also be used, e.g., in the social sciences).

In the final section, I offer a few general reflections on the attempts to devise general "logics of generation."

3.4   IS THERE A LOGIC OF GENERATION?

In the previous sections, I have critically examined three attempts to find the principles governing the generation of new biological theories. The first attempt was Kenneth Schaffner's proposal that Jacob and Monod's repressor model of enzyme induction in bacteria was generated by deductive reasoning (Section 3.1). I found this proposal not to be supported by the historical evidence. A more plausible thesis is that Jacob and Monod reasoned from an *analogy* with known cases of enzyme repression, thus employing a mode of reasoning that is not permissible in the context of justification. Schaffner's attempt to push his neat deductive argument into the context of discovery is unconvincing (as Schaffner acknowledges today).

The second attempt that I have examined was Lindley Darden's detailed study of the changes that occurred in the theory of classical genetics between 1900 and 1926 (Section 3.2). Darden devised a number of general "strategies" that could possibly have been followed by classical geneticists when they were diagnosing problems and introducing modifications into genetic theory. These strategies are not intended to be historically adequate; all that Darden aspires to is that the strategies are *general* and that they are *sufficient* for generating the theoretical changes extant in the (published) historical record. The problem with this account is not only that the relevance of Darden's strategies to the actual reasoning employed by the historical actors has not been established (as Darden recognizes), but also that it cannot even be shown that the classical geneticists used general procedures that are *similar in kind* to Darden's. But the early period of twentieth-century genetics is poorly documented in the first place. For this reason, I have examined the case of the urea cycle (Section 3.3), where a dense historical record is available and has been scrutinized from different angles by historians.

It turns out that there are several gaps in the historical reconstruction of the mental steps taken by Hans Krebs, the discoverer of the urea cycle. The reason is that the historical record reveals much about the experiments performed by Krebs on a daily basis, but much less about what Krebs was *thinking*. Krebs's own recollections are not entirely consistent, suggesting that the sequence of mental events was reconnected in his memory in a way that reflects Krebs's justificatory reasoning, rather than his generative reasoning. For example, doubts remain as to whether the concept of a catalyst actually guided Krebs's reasoning in devising the crucial steps of the urea cycle. Either way, Krebs's reasoning can be made intelligible by showing that he essentially solved a highly structured problem using some rules from organic chemistry. I have argued that the paradigm within which Krebs was working supplied some sort of problem-solving heuristic; however, this heuristic is highly specific to the field of intermediary metabolism. The only nonspecific principles that Krebs used were those of causal inference. But the latter were only involved in experimentally establishing the causal relevance of various substances for urea synthesis, not in generating the reaction scheme of the urea cycle itself.

Interestingly, all three cases exhibit the same basic dynamics. In all three cases, the historical actors were working with one or several initial hypotheses. These hypotheses guided the experimental strategies followed. In the case of Jacob and Monod's work on enzyme induction, the initial hypothesis was the generalized induction theory. In the case of Morgan and Bateson, the initial hypothesis was Mendel's law of segregation. In Krebs's case, the initial hypothesis was the assumption that urea nitrogen is derived from the amino groups of amino acids. In all three cases, anomalous findings eventually triggered revisions of the initial hypotheses.[24] In the enzyme induction case, the PaJaMo experiment violated the expectations from the generalized induction theory. In the classical genetics case, Bateson's and Morgan's linked characters violated Mendel's laws. In the urea case, the ornithine effect violated the expectation that amino acids are sources of urea nitrogen. In each of these cases, these anomalies defined a *problem*, the solution of which constituted an important new theoretical idea.

In addition to this common dynamical pattern, what else can be generalized from these cases? In particular, do the case studies examined provide evidence that there are any general problem-solving heuristics involved in the generation of new theories, something like a "deep logic," to use a term of Holmes's (2000a, 186)? With regard to this question, I am more skeptical than other friends of discovery, most of whom have claimed wide applicability for their heuristics or strategies (e.g., Kulkarni and Simon 1988, 171–173;

Darden 1991, 16–17; Schaffner 1993, 63; Grasshoff and May 1995, 58). In examining the three case studies, I cannot find much of *general* applicability that was really shown to contribute to the efficiency or rationality of scientific problem solving.

The best candidate, perhaps, is the use of *analogies* in devising new hypotheses or theoretical schemes.[25] Analogical reasoning seems to have been involved in the genesis of Jacob's and Monod's repressor model, in Morgan's later model of crossing-over (the "beads-on-a-thread" analogy), and possibly in Krebs's use of the concept of catalysis (if it was involved at all).[26] However, it is doubtful whether the use of analogies really qualifies as a powerful problem-solving heuristic. Analogies are too easy to find. The problem is always what the process in question should be analogous *to*, and in what respects the analogy should consist. If we take the case of Jacob and Monod (see Section 3.1), they compared the mechanism of enzyme induction in the *lac* genes to the known cases of enzyme repression. This may be an analogy, but it seems to me that this at best describes the *conceptual relationship* between the *lac* repressor model and the mechanisms for enzyme repression. The recognition that Jacob and Monod used an analogy does not really show what reasoning they actually used in solving the theoretical problem before them; it rather *describes the solution itself* in relation to extant theories of enzyme repression. Jacob and Monod had to devise a mechanism that accorded with the results of the PaJaMo experiments and that qualified as an acceptable problem solution according to the standards set by the exemplars of their discipline. But as soon as the situation is described in this way, we are back to very discipline-specific constraints and background information that are part of a particular scientific tradition.

It seems to me that Lindley Darden's strategies suffer from the same general problem. That Morgan used interrelations to another body of knowledge (when constructing his theory of linkage and crossing-over) is an abstract characterization of the relationship between the theory of linkage and the cytological knowledge of chromosomal mechanics. But how did Morgan know *what* other body of knowledge to use? Which parts of that body of knowledge? And which interrelations? Darden's abstract strategy hardly shows that Morgan used an *efficient* (i.e., rational) procedure to generate his solution. It is my contention that as soon as we start to fill in additional constraints and information as to how exactly Darden's strategies should be deployed (as, for example, Darden and Rada (1988a, 1988b) have done) we lose the general character of the strategy and end up once again with highly discipline-specific stratagems.

There are, however, problem-solving strategies that are part and parcel of a specific research tradition. This is most evident in the case of Krebs, who used rules for combining and stoichometrically balancing metabolic reactions that were essentially derived from organic chemistry, in combination with some theoretical principles from biochemistry, such as the rule that a metabolic intermediate must be transformed at least at the rate of the overall process. But these rules are hardly transferable to other scientific problems in or outside of experimental biology.

To conclude, the thesis of this chapter is that problem-solving heuristics, at least the more powerful ones, are *specific to scientific disciplines*. Such heuristics differ greatly in scope from certain principles of causal inference (or of theory testing; see Chapter 4) that are applicable to a great variety of scientific areas. Especially, the case of Krebs shows that the genesis of a novel problem solution (in this case constituting a new theory) can be made "intelligible to reason" by showing that the problem solved was highly structured by the rules of a specific scientific discipline. With the more generally applicable heuristics or strategies, such as the ones devised by Lindley Darden in her 1991 book, it is doubtful whether they can really explain the efficiency of scientific problem solving (in addition to the problem that their historical adequacy has not been demonstrated). I suspect that the more *abstract* and *general* a heuristic is, the less *effective* it will appear to be in solving specific scientific problems.[27]

Most recently, Carl Craver and Lindley Darden have drawn attention to the role of *mechanisms* (see also Section 2.1) in the elucidation of cellular and molecular processes (Craver and Darden 2001). On their account, the need to fill in "blanks" or missing causal links in schematic mechanisms provides a powerful heuristic for problem solving in experimental biology. I find Craver's and Darden's work on the discovery of mechanisms insightful (see Weber 2001c); however, I think this account also supports my thesis. For the mechanisms that Craver and Darden describe are quite specific to cellular and molecular biology. If there is a general problem-solving heuristic involved in the cases discussed by Craver and Darden, it is a very abstract and weak one, for example, "fill in missing causal links in mechanistic model," which, again, hardly explains the efficiency of scientific problem solving. As in the cases discussed in this chapter, the problem-solving strategy can be specified in more detail to become more effective (or even sufficient). However, then the *general* character is lost again.

What I am proposing, then, is not a return to the older view that the generation of theories is an irrational process that is not open to philosophical analysis, or inaccessible altogether. For to show that a kind of reasoning can be *rational* (in the sense of being efficient with respect to its goal) is not the same as showing that it employs *general* rules or procedures. That rational thought always involves general rules is one of the most deeply entrenched philosophical prejudices, which we should reject at least when analyzing what used to be called the "context of discovery."

SUMMARY

The genesis of biological theories is a form of reasoning that can be analyzed philosophically. I have first defined the goal of such analysis as trying to exhibit the *rationality* of the mental procedures used by scientists when generating new theories. To proceed rationally, in this context, does not mean to use methods that are conducive to truth (unlike in the context of justificatory reasoning). Instead, to show that scientists proceed rationally when generating new theories means to show that they deploy heuristic procedures or strategies that are *efficient* with respect to the goal of generative reasoning. The goal of generative reasoning is the production of theories, hypotheses, or problem solutions that are plausible or fruitful.

I have then examined three attempts to reconstruct the reasoning involved in the genesis of biological theories. The first attempt, Kenneth Schaffner's account of the genesis of Jacob's and Monod's 1959 repressor model of enzyme induction in bacteria (Perdee et al. 1959), examined the question of whether the reasoning involved in generation is different from that governing justification. Schaffner's conclusion was that generative and justificatory reasoning are identical, since the repressor model can be *deduced* from experimental results in conjunction with theoretical background beliefs. I have shown this account to be historically inadequate, since Jacob and Monod are more likely to have generated their model by *analogy* with known cases of enzyme repression in bacteria. Schaffner's deductive argument was at best part of the initial *evaluation* of the model.

The next attempt examined was Lindley Darden's detailed study of the development of the theory of classical genetics between 1900 and 1926. Darden has proposed a number of strategies that are not intended to reflect the *actual* reasoning used by classical geneticists such as W. Bateson or T. H. Morgan (which is unknown due to an insufficient historical record). Instead, Darden's criteria of adequacy are that her strategies are both *general* and *sufficient* for generating the changes in genetic theory extant in the published historical record. I have argued that Darden's strategies, as well as some other heuristics proposed by discovery researchers, are *either* sufficient *or* general, but not both. As soon as it is spelled out exactly how Darden's abstract strategies are to be deployed (thus rendering them sufficient), they lose their general character.

The third case examined is that of the urea cycle first described by the biochemist Hans Krebs in 1932. This case has been examined in much more historical detail than the other cases, not least because the historical record available is much richer (laboratory notebooks, correspondence, interviews). In spite of this, there remain gaps in the historical reconstruction of this research process, especially with respect to the exact sequence of Krebs's mental steps. Nevertheless, the genesis of the reaction scheme devised by Krebs in 1932 can be made intelligible to reason. This reconstruction showed that Krebs probably used a problem-solving heuristic that is specific to organic chemistry and intermediary metabolism.

The three cases discussed all follow a common dynamic pattern of initial hypotheses, experimental anomalies, and anomaly resolutions that introduced important new theoretical ideas. By contrast, the most efficient problem-solving heuristics used are probably less general; they seem to be part of specific scientific traditions or paradigms and are not applicable outside of these disciplines.

4

Scientific Inference: Testing Hypotheses

Under what conditions do experimental data or observations support a scientific hypothesis? Initially, philosophers sought the answer to this question in the relationship between theory and evidence. According to the hypothetico-deductive (H-D) account (Hempel 1945), empirical evidence supports or *confirms* a theory exactly if that evidence is *logically entailed* by the theory conjoined with additional assumptions, for example, about initial and boundary conditions or about the relationship between measurable quantities and theoretical magnitudes. The H-D account is plagued with various difficulties, which I am not going to review here (see Glymour 1980).

Clark Glymour has replaced the H-D account with an approach that came to be known as "bootstrapping." According to this account, scientists are allowed to use a theory subject to test in analysis of data (as they often do). A test of a theory, then, consists in using the data and the theory to derive an *instance* of the laws that make up the theory. An instance is a statement that the various parameters and variables that feature in the theory take a definite value. The theory is confirmed by an instance if the instance *satisfies* the theory. Glymour's bootstrap theory is rich and interesting, but I have no idea how it could be applied to experimental biology. What would an "instance" of a theory describing, for example, some molecular mechanism be? It seems to me that the bootstrap theory is more germane to physics than to experimental biology.

Probably the most influential account of scientific inference in recent years has been Bayesian confirmation theory (e.g., Howson and Urbach 1989; Earman 1992). The basic idea of this approach is that a good empirical test should increase a theory's *probability*. Thus, testing a theory amounts to calculating the theory's posterior probability given the evidence by using Bayes's theorem:

$$P(H|e) = \frac{P(e|H)P(H)}{P(e|H)P(H) + P(e|\neg H)P(\neg H)}$$

Bayes's theorem gives the conditional probability of a hypothesis H given some evidence e. In order to calculate this probability, the Bayesian needs the probability of the evidence given that H is true, $P(e|H)$. She also needs to know the *prior probability* of H, *P(H)*. This is the probability that H is true before the evidence is taken into consideration. In addition, this form of the theorem contains the conditional probability of the evidence given that H is false $P(e|\neg H)$. How does the Bayesian hope to calculate these probabilities? In order to see this, it must be understood how Bayesians interpret probability. On the standard Bayesian account, probabilities are interpreted as *subjective degrees of belief*. A statement's probability basically gives the betting rate that a rational agent would accept in a bet that the statement is true. In fact, this is how the Bayesian approach is justified; it is argued that anyone who does not calculate her probabilities in accordance with Bayes's theorem is acting irrationally in the sense that she will accept "Dutch books."

Prior probabilities express the confidence that a scientist has in a theory before the evidence. These probabilities are thus subject to interpersonal variation, and Bayesians argue that this does not matter, because prior probabilities "swamp out" after a sufficient number of tests. A difficult question is how to calculate the probability of the evidence given that the hypothesis is false, $P(e|\neg H)$, also known as the "catchall." Salmon (1991) argues that this is impossible, because it would involve predicting the future course of science; however, Bayesians have suggested some clever tricks for getting around this problem.

It is beyond dispute that the Bayesian approach has interesting and important applications to certain very specific problems. However, disagreement prevails on whether this approach is useful as a *general theory of scientific inference*, as some of its supporters maintain.[1] Some critics see the main problem with the standard Bayesian approach in its subjective interpretation of probability, which, in their view, leaves too much room for the individual scientist's subjective opinions (which enter as prior probabilities). I critically examine how the Bayesian approach fares in experimental biology in Section 4.5.

Deborah Mayo (1996) has developed a non-Bayesian theory of scientific inference. The important step that she has taken in the study of scientific methodology is to abandon the idea that we can give a full-blown account of how theories are tested by trying to analyze the relationship between theory and evidence. As I already mentioned, virtually all traditional accounts of scientific method share this idea. Inductivists used to think that there is some probabilifying relation from evidence to theory. Hypothetico-deductivists think that the relationship between theory and evidence is logical entailment. Bootstrappists argue that it is a satisficing relation (a special kind of logical entailment). Bayesians maintain that theory is related to evidence by Bayes's theorem via conditional probabilities. What all these philosophies of scientific method have in common is the assumption that, once the relationship between theory and evidence is understood, the methodologist's job is done. Not so in Mayo. According to her, the relationship between theory and evidence can basically be left open (it may depend on the form and content of a theory and of the evidence). But what is really relevant to understanding scientific methodology is how evidence is *generated* in the first place. In other words, evidence is not just *given*; if we want to know whether some experimental evidence tests or supports a theory, we need to investigate how this evidence was produced.

Mayo presents us with a rich methodological instrumentarium that tries to incorporate some of the insights of the "New Experimentalist" movement in history and philosophy of science, which considers the nuts and bolts of experimental practice to be of primary importance for understanding the nature of scientific knowledge (see also Chapter 5). The centerpiece of Mayo's account is her statistical concept of a *severe test*, but she offers a wealth of additional resources as well. As we shall see, Mayo's general approach is helpful in grappling with methodological issues in experimental biology, and I will later use her account in my case study (Sections 4.5 and 4.6). However, I am also going critically examine Mayo's formal notion of a severe test as well as her informal concept of "arguing from error."

Before I move on to investigate in detail how experimental biologists test theories, I would like to introduce some problems for the appraisal of scientific theories that have been discussed extensively by philosophers.

The first problem is known as *Duhem's problem*, after Pierre Duhem (1954, 185). Duhem pointed out that when a theory is in conflict with the evidence, this does not necessarily mean that the theory is false. A false prediction, for instance, could also be due to an error in the *auxiliary assumptions* that were used to predict the phenomenon. For this reason, it is not clear how to distribute praise and blame between the theory and whatever extra assumptions were made, for example, about the measurement apparatus used, about the way measurable quantities are related to theoretical magnitudes, and so on. Duhem's problem is often (and mistakenly) lumped together with a second problem, namely *underdetermination* of theory by the evidence. Underdetermination is really a whole set of problems, as Laudan (1990) has shown. We have to differentiate at least between *logical* and *methodological* underdetermination. The former is the problem that a theory may have infinitely many alternatives with the same empirical consequences (empirical equivalence). The latter arises if two theories are equally well supported by the evidence. A variant of methodological underdetermination is the "other hypothesis objection," which states that some evidence does not support a theory if an alternative theory can account for it as well. Recent accounts of scientific inference, including the Bayesian approach and Mayo's account, contain attempts to solve Duhem's problem as well as underdetermination. Therefore, it will be interesting to investigate whether these philosophical conundrums actually have instances in the history of experimental biology.

After having introduced some of the main topics in recent philosophical debates of the methodology of theory testing, I move on to consider these problems in the light of a historical case study. The next two sections present this case study, the oxidative phosphorylation controversy in biochemistry, focusing on the two major competing theories involved (Section 4.1) and some of the experimental tests done in the attempt of resolving the controversy (Section 4.2). Section 4.3 examines whether this case instantiates any of the problems that are associated with underdetermination and Duhem's problem. In Section 4.4, I discuss the crucial evidence that led to the final resolution of the controversy. Section 4.5 takes a critical look at the Bayesian approach to scientific inference by investigating to what extent this approach illuminates my historical case. In Section 4.6, I turn to Deborah Mayo's "error statistical" approach. Finally, in Section 4.7 I present what I take to be the correct approach.

4.1   THE OXIDATIVE PHOSPHORYLATION CONTROVERSY IN BIOCHEMISTRY

The case I want to examine more closely is a fascinating episode from the history of biochemistry, known as the "oxidative phosphorylation controversy" (1961–1977).[2] Oxidative phosphorylation is a central part of the cell's energy metabolism. Cells break down carbohydrates (sugars and similar energy-rich compounds) in a complex set of biochemical pathways, which can be roughly grouped into two phases. The first phase, which is known as "glycolysis," requires no oxygen. The six carbon atoms of the sugar molecule are first split into a compound made of three carbon atoms (pyruvic acid). One of these carbon atoms is released in the form of carbon dioxide (CO_2). The remaining two carbon atoms end up as acetic acid residues bound to a coenzyme (acetyl-coenzyme A or acetyl-CoA). Fermenting yeasts release these two-carbon units in the form of ethanol. By contrast, all cells that utilize oxygen feed acetyl-CoA into a second phase of metabolic breakdown reactions. One part of this second phase is the so-called Krebs cycle, named after its discoverer Sir Hans Krebs (whom we have already met in Section 3.3). The other part is respiration. The Krebs cycle completely breaks down the two-carbon units of acetyl-CoA into carbon dioxide. The hydrogen atoms contained in the acetic acid moiety are removed from the Krebs cycle by a compound called NADH (nicotinamide adenine dinucleotide; the H stands for the extra hydrogen atom of the reduced form). Respiration then transfers these hydrogen atoms to molecular oxygen (O_2), producing water (H_2O).

This last step in the metabolic breakdown of foodstuffs provides most of the cell's energy supply. The reason is that the formation of water from chemically bound hydrogen atoms and molecular oxygen releases an enormous amount of energy (as witnessed by the loud bang produced by a mixture of hydrogen and oxygen when it is ignited). With the process of respiration, evolution has invented an extremely efficient and gentle way of harnessing this energy and making it available for all the chemical processes that constitute life.

While all that I have told so far was known to biochemists by 1940, the question of how the cell can actually utilize the energy derived from respiration remained a mystery for many years to come. This question was the very heart of what came to be known as the oxidative phosphorylation controversy.

In the 1920s, the British biochemist David Keilin discovered a set of enzymes that changed their spectroscopic properties in the presence of oxygen. He named these enzymes "cytochromes" (cell pigments). The cytochromes turned out to be part of a group of sequentially acting enzymes that are located in the inner membrane of mitochondria and that are responsible for respiration (Keilin 1966). The main substrate for this "respiratory chain" was shown to be NADH derived from the Krebs cycle. The respiratory chain extracts two electrons from each NADH molecule and transfers them to molecular oxygen in a series of enzymatic reduction–oxidation steps. The main steps of this process were elucidated in the 1940s and 1950s. But there was still an important part of the story missing.

What was missing was the link from respiration to the cell's main energy stores. The universal carrier of biological energy is a substance called

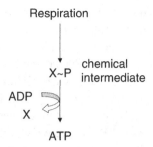

Figure 4.1. Slater's chemical mechanism. The squiggle symbol in the chemical intermediate indicates its high potential to transfer the phosphate group (P) to an acceptor molecule, in this case ADP.

ATP (adenosine triphosphate). ATP is thermodynamically unstable in water; it tends to hydrolyze to give ADP (adenosine diphosphate) and inorganic phosphate ($H_2PO_4^-$). As Fritz Lipmann (1941) proposed, cells use the so-called phosphate bond energy that is released in the hydrolysis of ATP to ADP and phosphate to drive most biochemical reactions that require energy, such as muscle contraction, biosynthesis, regulation, and DNA replication. To understand how ATP is regenerated from ADP and inorganic phosphate is to understand the main principle of energy metabolism. For this reason, the regeneration of ATP was one of the main questions that was on the biochemists' minds. Because this regeneration is a phosphorylation step (ADP + $H_2PO_4^- \rightarrow$ ATP), and because the energy for this step largely comes from the oxidative breakdown of foodstuffs, the process was termed "oxidative phosphorylation." It was first observed in 1939 by the two Russian biochemists V. A. Belitser and E. T. Tsybakova in rat liver and by Severo Ochoa in brain tissue (Kalckar 1969).

In 1953, the Australian biochemist E. C. Slater, working at the University of Amsterdam, proposed a mechanism for oxidative phosphorylation (Slater 1953). He suggested that the energy released in respiration is first stored in a so-called high-energy intermediate directly generated by the respiratory enzymes. This intermediate would then acquire a phosphate group. This phosphate group would finally be transferred to ADP to give ATP (see Figure 4.1). Slater's conjecture came to be known as the "chemical hypothesis." It was closely modeled after the one ATP-yielding step of glycolysis, which had been described by Racker and Krimsky (1952).

The chemical hypothesis was widely acclaimed and completely dominated experimental investigations for several years. At the Fifth International Congress of Biochemistry, held in 1961 in Moscow, most of the experimental findings presented were interpreted in terms of Slater's mechanism (Slater

1963). An especially noted set of findings concerned so-called energy-linked functions, which involved the redirection of energy from the respiratory chain. Apparently, the hypothetical high-energy intermediates were able to deliver their energy to other processes (i.e., processes other than ATP synthesis). Slater had predicted this effect in his 1953 *Nature* paper, and this was widely seen as a success for the chemical hypothesis. Like phlogiston 200 years earlier, the hypothetical chemical intermediates seemed to reveal their existence in various processes.[3] Hopes were rising that they would soon be identified (Chance 1963). The epistemic values operative in biochemistry at that time required that the intermediate be *chemically isolated*, as Racker and Krimsky had done for the ATP-yielding step in glycolysis. The search was thus on for high-energy intermediates of oxidative phosphorylation.

Between 1956 and 1972, there were at least seventeen published claims of the isolation of an intermediate (Allchin 1997). However, not a single one of these claims held up to critical scrutiny. One of the most promising candidates (phosphohistidine), found in 1962 in Paul Boyer's laboratory in Minnesota, turned out to be an intermediate of the Krebs cycle. The intermediates, if they existed, seemed to be extremely elusive. As the years passed, they increasingly looked like a "twentieth-century phlogiston."

In 1961, the British biochemist Peter Mitchell published an article in *Nature* that radically challenged orthodox thinking in biochemistry at that time (Mitchell 1961). Specifically, Mitchell proposed to seriously consider the possibility that the chemical intermediates did not exist. He proposed a mechanism that did not require such an intermediate. According to Mitchell, the energy released by respiration is used by the respiratory enzymes to transport protons across the mitochondrial membrane. Thus, a gradient would build up across the membrane, composed of an electric potential (because protons are positively charged) and of an osmotic component (because of the concentration difference between the two sides of the membrane). This gradient would then be used by another enzyme located in the membrane, the so-called ATPase, to phosphorylate ADP. Mitchell termed his idea the "chemiosmotic hypothesis" (see Figure 4.2).

Mitchell's hypothesis was largely ignored at first. When it was published, the biochemical community was immersed in hypothetical chemical intermediates, hyped by the "energy-linked functions." They had no need for an alternative mechanism.[4] Only the American biochemist Albert L. Lehninger, for example in his textbook published in 1964, praised the chemiosmotic hypothesis for providing "some rationale for the fact that the respiratory and phosphorylating enzymes are located in the membrane" (Lehninger 1964, 129).

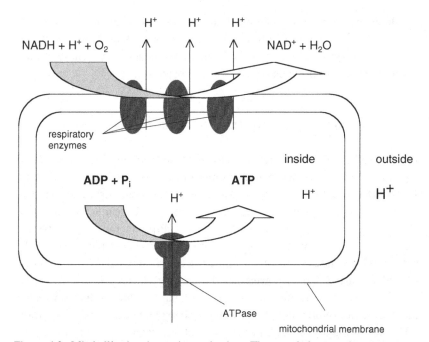

Figure 4.2. Mitchell's chemiosmotic mechanism. The rounded rectangle represents a mitochondrion. The respiratory chain enzymes reduce NADH (nicotinamide adenine dinucleotide) using oxygen (O_2) as a final electron acceptor. This process, which is called "respiration," involves many steps but has been simplified here (details can be found in any biochemistry textbook). The energy released by respiration is used by the respiratory enzymes to transport protons (H^+) across the mitochondrial membrane. This creates a proton gradient across the membrane, which is depicted by the small and large H^+ symbols to the right of the diagram. These protons then follow their concentration gradient and electric potential, flowing back inward through the ATPase. This proton flow powers the synthesis of ATP from ADP and inorganic phosphate (P_i).

In 1965, experimental evidence became available that forced biochemists to take notice of Mitchell's theory (discussed in the next section). From then on, the chemiosmotic hypothesis was highly present in scientific discourse. Most biochemists initially wanted to disprove it, but the hypothesis was clearly not ignored any more. However, it would take another decade of painstaking research and vigorous debate until Mitchell's hypothesis was accepted by a majority of the main investigators in the field. It eventually was accepted, and Mitchell was awarded the Nobel Prize for chemistry in 1978.[5]

Why did it take so long to resolve this controversy? Why was all the experimental evidence produced in the 1960s considered to be insufficient to adjudicate between the chemical and chemiosmotic theories? And how could the controversy be resolved after all? What kind of evidence warranted a choice in favor of Mitchell's theory in the 1970s and why? These are the questions that concern us in the rest of this chapter.

4.2   EXPERIMENTAL TESTS OF MITCHELL'S THEORY

At the 1965 Gordon Conference in Boston, the German biochemist André Jagendorf presented experimental findings that commanded attention. Oddly, Jagendorf was not working on oxidative phosphorylation itself, but on photosynthesis in green plants. However, these processes have much in common. The chloroplasts of green plants contain an ATP-producing enzyme system that resembles the respiratory chain of mitochondria considerably (they probably have the same evolutionary origin). The only functional difference is that the chloroplast system is powered by light energy, while the mitochondrial system is driven by respiration. In his 1961 paper, Mitchell had suggested that the chemiosmotic hypothesis would also explain how chloroplasts form ATP. According to Mitchell, the photosynthetic reaction chain forms a proton gradient across the innermost membrane of chloroplasts. A chloroplast homologue of the mitochondrial ATPase would then use this proton gradient to synthesize ATP.

Jagendorf and his co-worker Ernest Uribe incubated isolated spinach leaf chloroplasts in a weakly acidic solution. Then they rapidly increased the pH of the solution surrounding the chloroplasts in order to generate an artificial proton gradient across the membrane. Remarkably, the chloroplasts synthesized ATP in the dark, as Mitchell's theory predicted (Jagendorf and Uribe 1966). Even some of Mitchell's staunchest opponents like E. C. Slater agreed that this finding supported the chemiosmotic theory (e.g., Tager, Veldsema-Currie, and Slater 1966, 376). Several laboratories, including Slater's in Amsterdam, Britton Chance's in Philadelphia, and Efraim Racker's in New York, started to do experiments on the proton gradient. Even though they tried to obtain experimental evidence *against* Mitchell's theory, the fact that some of the most prestigious laboratories in the field invested resources in investigating the role of proton gradients indicates that the chemiosmotic theory's standing had been greatly increased. Scientists do usually not invest in maverick theories, not even in order to refute them.

Jagendorf's experiment, which came to be known as the "acid bath experiment," seems to have been widely accepted as support for Mitchell's theory (Robinson 1986). However, the experiment by no means ended the ox-phos controversy. In fact, the real historical significance of the acid bath experiment is that it *started* the controversy. Before it become known, there was no controversy, because the chemical theory ruled supreme.

This raises the question of why the acid bath experiment was not considered to be decisive or crucial evidence for the chemiosmotic theory. The reason is that there were alternative explanations for the effect observed by Jagendorf – explanations within the framework of the chemical theory (see below). Jagendorf had not established that ATP synthesis after an acid–base transition was *relevant* to the mechanism of photosynthesis; it could have been an irrelevant side effect. Thus, there were no strong reasons for Mitchell's opponents to accept this as crucial evidence and reject their preferred chemical explanation.

Mitchell himself was encouraged by Jagendorf's findings to obtain experimental evidence for his own theory. He devoted his newly founded laboratory to this task. Using extremely simple equipment, Mitchell and his co-worker Jennifer Moyle performed a series of experimental tests of the chemiosmotic hypothesis. I now review these tests in a condensed form, followed by the objections raised by Mitchell's opponents.

Test 1: Isolated mitochondria eject protons after an oxygen pulse.

Mitchell and Moyle were able to show with very simple pH and oxygen electrodes and a suspension of isolated mitochondria that when respiration sets in after the addition of an oxygen-saturated solution, the mitochondria release protons into the surrounding medium (Mitchell and Moyle 1965b, 1967). This, of course, was a central prediction of chemiosmotic theory that appeared to be borne out by the evidence. However, Mitchell's opponents quickly raised the following objection:

Objection to Test 1: Mitochondria have other "energy-linked functions." Proton movements could be a side effect of ox-phos.

Remember that it was considered to be an empirical success of Slater's chemical theory that it predicted "energy-linked functions," which were thought to be powered by the hypothetical chemical intermediates. One of the proposed energy-linked functions was ion transport across the mitochondrial membrane, so what Mitchell and Moyle were observing could be an energy-linked function instead of oxidative phosphorylation proper. Slater (1967) proposed three alternative schemes of the relationship between respiration, proton transport, and ATP synthesis (Figure 4.3). According to these alternative schemes, proton transport is a secondary or even a tertiary effect of oxidative phosphorylation and is not directly involved in ATP synthesis. The experimental

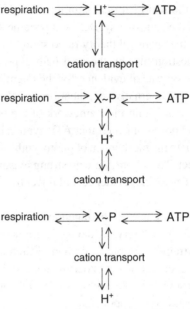

Figure 4.3. Slater's alternative schemes for the relationship between respiration, proton transport (H^+), ATP synthesis, and cation (e.g., Ca^{++}) transport. The scheme on the top is Mitchell's mechanism, but the two alternatives schemes below (involving a chemical intermediate, X~P) were equally well supported by the experimental evidence available at that time.

evidence available at that time did not allow discrimination between these alternative schemes. The same problem beset the acid bath experiment, which was discussed at the beginning of this section.

The next experimental test attempted in Mitchell's lab made use of an effect that had been observed early on in the history of oxidative phosphorylation: Certain chemicals such as dinitrophenol (DNP), when added to respiring mitochondria, inhibit ATP synthesis while allowing respiration to proceed at an increased rate (Loomis and Lipmann 1948). These chemicals were named "uncouplers" because they seemed to sever the coupling of respiration and phosphorylation. They constitute a chemically quite heterogeneous class of compounds. Uncoupling was considered to be a central phenomenon that a theory of oxidative phosphorylation would have to explain. Slater explained it by proposing that uncouplers destroy the hypothetical chemical intermediates. By contrast, Mitchell suggested that uncouplers collapse the proton gradient of the chemiosmotic mechanism. Indeed, Mitchell and Moyle (1967) obtained the following experimental result:

Test 2: Uncouplers increase the proton permeability of mitochondrial membranes.

Mitchell and Moyle showed this by measuring the time it takes for a suspension of mitochondria to reach its equilibrium pH after the addition of an acidic solution. Acids release protons. As long as the membranes of the mitochondria have low proton permeability, they will be slow to take up the additional protons added from the outside. But in the presence of DNP, proton uptake was a lot faster, suggesting that DNP made the membrane more permeable to protons. This corresponded to a prediction made by Mitchell in his 1961 *Nature* paper, namely that uncouplers act by making the membrane permeable to protons, thus collapsing the proton gradient, which, according to Mitchell's theory, is necessary for oxidative phosphorylation to work. Thus, what Mitchell suggested is that the chemically heterogeneous class of uncoupling agents share a *physical* property; they affect the proton conductance of membranes. This not only was a *novel prediction*,[6] but offered the first explanation of how such a chemically heterogeneous class of compounds could have such similar effects on oxidative phosphorylation. For these reasons, this experimental test was widely acknowledged as providing strong support for the chemiosmotic theory. However, the following objection was raised (Greville 1969, 58):

Objection to Test 2: Uncouplers could act by causing a proton current that destroys the chemical intermediate.

Thus, although the chemical theory would not have *predicted* the physical effect of uncouplers on membranes, it was at least able to *explain* this phenomenon. However, this explanation differed from the explanation for uncoupling that Slater gave in 1953. As already mentioned, he had suggested that uncouplers act by *chemically* destroying the high-energy intermediate. At any rate, the existence of an alternative explanation for the effect of uncouplers on membrane permeability seems to have damaged the evidential force of Mitchell's successful prediction.

The next test I would like to discuss was a *negative* result, which was not obtained in a single experiment or in a particular laboratory. Instead, this result was gradually established over the years.

Test 3: Oxidative phosphorylation requires an intact membrane.

Traditionally, biochemists had proceeded by grinding up some biological material until it was completely homogenized and soluble. They then measured the activity of enzymes in these homogenates and subsequently tried to isolate the enzymes. This approach had been successful in unraveling most of the biochemical pathways known, including glycolysis, the urea cycle, and the Krebs cycle.[7] From the 1940s on, several laboratories applied this approach to oxidative phosphorylation and some even claimed success (e.g., Cooper and Lehninger 1956). However, during the 1960s it became clear that oxidative phosphorylation only works in vitro if intact mitochondrial membranes are present, as Mitchell's theory predicts. This point was somewhat controversial and several laboratories claimed to observe ox-phos in completely solubilized systems, that is, in experiments in which all membranes had been dissolved by detergents. However, these results were not generally accepted, because there were doubts whether the mitochondrial membranes had been solubilized completely (Racker 1976, 60). Thus, there was at least a partial consensus that oxidative phosphorylation requires intact membranes. However, opponents of chemiosmotic theory did not accept this as decisive evidence, because of the following objection:

Objection to Test 3: Respiratory enzymes might need membranes for mechanical support or structural integrity.

It was perfectly plausible to argue that if enzyme complexes as intricate as the respiratory chain are subjected to detergents in order to dissolve the membrane in which they are embedded, they simply "die"; in other words, they lose their functional integrity. Thus, another attempt to establish Mitchell's theory failed because of the existence of an alternative explanation that was fully compatible with the chemical theory.

Another opportunity for testing Mitchell's theory was provided by so-called submitochondrial particles. Such particles were produced by treating isolated mitochondria with ultrasound. This treatment generated small, sealed vesicles of mitochondrial membrane that showed oxidative phosphorylation activity (Penefsky et al. 1960; Racker et al. 1963).[8] At a conference in Bari in 1965, Lee and Ernster (1966) suggested that the membrane of submitochondrial particles displays an inside-out orientation compared to intact mitochondria. Indeed, Mitchell and Moyle (1965a) showed that the movement of protons across the membrane is reversed in inside-out particles. Furthermore, if the respiratory chain was forced to operate in reverse, the flow of protons in submitochondrial particles was also reversed. This suggested the following test:

Test 4: The inner mitochondrial membrane is functionally asymmetric.

The asymmetry of the membrane was a basic postulate of chemiosmotic theory. However, the following objection was raised:

Objection to Test 4: Energy-linked functions also require an asymmetric membrane.

Clearly, nothing in the chemical theory ruled out the membrane being asymmetric. In fact, since defenders of the chemical theory thought that the hypothetical chemical intermediates also mediate ion transport processes, they would also expect the membrane to be asymmetric.

After having discussed four attempted experimental tests of Mitchell's chemiosmotic theory, it is time now to take stock and examine the possible reasons that, so far, prevented a resolution of the controversy.

4.3 UNDERDETERMINATION AND DUHEM'S PROBLEM

Underdetermination of theory by evidence is widely viewed as a major problem for theory appraisal. It is therefore worth examining whether underdetermination could be the reason for the difficulties that biochemists experienced in comparing the chemical and chemiosmotic theories. As already mentioned in the introduction to this chapter, underdetermination comes in at least two forms, which Laudan (1990) has termed *logical* and *methodological* underdetermination, respectively. Logical underdetermination is sometimes referred to as *empirical equivalence*. The difference between the two concepts is that logical underdetermination applies to sets of theories that have the same empirical consequences purely on the grounds of logical entailment. By contrast, methodological underdetermination arises if two theories are equally well supported by the evidence. As Laudan has emphasized, two theories could be empirically equivalent and yet some rule of scientific inference could still distinguish one of the two theories as better supported. Good examples of empirical equivalence are typically found in physics, where, in some cases, infinitely many empirically equivalent alternatives to a theory can be constructed (e.g., physical geometry).

I now examine whether the chemical and chemiosmotic theories were underdetermined by the evidence in either of these two senses.

First, logical underdetermination. In order to decide whether the theories were empirically equivalent, we must identify the empirical consequences of both theories. The chemiosmotic theory predicted (1) that mitochondria move protons across the membrane, (2) that uncouplers increase the proton permeability of the membrane, (3) that oxidative phosphorylation requires a membrane, and (4) that the membrane is asymmetric. Three of these four facts, namely (1), (3), and (4), are *constitutive* for the chemiosmotic theory, that is, part of its core postulates. (2) is an additional assumption that accounts for the phenomenon of uncoupling and explains why a chemically heterogeneous class of chemicals have the same effect on ox-phos. The chemical theory was also able to account for these four experimental facts; however, none of these facts correspond to a core postulate of the chemical theory. In all of these four cases, the defenders of the chemical theory had to introduce additional theoretical assumptions, namely (1) and (4) that proton transport is an energy-linked function, (2) that an uncoupler-induced proton current destabilizes the chemical intermediate, and (3) that respiratory enzymes are inactive outside a membrane. These additional assumptions are *consistent* with the chemical theory, but they are not *entailed* by it; that is, they are logically independent. By contrast, facts (1), (3), and (4) are clearly logically dependent on the chemiosmotic theory. In other words, if any one of them is false, the chemiosmotic theory is false, too. From these considerations, we see that the two theories were not empirically equivalent. Each theory made predictions for which there was *no corresponding prediction* of the other theory.

Second, methodological underdetermination. In theory, whether two theories are methodologically underdetermined by the evidence depends on a particular account of how theories are supported by the evidence. However, I suggest that we do not need to refer to a particular account of scientific inference in order to see that the chemical and chemiosmotic theories were not equally well supported by the evidence produced in the 1960s, because such a relation may obtain under any (reasonable) account of scientific inference.

It was generally accepted that some of the experimental findings supported the chemiosmotic theory. For example, Greville (1969, 55) called Mitchell's prediction that uncouplers would increase the proton permeability of the membrane, which was experimentally verified in Mitchell's own and in other laboratories, "an outstanding achievement" of the chemiosmotic theory. As we have seen, this finding did not contradict the chemical theory. However, it was clearly not *evidence* for the chemical theory. The same is true of the membrane requirement and the asymmetry of the membrane: It was possible to explain these findings within the framework of the chemical theory, but they did not support the latter, because the defenders of the chemical theory had to introduce *ad hoc* assumptions in order to account for these facts (e.g., that the chemical intermediate is destroyed by uncoupler-induced proton currents or that respiratory enzymes are inactive when dissociated from the membrane). Although it was pointed out by several reviewers that the chemiosmotic theory offered a more *parsimonious* explanation of these facts, such a philosophical argument was unlikely to impress biochemists.[9]

Table 4.1. *Racker's Scoreboard of Experimental Evidence*

	Chemical	Chemiosmotic
Role of the membrane	−	+
Ion transport	+	×
Action of uncoupling agents	−	+
Isolation of high-energy intermediates	−	±
^{32}P–ATP exchange		
ADP–ATP exchange	+	−
$H_2{}^{18}O$ exchanges		

Source: Redrawn after Racker (1970).

But even when parsimony is ignored, we cannot say that the two theories were equally well supported by the evidence, no matter what particular account of evidence or empirical support is used. There were several findings that clearly supported the chemiosmotic theory better. However, there were also findings that supported the chemical theory, for example, the prediction of energy-linked functions by Slater (1953). Some investigators argued that the chemiosmotic theory was better supported by kinetic measurements and by the so-called P:O ratios, that is, the numbers of phosphate groups incorporated per oxygen atom consumed (Chance, Lee, and Mela 1967); however, the reliability of these measurements was dubious.

In biochemistry, quantitative predictions clearly do not play the important role that they do in physics (see Section 9.2). The obvious explanation for this fact is that the measurement techniques are generally too "dirty" to permit scientists to determine the precise value of some measurable quantity. In other words, measurement error is often too large to allow using quantitative predictions to discriminate between different hypotheses.

It thus seems that each of the two theories fared better with respect to some evidence, but that no piece of evidence supported the two theories equally. Tables 4.1 and 4.2 show two "scoreboards" of experimental evidence that were tabulated by Racker (1970) and Slater (1971), including some evidence that I have omitted from my account. These tables show two things. First, two of the leading scientists in the field acknowledged that there was evidence for and against the two hypotheses. Second, not all of the evidence tabulated was equally relevant to the two theories. The tables include effects concerning which one of the two theories made no specific predictions.

I conclude that the theories were not underdetermined by the evidence in any sense. The real problem was that both theories had empirical advantages *and* disadvantages and that there were no generally accepted methodological principles that would have allowed the scientists to weigh these different

Table 4.2. *Slater's Scoreboard of Experimental Evidence*

Chemical Hypothesis	Chemiosmotic Hypothesis
1. There is no evidence for the existence of the hypothetical A~C compounds in state-4 mitochondria	1. There is no evidence for the existence of a membrane potential of sufficient magnitude in state-4 mitochondria
2. A high-energy compound with a $\Delta G'_0$ value of hydrolysis of 17 kcal/mole is unlikely	2. A membrane potential of 370 mV is unprecedented in either artificial or natural membranes
3. No explanation is given for the multiplicity of electron carriers in the respiratory chain	3. There is no experimental support for alternate hydrogen and electron transfer in the respiratory chain
4. An *ad hoc* hypothesis (the proton pump) is necessary to explain energy-linked cation uptake	4. There is no experimental evidence for the translocation of H^+ in the absence of cation
5. This hypothesis takes insufficiently into account the fact that the energy-transducing reactions take place in membranes	5. This hypothesis takes insufficiently into account recent advances in our knowledge of the chemical properties of haemoproteins
6. No explanation is given for the fact that uncouplers increase the electrical conductivity of artificial membranes	6. No explanation is given for the fact that some uncouplers are not proton conductors
7. An oligomycin- and uncoupler-sensitive ATP-P_i exchange reaction is found in pro-mitochondria lacking a respiratory chain	7. There is no experimental support for the postulated diffusible X^- and IO^-
8. There is no site specificity for reaction with ADP or for the action of uncouplers or inhibitors of oxidative phopshorylation	8. No explanation is given for kinetics of ADD-induced oxidation of ubiquinone

Source: Redrawn after Slater (1971).

empirical merits against each other. This is not underdetermination.[10] Although the present discussion has taught us what the real problem was in comparing the two theories, we can dismiss underdetermination as a methodological problem that is relevant to our case. Underdetermination is an intriguing theoretical possibility, but it seems that real-life theories in experimental biology do not necessarily exhibit this relation.[11]

As I mentioned in the introduction to this chapter, underdetermination should be distinguished from the so-called Duhem problem. In contrast to the more general thesis of underdetermination of theory by the evidence, Duhem's problem specifically addresses situations where a mismatch between theory and experimental results is observed. The problem is which should be blamed for this mismatch: the theory or the additional assumptions that were made in interpreting the results, connecting the evidence with the theory, and so forth. I briefly discuss an instance of Duhem's problem from the ox-phos story.

The laboratory of Britton Chance employed sophisticated spectroscopic techniques in order to measure the kinetics of proton transport in mitochondria in different states (Chance and Mela 1966a, 1966b, 1966c). They used the dye bromthymol blue to measure the pH (proton concentration) inside isolated mitochondria (the spectroscopic properties of this dye depend on the pH of the medium; it is a so-called pH indicator). What they claimed to have found was that, when mitochondria activate respiration, the proton gradient across the membrane builds up too slowly, that is, only after energy coupling is already engaged. This, of course, would be bad news for Mitchell's theory. However, Mitchell, Moyle, and Smith (1968) claimed that this result was an experimental artifact.[12] They were able to show that bromthymol blue could not be used for measuring the pH gradient across mitochondrial membranes, because the dye molecules were themselves unevenly distributed across the membrane. Chance's technique only would have worked if the concentrations of the indicator inside and outside the mitochondrion had been the same. It seems that these arguments were generally accepted, since nobody claimed that Chance and his co-workers had successfully refuted Mitchell's theory.

This example suggests that biochemists can solve Duhem's problem at least in some situations. In this example, an anomaly of Mitchell's theory was detected. Mitchell was able to locate the culprit in an auxiliary assumption made by Chance and his co-workers, namely the assumption that the absorption spectrum of bromthymol blue in respiring mitochondria reflects the pH inside the mitochondria. Thus, Duhem's problem was solvable because the auxiliary assumptions could be checked independent of Mitchell's theory.

I conclude that neither logical nor methodological underdetermination nor Duhem's problem was responsible for the difficulties in resolving the ox-phos controversy in the 1960s. The real problem was that both theories had advantages and disadvantages in accounting for the experimental evidence. Nobody knew how to weigh these advantages and disadvantages. And I do not think that there exist sound methodological principles that would allow this. Parsimony or "Occam's razor" did not conform to the methodological standards of mid-twentieth-century biochemistry. Mitchell's hypothesis had to wait for a crucial test until it was acceptable to a majority of biochemists. In the next section, I show that such a crucial test was eventually accomplished with the help of new experimental techniques.

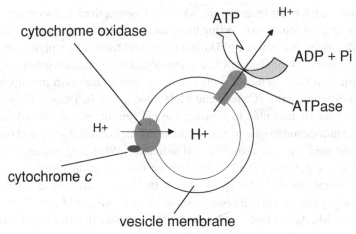

Figure 4.4. Reconstitution of respiratory enzymes and ATPase in artificial vesicles. Cytochrome oxidase reduces cytochrome c and transfers the electrons received to molecular oxygen (not shown). The energy released by this oxidation/reduction reaction is used to build up a proton (H^+) gradient across the membrane, which drives ATP synthesis. The diameter of the vesicles is approximately 0.1 μm.

4.4   RECONSTITUTION AND CLOSURE

Around 1970, Efraim Racker's laboratory at Cornell University took a novel approach to the study of oxidative phosphorylation. While all the previous experiments had been done on whole, isolated mitochondria or submitochondrial particles, Racker and his co-workers started to *reconstitute* oxidative phosphorylation from purified components. In one series of experiments, they took purified ATPase and cytochrome oxidase (the terminal enzyme complex of the respiratory chain) and inserted them into artificial phospholipid membrane vesicles (see Figure 4.4).

When these vesicles were fed with cytochrome c (the electron donor for cytochrome oxidase) as well as oxygen, they synthesized ATP from ADP (Racker and Kandrach 1971). Thus, they performed the following test:

Test 5: A reconstituted system containing ATPase and cytochrome oxidase synthesizes ATP in the presence of cytochrome c, oxygen, inorganic phosphate, and ADP.

This was the first time that one of the energy-coupling reactions of oxidative phosphorylation was directly observed outside an intact mitochondrion. In a similar way, Racker's laboratory obtained vesicles containing another main

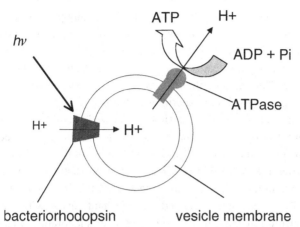

bacteriorhodopsin   vesicle membrane

Figure 4.5. Reconstitution of bacteriorhodopsin and the ATPase in artificial vesicles. Bacteriorhodopsin uses light energy ($h\nu$) to pump protons across the membrane. The mitochondrial ATPase can use the resulting proton gradient to phosphorylate ADP.

enzyme complex of the respiratory chain, namely NADH dehydrogenase (Ragan and Racker 1973).

Using the same artificial vesicle system, Racker and Stoeckenius (1974) published a key experiment which has been mentioned in textbooks (e.g., Alberts et al. 1994, 673) ever since as the crucial experiment that solved the oxidative phosphorylation controversy (cf. Allchin 1996). The experiment was made possible by D. Oesterhelt and W. Stoeckenius's work on the photosynthetic *Halobacterium halobium*. This bacterium is equipped with a very simple photosynthetic system that allows it to generate ATP from light (without the complex electron transport system found in green plants). It had already been shown in Stoeckenius's laboratory that photosynthesizing *Halobacterium* forms a proton gradient across its cell membrane, suggesting that bacteriorhodopsin functions as a light-driven proton pump. Stoeckenius and Racker set out to test this idea. They reconstituted both bacteriorhodopsin and the mitochondrial ATPase into liposomes and added ADP and inorganic phosphate (see Figure 4.5).

When these "chimeric" vesicles were illuminated, they formed ATP:

Test 6: A reconstituted system containing bacteriorhodopsin and ATPase synthesizes ATP in the presence of light, ADP, and inorganic phosphate.

The reconstitution experiments done in Racker's laboratory seem to have been a major factor in the closure of the ox-phos controversy.[13] By 1977, most of the leading scientists involved in the controversy had accepted the basic postulates of Mitchell's theory, as witnessed by a jointly authored review article published in the *Annual Review of Biochemistry*[14] and, one year later, by Mitchell's Nobel Prize.

We may now ask what features of the reconstitution experiments made them so crucial. What methodological properties did these experiments exhibit that were absent in all the experiments done up to 1970?

What the first reconstitution experiments showed was that a system that contained only the components that, according to Mitchell, were necessary and sufficient for oxidative-phosphorylation-synthesized ATP. These components included the ATPase, a respiratory enzyme (cytochrome oxidase), and a sealed membrane. A more direct test of Mitchell's theory is hardly conceivable. However, there remained an important caveat with the initial series of reconstitution experiments done in Racker's laboratory. In the vesicles containing respiratory enzymes (e.g., cytochrome oxidase), it was still possible that there was a direct chemical interaction (i.e., via a chemical intermediate) between the respiratory enzyme and the ATPase. This is where the bacteriorhodopsin experiment came into play. Because bacteriorhodopsin is devoid of any oxidation–reduction properties (*Halobacterium*, unlike the chloroplasts of higher plants, lacks a photosynthetic electron transport chain) it was highly unlikely that there was a direct chemical interaction between bacteriorhodopsin and the ATPase. In other words, in the chimeric vesicles it could finally be ruled out that a chemical intermediate was involved, yet these vesicles made ATP.

It was thus the *combination* of all the reconstitution experiments done in Racker's laboratory that provided the crucial evidence for Mitchell's theory. The first series of experiments, which co-reconstituted respiratory enzymes and the ATPase, directly demonstrated the causal sufficiency of Mitchell's mechanism, and the Racker–Stoeckenius experiment with chimeric vesicles ruled out a possible error in the chemiosmotic interpretation of these experiments.

I now turn to a methodological analysis of the experimental tests presented, beginning with a critical examination of the Bayesian approach.

4.5   WHY BIOCHEMISTS ARE NOT BAYESIANS

It is worth examining whether Bayesian confirmation theory is any help in the methodological analysis of the ox-phos case. How would a Bayesian reconstruct the experimental tests that I have reported in Sections 4.2 and

4.4? I first attempt such a reconstruction. Then I show that such a Bayesian account faces considerable problems.

Because we are comparing two hypotheses, we can use the so-called odds form of Bayes's theorem (which spares us the difficulty of estimating the "catchall" mentioned in the introduction to this chapter):

$$\frac{P(H_{co}|e)}{P(H_c|e)} = \frac{P(H_{co})P(e|H_{co})}{P(H_c)P(e|H_c)}$$

I use the terms H_{co} and H_c for the chemiosmotic and chemical hypotheses, respectively. This form of Bayes's theorem gives us the ratio of the posterior probabilities of the two hypotheses given some evidence e. To calculate this ratio, we need the prior probabilities of the two hypotheses, $P(H_{co})$ and $P(H_c)$. On the standard Bayesian account, these priors basically express how much a scientist would bet on the hypothesis before the evidence is taken into consideration. Clearly, this value will differ from person to person. Bayesians also allow that scientists use background beliefs to set their priors. Thus, the initial prior probability of the chemical hypothesis was high for most biochemists because of the successes of this type of hypothesis in elucidating other biochemical pathways such as glycolysis (where high-energy intermediates are actually involved). For the same reason, the initial prior probability of the chemiosmotic hypothesis was probably low, except for Mitchell himself. At any rate, these probabilities are subject to revision in light of evidence. In addition, we need to know the likelihood of a particular piece of evidence given the truth of the two hypotheses, $P(e|H_{co})$ and $P(e|H_c)$.

Let us now try to determine how the relative probabilities of the two hypotheses could have evolved as new experimental evidence came in. I begin with the evidence from proton translocation in isolated mitochondria obtained by Mitchell and Moyle in 1965 (Test 1). What we need to know is how likely the effect of proton translocation was given the truth of the two hypotheses. We could say that, on the chemiosmotic theory, this likelihood was high. In fact, it could be suggested that the chemiosmotic hypothesis *entails* that mitochondria transport protons, thus rendering $P(e_{pt}|H_{co}) = 1$, where e_{pt} is the evidence that mitochondria translocate protons. Alternatively, this probability could be set to a value close to 1, taking into account some uncertainty as to whether proton translocation would actually be measurable. How about the likelihood of e_{pt} under the chemical hypothesis? Since the hypothesis does not directly entail this evidence, we should set $P(e_{pt}|H_c)$ to a value <1. However, because of the postulate of energy-linked functions (one of which was thought to be ion transport), we cannot set it to a very low value.

Given these assumptions, a Bayesian can now argue as follows. Since the prior probability of the chemical hypothesis, in 1965, exceeded that of the chemiosmotic hypothesis by far for most biochemists at that time, the ratio of the likelihoods of the proton transport evidence given the two hypotheses was not large enough to trip the ratio of the probabilities given this evidence in favor of Mitchell's hypothesis. And this, in fact, is what we find: The proton translocation data were not considered to be decisive evidence.

We could now repeat this game for all the other experimental tests discussed in Section 4.1; however, I am going to spare the reader this boring exercise (Test 2 will be discussed later). Instead, I jump directly to the crucial evidence obtained in the 1970s, the evidence from reconstitution experiments (Test 6).

Let us consider the likelihood that artificial vesicles containing bacteriorhodopsin and the mitochondrial ATPase would synthesize ATP in the presence of light given the two hypotheses. Can we say that Mitchell's hypothesis entailed this evidence, thus rendering $P(e_{RS}|H_{co}) = 1$? (I use e_{RS} to refer to the evidence from the Racker–Stoeckenius experiment with bacteriorhodopsin and artificial vesicles.) I think this would be wrong. Mitchell's hypothesis does not directly entail that biochemists can grind up cells, isolate bacteriorhodopsin and the ATPase, stick them in artificial phospholipid vesicles, add ADP and phosphate, switch on a lamp, and observe the synthesis of ATP. In fact, it is still a cause for amazement today that this intricate experiment worked, even given the truth of Mitchell's hypothesis. A large number of additional assumptions have to be plugged into Mitchell's theory in order to make such a prediction. Therefore, we must set $P(e_{RS}|H_{co})$ to a value below 1. However, what seems clear is that the bacteriorhodopsin experiment was very unlikely to work given the *chemical* hypothesis, thus rendering $P(e_{RS}|H_c)$ very low. Now, the Bayesian could argue as follows. The prior probability of Mitchell's hypothesis was much higher in 1974 compared to 1965. The hypothesis had successfully explained several experimental effects in the meantime. By the same token, the prior probability of the chemical hypothesis had deteriorated over the years because of the failed attempts to detect the hypothetical chemical intermediates. Since the likelihood $P(e_{RS}|H_c)$ is considerably lower than $P(e_{RS}|H_{co})$ by any standards, the bacteriorhodopsin experiment finally pushed the ratio $P(H_{co}|e)/P(H_c|e)$ well above unity – and Lo! The chemiosmotic theory was accepted. Thus, it seems that the Bayesian approach correctly predicts the scientists' actual decisions.

I now turn to a critical evaluation of the Bayesian approach as applied to my case study. I find at least two major difficulties, which I discuss in turn.

The first problem I want to call attention to is that the Bayesian approach could be used to show that chemiosmotic theory ought to have been accepted *much earlier*. Take, for example, Mitchell's prediction that uncouplers increase the proton permeability of membranes (Test 2). It could be argued that the likelihood of this evidence given Mitchell's hypothesis exceeded the likelihood given the chemical theory by far. It was considered to be a great virtue of Mitchell's theory that he could explain why such a chemically heterogeneous class of chemical compounds had the same effect on oxidative phosphorylation from the central postulates of his theory.[15] Thus, $P(e_{uc}|H_{co}) = 1$ or close to 1 (where e_{uc} is the evidence from the effect of uncouplers). By contrast, $P(e_{uc}|H_c)$ was arguably very low, since the chemical theory initially explained uncoupling by hydrolysis of the hypothetical chemical intermediates. Why was this not sufficient to increase the probability of Mitchell's theory above the chemical theory? Bayesians will say that, obviously, the prior probability of Mitchell's theory was still too small compared to the chemical theory. (Bayesians can always make this move, which is worrying.) But is *conservatism* on the part of the scientists really the explanation of why the chemiosmotic theory was not accepted in 1966 or 1967? Even if conservatism did play a role in this controversy, the philosopher of science is well advised to take the actual judgments of the scientific community very seriously. And their judgment was that the evidence for Mitchell's theory was *not* sufficient at that time. Good scientific judgment in this case said that the evidence from the mode of action of uncouplers, although it clearly did favor Mitchell's theory, was somehow *indirect*. It did not show that a chemiosmotic mechanism was, in fact, operating; it merely showed that one of Mitchell's central predictions obtained. Bayesianism gives us no resources to differentiate between direct and indirect evidence.

If it is true that the Bayesian approach renders some of the early evidence for Mitchell's theory decisive, and given that more than just conservatism was involved in the scientific community's judgments, then we are forced to conclude that the Bayesian inference scheme licenses unwarranted inferences. In other words, Bayesianism is *unsound*; it allows too much compared to the standards that scientists apply to experimental tests.

It must be stressed that my point is not just that the Bayesian account does not correctly represent how the biochemists *actually* reasoned. For this would be consistent with the possibility that Bayesianism provides a correct *normative* (prescriptive) methodology for science. But what my analysis shows is that the prescriptions issued by Bayesians are not *good* prescriptions. If scientists followed them, they would make bad decisions (even if, in this example, the Bayesian calculus picked out the right theory as being better supported). Had the majority of players in the ox-phos controversy been Bayesians, they would have accepted Mitchell's theory too early, at a time when its empirical superiority had not really been exhibited. Bayesianism licenses inferences that are not acceptable according to the highest scientific standards of evidence.

A second problem, I suggest, lies in the amount of *guessing* we had to use in order to estimate the likelihoods $P(e|H)$. Bayesians seem to assume that these likelihoods are easy to calculate. In statistical tests, this is sometimes the case. Statistical hypotheses frequently directly entail probability distributions that a certain outcome will be observed. And indeed, statistical testing is where the Bayesian approach may have its merits. For example, Kenneth Schaffner has presented a Bayesian approach to statistical testing in Mendelian genetics (Schaffner 1993, 256–260). However, we are dealing here with nonstatistical tests, which are more frequent in biochemistry and molecular biology.

In cases where the evidence is directly *entailed* by a hypothesis, it can of course be argued that $P(e|H) = 1$ or close to 1 (most of the nonstatistical examples treated by Bayesians are of this type). But how shall we deal with those cases where the evidence is not *directly* entailed by the hypothesis, for example, the reconstitution experiments?

In cases like these, an additional *subjective element* seems to be introduced. The likelihood of the evidence takes on the character of a *prior* probability itself; it expresses the confidence that an individual scientist has that a hypothesis accounts for some evidence. In fact, this problem has been noted by Bayesians (e.g., Schaffner 1993, 239–240). I suggest that the ox-phos case provides a clear example for this difficulty. For a supporter of Mitchell's theory, the evidence from the mode of action of uncouplers indeed seemed to be strong evidence, since a central prediction of Mitchell's theory was confirmed. However, for the opponents of chemiosmotic theory, this evidence was not considered to be very strong, since explanations for this effect within the framework of the chemical theory could be given. Thus, how well a hypothesis accounts for some evidence seems to be a matter of belief. Of course, a Bayesian could simply shrug her shoulders and point out that subjective beliefs are already part and parcel of the Bayesian approach (due to the prior probabilities). But the extent of subjective belief that the Bayesians allow in scientific reasoning is beginning to look intolerably large.

Adherents of Bayesianism are generally confident that problems such as the ones I have presented can be adequately dealt with from within the Bayesian framework. For those of us who are less sanguine about this approach, my example may serve to reinforce the view that Bayesianism poorly reflects the actual practice of science, no matter how plausible it may seem philosophically. For the present analysis, I have found the Bayesian approach unhelpful for illuminating the actual choices that the historical actors have made. It seems that Bayesians can always finagle their priors and evidential likelihoods until the Bayesian machinery returns the correct posterior probabilities, that is, the ones that we expected anyway from our pretheoretic intuitions. But what we want to do is to advance *beyond* these pretheoretic intuitions and to understand what constitutes a good test of a scientific theory or hypothesis.

In the next section, I examine a different approach to theory testing, namely Deborah Mayo's so-called error-statistical approach, which is intended as an alternative to the Bayesian account.

4.6   ARGUING FROM ERROR

As we have seen in Section 4.4, what turned out to be essential to resolve the ox-phos controversy was an experimental inquiry that directly demonstrated that the causal transmission from oxidation to phosphorylation flows through the proton gradient and not through a chemical intermediate. I have also argued that a Bayesian approach fails to illuminate why this evidence was better than, for example, the confirmation of Mitchell's successful predictions from the 1960s. In this section, I examine whether a different approach to theory testing does a better job in analyzing the present case study, namely the "error-statistical" approach that has been developed by Deborah Mayo (1996, 1997, 2000).

Central to Mayo's account is the notion of a *severe test*. Severity is a technical concept from the Neyman–Pearson theory of statistical testing. It is related to the probability that some test procedure passes a hypothesis if this hypothesis is false. Statisticians call this a type I error, and the probability of committing this error is called "error probability." A test procedure earns the label "severe" exactly if this error probability is very low.[16] In contrast to the Bayesians, Mayo does not interpret probabilities as subjective degrees of belief. Instead, the error probabilities are relative frequencies in a suitable reference class of tests. Thus, Mayo adopts an objective interpretation of probability. Furthermore, error probabilities do not attach to statements or hypotheses. They are a property of tests.

Far from just expounding a theory of statistical testing, Mayo suggests that the kind of reasoning employed in the Neyman–Pearson approach underlies a broad variety of test situations in the empirical sciences. It is also supposed to be applicable, in slightly modified form, to cases where scientists do not use formal statistics in order to test a theory or hypothesis on the basis of some experimental evidence. In such cases, Mayo argues, scientists must also conduct a suitable inquiry into the possible errors in inferring from some evidence to a hypothesis. Here, the term "error" presumably does not necessarily have the technical meaning that Neyman–Pearson statistical test theory gives it (type I statistical error), but the more commonsensical meaning of "mistake." Canonical mistakes in experimental inquiries include, for example (Mayo 1996, 18): (1) mistaking experimental artifacts for real effects, or mistaking chance effects for genuine correlations or regularities; (2) mistakes about a quantity or value of a parameter; (3) mistakes about a causal factor; and (4) mistakes about the assumptions of experimental data. In spite of this diversity of possible errors, Mayo thinks that there is a common argument schema underlying tests that have some means of controlling these possible errors, namely the so-called argument from error (the designations A_1 and A_2 are mine):

(A_1) It is learnt that an error is absent when [...] a procedure of inquiry [...] that has a very high probability of detecting an error if (and only if) it exists nevertheless detects no error (Mayo 1996, 184).

(A_2) It is learnt that an error is present when a procedure of inquiry that has a very high probability of not detecting an error if (and only if) none exists nevertheless detects an error (Mayo 1996, 185).

The reader should note that these arguments still make use of the concept of probability. Without further ado, I now examine how Mayo's account handles the ox-phos case.

Consider the following two hypotheses, which I take to be the cores of the chemiosmotic and chemical theories, respectively:

(H_{co}) Coupling of ATP synthesis and electron transport is caused by a chemiosmotic gradient

(H_c) Coupling of ATP synthesis and electron transport is caused by a chemical intermediate

Now, I examine the reconstitution experiments discussed in Section 4.4. Did these experiments constitute a good test of the hypotheses H_{co} and H_c in light of Mayo's account?

As we have seen, the point in Racker's reconstitution experiments seems to have been that they directly demonstrated the operation of a chemiosmotic mechanism in a simplified in vitro system. However, in the first series of reconstitution experiments (an example of which is depicted in Figure 4.4), coupling could, in fact, have been mediated by a unintentionally co-purified chemical intermediate. We could now argue with Mayo that the bacteriorhodopsin experiment (Figure 4.5) had a high probability of detecting this possible error, if it had been present. In the hybrid vesicles prepared by Racker and Stoeckenius, no chemical intermediate could have been co-purified, because that system contained no respiratory enzymes (only bacteriorhodopsin, which has no oxidation–reduction functions). The test detected no error in H_{co}, but it was likely to find an error if there had existed one. This is an argument from error of type A_1. The error that was controlled by this experiment appears to be a mistake about a causal factor.

Conversely, the same experiments can be seen as detecting an error in H_c by a procedure that had a high probability of *not* detecting in error if none existed. According to H_c, the bacteriorhodopsin experiment was unlikely to work because the cause of phosphorylation postulated by H_c (the chemical intermediate) was absent in this experiment. Thus, had H_c been true, the experiment should not have detected an error. Since the experiment did work, it found an error in H_c. Thus, an argument from error of type A_2 applies.

Thus, the crucial experimental evidence that contributed greatly to resolving the ox-phos controversy can be represented as Mayo-style arguments from error. If we consider Mayo's methodology to be an adequate model of experimental reasoning in science, we can now understand why the reconstitution experiments warranted a choice between the two competing theories. At the same time, we are in a better position to understand why the experimental evidence available before 1970 did *not* warrant such a decision: Before the in vitro reconstitution system was developed, it was not possible to control for a major possible error in the chemiosmotic interpretations of the experimental data, namely to attribute energy coupling to the proton gradient even though it was, in fact, mediated by a chemical intermediate. If the intermediate existed, it would always be present in the experimental systems available at that time, namely intact, isolated mitochondria and submitochondrial particles. Only by simplifying the experimental system did it become possible to conduct an experimental inquiry that allowed biochemists to subject the basic assumptions of the two competing theories to a highly severe experimental test.

After having shown that the crucial evidence for Mitchell's theory can be analyzed using Mayo's methodology of theory testing, I am now going to point out a problem with this account.

As so often in the philosophy of science, problems loom whenever we encounter the term probability. According to Mayo, error probability is not to be interpreted as a subjective probability or degree of belief. In other words, error probabilities do not attach to hypotheses or statements. They are *objective properties* of experimental test procedures. Thus, the relevant interpretation of probability is a frequency interpretation. On this interpretation, "[t]he probability of an experimental result is understood as the relative frequency of that result in a hypothetical long run of applications or trials of the experiment" (Mayo 1996, 164). In the context of certain statistical tests, this notion does indeed make sense. Suppose that we want to test the hypothesis that an English lady can tell whether tea or milk has been poured into the cup first (unsurprisingly, this example is originally due to an Englishman, namely R. A. Fisher). The obvious way of testing this hypothesis is to blindfold the lady and offer her a random series of cups of tea for tasting. According to statistical methodology, the hypothesis passes such a trial if an appropriate null hypothesis can be rejected. The best null hypothesis, in this example, says that the lady does no better than chance in telling whether tea or milk has been poured first. Now suppose that a trial returns the result that the lady had the order of milk and tea right in 60% of the cases. If the sample size of the trial is 100, statistical theory predicts that the probability that the lady guessed correctly in 60 or more out of 100 cases is 0.03. In other words, even if the lady does no better than chance, we would expect her to get at least 60 out of 100 tea cups right in three out of a hundred trials. Thus, the probability of erroneously accepting the hypothesis under test amounts to 0.03.

In this example, the error probability can be interpreted as the frequency of trials in which the test passes the hypothesis given that it is false. Thus, this interpretation of error probability requires that there be a class of trials – called the reference class – that includes both trials that pass the hypothesis and trials that fail it. The error probability gives the frequency of trials that falsely return a passing result within the reference class of trials. In this example, the reference class is a class of simple repetitions of the same test.

So far, so plausible. The question is what happens if we transfer the error statistical approach to nonstatistical tests. In experimental biology, for example, in oxidative phosphorylation, experimental tests are often nonstatistical. That is, even though statistical methods play an important role in population biology (e.g., population genetics, ecology), the same is not true of molecular biology or biochemistry. With some exceptions, experimental biologists in these disciplines do not apply statistical tests when they are testing a hypothesis such as Mitchell's chemiosmotic hypothesis. For example, Racker's laboratory, when they were reconstituting oxidative phosphorylation in artificial vesicles, simply observed whether or not the vesicles synthesized a detectable amount of ATP. Of course, they performed many control experiments in order to make sure that the ATP synthesis observed was really due to oxidative phosphorylation and not some experimental artifact. For example, they tested whether oxidative phosphorylation in the reconstituted system displayed the known characteristic properties of the process, such as proton translocation, uncoupling, or sensitivity to specific inhibitors of the enzymes involved (see the next section). We could say with Mayo that what they did in these control experiments was to rule out possible errors. However, what they did *not* do was statistical testing. How, then, can Mayo's account of theory testing be relevant to our ox-phos example?

Mayo's account of experimental testing contains a distinction between a formal and an informal approach. The formal approach consists in actually conducting a statistical test, that is, calculating the error probability or severity of a test according to Neyman–Pearson theory. This approach is applicable in cases where actual test statistics are available, an appropriate null hypothesis can be constructed, and expected experimental distributions can be modeled under the assumptions of the null hypothesis. In the tea-tasting example, this is possible because the expected outcome of trials under the assumptions of the null hypothesis can be modeled by the binomial distribution (assuming that the lady is merely guessing, i.e., that she can only get the order of tea and milk right by chance). In more elaborate cases such as the examples from experimental physics discussed by Mayo (1996, 92–99), the expected experimental distributions given the truth of the null hypothesis can be modeled using Monte Carlo methods.

But there is also an *informal* use of error statistical reasoning according to Mayo. In the informal mode, an experimental test involves constructing *arguments from error* as I have done for the crucial evidence for Mitchell's theory in this section. Arguments from error need not contain formal statistical tests; they can be constructed from any experimental investigation that inquires into the possible errors in inferring to a hypothesis. In my account of the crucial test for Mitchell's hypothesis, I have shown that the famous bacteriorhodopsin experiment in the reconstituted system can be nicely reconstructed as an investigation into the possible errors in the chemiosmotic interpretation of the reconstitution experiments. However, in light of the aforementioned considerations on the interpretation of probability, we must face the following question: What does "probability" mean in an *informal* argument from error?

There are two options for answering this question. First, it could be suggested that "probability" means the same as in a formal error-statistical test, namely a frequency of passing (i.e., false positive) results in a series of repetitions of the experiment. But this immediately raises the question of what the relevant reference class is in an example such as Racker's reconstitution experiments. When we say that the reconstitution experiments had a low probability of passing the chemiosmotic theory had it been false, what is the relevant class of experiments out of which the frequency of passing results is to be calculated? Obviously, it is not a simple repetition of the same experiment, since such a repetition yields more or less the same result if the experiment is done properly (as ascertained by the controls), namely that ATP synthesis occurs. But what else could the reference class be? Is it a class of hypothetical similar experiments, some of which will pass Mitchell's hypothesis while others will fail it? But what are these hypothetical experiments? Unless this question can be given a satisfactory answer, we do not really understand what "probability" means in the informal argument from error. Thus the question becomes what exactly the error-statistical reconstruction of the episode illuminates.

The second option is to deny that "probability" is understood in the frequentist way as in a formal error-statistical test. Perhaps it should just be viewed as an undefined theoretical term of scientific methodology. But this move is not satisfactory either. It immediately prompts the question of what explanatory work the concept of probability does here.

My attempt to apply Mayo's account of theory testing to the ox-phos case has led me to a somewhat skeptical conclusion. Although this account can be used to provide what seems to be a plausible analysis of the crucial test of Mitchell's theory, this analysis leaves us with the question of exactly what the probability of committing an error in an experimental test is. What might be the way out of this conundrum? There are several options. One option is to shrug our shoulders and say: Well, this happens in philosophy. You answer one question and you end up with new ones. We will never answer all of our philosophical questions. This option is cheap, and therefore unattractive. But there is an alternative, which will be explored in the final section of this chapter. This alternative will involve dropping probability altogether.

4.7   THE CONTROL EXPERIMENT

In general, Mayo's approach seems more promising than the Bayesian one for analyzing my case study, because it correctly identifies what biochemists considered to be a good test of the chemiosmotic hypothesis. Furthermore, this account has the advantage that, according to Mayo, how good a test is does not depend on subjective beliefs; it is an objective feature of the test procedure itself. However, in the previous section I have identified a major problem in applying Mayo's account – which is intended as a general methodology for all the sciences – namely that error probabilities are not definable in typical test situations in experimental biology. This problem notwithstanding, there might be aspects of Mayo's overall account that we can use.

Mayo has convincingly argued that a viable account of theory testing needs to pay attention how evidence is *produced* in the laboratory. In particular, it

118

must reflect how scientists scrutinize data, experimental arrangements, possible sources of error, and so forth in order to generate reliable inferences to theoretical claims. Mayo's philosophy of science is precisely an attempt to bring these aspects of experimental practice into focus. The problem with her account, as I have shown in the previous section, is her reliance on statistical methods, in particular the concept of error probability. I shall therefore try to develop an alternative.

The key to an adequate account of experimental testing that works in biology, I suggest, is to accept Mayo's central insight that successful experimentation involves *controlling error*. However, "error" should not be understood in a statistical sense, nor should such an account rely on error probabilities. Biologists have developed strategies for controlling error that do not involve statistical reasoning, nor do they involve probabilities. These methods are *qualitative*; however, they are no less reliable. In order to develop this point, I go back to my case study. Specifically, I examine a particular set of experiments in greater detail.

The experiments I want to examine more closely are the first reconstitution experiments with artificial vesicles done in Racker's laboratory beginning in 1970 (Test 5). Racker's group purified the hydrophobic proteins from bovine heart mitochondria by extracting submitochondrial particles with the detergent cholate and precipitating the resulting solution with ammonium sulfate. After centrifugation, they suspended the pellet in a solution of soybean phospholipids. Electron microscopy showed that this preparation yielded phospholipid vesicles with a diameter of about 0.1 μm. When they added "coupling factors" (a soluble fraction that contained parts of the ATPase that had been previously been removed), these vesicles catalyzed the ^{32}P-ATP exchange reaction, that is, the incorporation of radioactively labeled phosphate into ATP. This was the first time that any partial reaction of oxidative phosphorylation (other than ATP hydrolysis) was measured in a system that contained no intact mitochondria or submitochondrial particles. Furthermore, Racker and co-workers were able to demonstrate that their reconstituted ATPase can function as a proton pump; that is, it generated a pH gradient across the liposome membrane upon addition of ATP. All this spawned the promise that the new reconstitution system with the artificial vesicles would eventually lead to the reconstitution of full-blown oxidative phosphorylation (which it did; see Section 4.4).

What I would like to draw attention to is the complex set of *control experiments* that accompanied the demonstration of ^{32}P-ATP exchange in reconstituted vesicles by Kagawa and Racker (1971). One set of control experiments is shown in Table 4.3. It shows the effect of various chemicals on the

Table 4.3. *Control Experiment I Showing the Response of Whole Mitochondria,*
Submitochondrial Particles (SMP), and the Reconstituted Vesicles (33P and
33–50P) to Various Chemicals

Additions	Mitochondria	SMP	33P	33–50P
None	2040.0	1516.0	145.2	203.0
Nigericin	2206.0	1375.0	153.6	176.4
TPA	182.4	1588.0	103.8	105.2
TPA + nigericin	60.8	1296.0	88.2	94.8
TPB	1988.0	508.0	27.2	12.3
TPB + nigericin	2080.0	14.8	2.6	2.9
FCCP	46.0	84.4	2.4	0.0

Note: The numbers express the rates by which the preparations exchanged radioactive phosphate with ATP (in nmoles/mg/10min). Nigericin is an antibiotic that makes membranes permeable to ions. TPA and TPB are synthetic, lipid-soluble ions that affect the transmembrane electric potential. FCCP is an uncoupling agent. Note that the response profile of the reconstituted vesicles is more similar to that of submitochondrial particles than to that of whole mitochondria, which is explained by their similar structure (inside-out orientation, only a single membrane).
Source: Redrawn after Kagawa and Racker (1971).

reconstituted vesicles as well as on whole mitochondria and submitochondrial particles. For example, the uncoupler p-trifluoromethoxyphenyl hydrazone (FCCP) completely abolished the exchange reaction in whole mitochondria, in submitochondrial particles (SMP), and in the artificial vesicle preparations. The response of the reconstituted vesicles to a series of other reagents such as the antibiotic nigericin was similar to that of submitochondrial particles, which was to be expected due to their inside-out orientation. Kagawa and Racker also tested the response of the reconstituted system to antibodies[17] specific for the ATPase (Table 4.4). In comparison with an unspecific control antiserum, these antibodies inhibited both the ^{32}P-ATP exchange reaction and the ATPase activity, showing that the ATPase itself was responsible for the observed ^{32}P-ATP exchange. The lack of a significant effect with the control antiserum made sure that the inhibition observed was not caused by some unspecific factor present in the antiserum.

Methodologically, what we have here is an instance of John Stuart Mill's "method of difference." In essence, Mill suggested that some factor C can be inferred to be a cause of some effect E if C is the only difference between a situation in which E is observed and a situation in which E is not observed (Mill 1996 [1843], 391). This principle is frequently employed in control experiments in biochemistry. By its application to our present control experiment, it was inferred that a specific anti-ATPase antibody caused the inhibition because the presence of this antibody was the only difference

Table 4.4. *Control Experiment II Showing the Effect of Antibodies Specific for ATPase (Anti-F_1 Serum) on Submitochondrial Particles (SMP) and Reconstituted Vesicles (33P and 33–50P)*

Additions	SMP	SMP + P-lipids	33P	33P–50P	F_1
$^{32}P_i$-ATP Exchange:					(nmoles/μg/10min)
Control serum	364	443	63.3	173	
Anti-F_1 serum	82	61	38.4	61	
ATPase:					
Control serum	2150	2200	295	455	1170
Anti-F_1 serum	1250	1150	155	355	350

Note: F_1 is a water-soluble portion of the ATPase enzyme, which can catalyze only ATP hydrolysis. The numbers in the upper two lines give the rate of exchange of radioactive phosphate in ATP (in nmol/500 μg/10min, except in the far right column). In the lower two lines, the numbers give the rate of ATP hydrolysis (in nmol/500 μg/10min). The control with submitochondrial particles and phospholipid was done in order to rule out that the phospholipids effect the binding of the antibodies.

Source: Redrawn after Kagawa and Racker (1971).

between the control antiserum and the anti-ATPase antiserum (the control antiserum also contains antibodies, but none that specifically bind ATPase).

These control experiments were necessary to rule out the possibility that the reactions observed in the reconstituted vesicles were some unspecific reactions that were unrelated to oxidative phosphorylation. In other words, such control experiments are designed to eliminate *experimental artifacts* (see Section 9.5). These experiments showed that the reactions in the reconstituted system shared the main characteristics of oxidative phosphorylation known from studies of intact mitochondria and submitochondrial particles and, therefore, the reactions were no experimental artifact.

Reconstitution of partial reactions of oxidative phosphorylation using artificial phospholipid vesicles proved to be a powerful experimental system. However, before it became possible to reconstitute oxidative phosphorylation itself, a major difficulty had to be resolved: As Mitchell's theory predicts, coupling between electron transport and phosphorylation crucially depends on the orientation of the respiratory chain and ATPase enzyme complexes. This is because of their spatially asymmetric mode of action. The respiratory chain was thought to translocate protons from the inside of the mitochondrion to the outside. The ATPase uses an inward proton current to generate ATP. In order to reconstitute this process or parts of it in artificial phospholipid vesicles, it was necessary to incorporate the enzyme complexes into the artificial membrane in the right orientation.

The first attempts to reconstitute cytochrome oxidase (the terminal respiratory chain enzyme) together with the ATPase failed because the reconstitution procedure used incorporated the cytochrome oxidase complex randomly in both orientations (Racker 1976, 121). Obviously, the enzyme cannot generate an electrochemical potential that way. This difficulty was resolved by applying a procedure to remove external cytochrome c. Cytochrome c is the electron donor for cytochrome oxidase and, apparently, it must be available for the oxidase on the right side of the membrane. If cytochrome c is completely depleted from the outside of the vesicles, the enzyme molecules sitting in the membrane in the wrong orientation cannot receive electrons from cytochrome c, allowing those molecules with the right orientation to generate a proton gradient (using the cytochrome c available from the inside of the vesicles). The ATPase, it seems, inserts preferentially in one orientation. Thus, Racker and Kandrach (1971, 1973) obtained vesicles containing cytochrome oxidase, cytochrome c, and ATPase that consumed oxygen and catalyzed ATP synthesis with a P:O ratio of 0.5 or higher.[18] The reconstituted system was uncoupled by externally added cytochrome c. This was readily explained by the fact that the vesicles contained cytochrome oxidase complexes in both orientations: externally added cytochrome c induces the molecules that were incorporated in the wrong direction to destroy the proton gradient. Racker (1976, 121) pointed out that uncoupling by cytochrome c was difficult to explain under the chemical coupling theory and that, therefore, the initial problems of reconstituting cytochrome oxidase turned out to be a "blessing." That is, an initial problem was successfully turned into another control experiment.[19]

I suggest that a very large part of the experimental work going into reconstitution had the purpose of containing possible experimental errors. In other words, a lot of experiments were performed simply in order to make sure that the data were not misinterpreted. If an artificial vesicle containing purified enzymes makes ATP, this *could* be because the vesicles perform some of the steps of oxidative phosphorylation. But the amount of ATP made was tiny, and it could be due to, for example, contamination by residual amounts of other enzymes present in the experiments. Thus, the kind of error the biochemists had to control for was falsely attributing the measured ATP synthesis to reconstituted ox-phos although it was, in fact, caused by contaminants. The strategy to control for such errors involved systematically varying the experimental conditions. For example, if the reconstituted vesicle system had still synthesized ATP in the presence of a strong detergent (which dissolved membranes), this would have indicated that the ATP synthesis was due to contamination. Similarly, had ATP synthesis in the reconstituted system proved

122

to be insensitive to known specific inhibitors of oxidative phosphorylation (such as oligomycin or uncouplers), this would have been bad news. Such control experiments serve to rule out false attributions of the experimental results, that is, thinking that the effects observed are due to some process while they are, in fact, due to some other process (e.g., a contaminant).

Thus, Mayo is right to place so much emphasis on the study of the nitty-gritty of experimental practice and to suggest that most of this work serves the purpose of controlling for possible errors. Only once all the possible ways in which the interpretation of experimental data could go wrong have been checked with suitable control experiments will an experimental biologist announce that he or she has, for instance, "demonstrated the occurrence of oxidative phosphorylation in a reconstituted system."

However, Mayo errs in thinking that controlling for error always involves calculating severity in the formal mode or arguing from error in her informal sense. Her informal arguments from error are still too formal because they invoke probabilities. I suggest that Mayo commits the same mistake as the Bayesians in seeking formal unity in scientific inference. As Ronald Giere (1997) and others have argued, there is no unity of this kind to be found in scientific methodology; there is only a set of different methods that are suited to different tasks. Thus, we need to examine the specific ways in which experimental results in biology are subjected to critical scrutiny.

I suggest that the main strategy used in "wet" biological research is to vary the experimental conditions systematically in ways that are designed to eliminate disturbing causal factors. In my example, such factors were eliminated by adding detergents, inhibitors, antibodies, and so forth and by showing that ATP synthesis ceased under these conditions. Experimental biologists do not need statistical methods because they usually have sufficient control over their experimental systems through causal interventions.

At this point, the following question arises: Since the earlier evidence from the 1960s was also accompanied by suitable control experiments, why did this evidence fail to resolve the ox-phos controversy? In order to answer this question, it is important to ask *what* precisely is being tested by a controlled experiment.

As we have seen, Mitchell and Moyle (1967) demonstrated the occurrence of proton transport (Test 1). In order to establish that their measurements supported this conclusion, they also performed control experiments. For example, they showed that proton transport was abolished when they added the detergent Triton X-100, which dissolves biological membranes (Mitchell and Moyle 1967, 67). This was an important control to make sure that the observed pH drop observed in the mitochondrial suspension after the oxygen pulse was, in fact, linked to respiration and not caused by some unrelated process, for example, the release of some acid. Therefore, Mitchell and Moyle's result was backed by control experiments; they did control for some possible errors. However, one potential error they could *not* control for in these experiments was the erroneous attribution of proton translocation to the energy-transfer reaction of oxidative phosphorylation although proton translocation was, in fact, a side-reaction of ox-phos (i.e., an energy-linked function, as Slater and others believed). For similar reasons, all the other evidence obtained in the 1960s did not strongly support the chemiosmotic theory, because this crucially important possible error could not be ruled out. Only Racker's artificial vesicle system had the capacity to control for this error, because it allowed the observation of ATP synthesis in the absence of respiratory enzymes (e.g., by substituting the respiratory enzymes by bacteriorhodopsin in reconstitution experiments). What this suggests is that experimental evidence only supports a theoretical claim to the extent that it "probes" the possible ways in which *this particular claim* could be in error.

According to Mayo, experimental testing of theoretical proceeds in a "piecemeal" fashion. She differentiates between three different kinds of hypothesis that are subject to experimental testing in a typical scientific investigation (1996, 129): (1) *Primary hypotheses* are the main theoretical questions that are ultimately the goal of the inquiry. Theories subject to testing are broken down into sets of primary hypothesis, which are tested individually. (2) *Experimental models* are hypotheses that link the primary hypotheses to the data. More precisely, experimental models link primary hypothesis not to raw data, but to so-called *data models*. (3) Raw data are always processed, and it is these processed data that are linked to the experimental models. Data models might include, for example, a hypothesis stating that a measurable quantity takes a certain value under some specified experimental conditions.

The crucial point in Mayo's methodology, then, is that the assumptions of the data models, experimental models, and primary hypotheses must be tested *independently*. In other words, scientists must try to eliminate errors in all the assumptions that were made in constructing the data models and experimental models, as well as possible errors in inferring to the primary hypothesis. This is also the key to solving Duhem's problem. The possibility of locating the culprit in the case of an anomaly then turns on the extent to which possible mistakes in these assumptions can be eliminated.[20]

How do Mayo's three levels of inquiry map onto our biochemical experiments? Let us begin with the data models. An example of a data model is Mitchell's and Moyle's result that, in a particular series of experiments with mitochondrial suspensions, the pH of the suspension dropped by a certain value. The exact value, of course, will vary from experiment to experiment. In the written report that was published, the authors both show results from individual experiments and averaged values from several experiments (e.g., Mitchell and Moyle 1967, 59). The latter is the *data model*. The *experimental model* states that mitochondria eject protons and that this effect is linked to respiration. In order to establish this claim, Mitchell and Moyle had to control for erroneous attribution of the measured pH drop to respiration although it was, in fact, due to some unrelated process. Thus, Mitchell and Moyle were able to control for some possible errors at the level of experimental models. However, Mitchell and Moyle were unable to control for a possible error in linking this experimental model to the primary hypothesis. The primary hypothesis, in this case, was the claim that proton transport couples respiration to phosphorylation. Mitchell and Moyle's experimental models obtained in 1967 were consistent with this hypothesis; however, these experimental models were not a crucial test.

On my analysis, the reason that it was not a crucial test lies in the fact that Mitchell and Moyle had no way of controlling for the possible error that energy coupling was attributed to the proton gradient although it was, in fact, caused by another process (e.g., a chemical intermediate). Thus, although the possible errors had been controlled for at the level of experimental models, a major error in inferring to the primary hypothesis could not be ruled out. Indeed, Mitchell and Moyle's experimental model of 1967 (which only says that there is proton transport, not what its role is in the mechanism of ox-phos) was widely accepted, whereas the primary hypothesis (which says that the proton gradient couples respiration to phosphorylation) was not.

By contrast, Racker's reconstitution experiments were able to control for possible errors at all three levels. Experimental controls such as the ones shown in Tables 4.3 and 4.4 served to rule out possible errors in the experimental model, which stated that a system composed of reconstituted cytochrome oxidase, cytochrome *c*, artificial membranes, and ATPase performed the last step of oxidative phosphorylation. This experimental model was still consistent with the chemical hypothesis; that is, it was still possible that a chemical intermediate was somehow involved. But the bacteriorhodopsin experiment finally allowed a controlled test of the primary hypothesis that had been the main bone of contention in the entire controversy, namely, that the chemiosmotic gradient is sufficient for coupling respiration to phosphorylation. The reason is that this experiment allowed the researchers to rule out the possible error of attributing energy coupling in the reconstituted system to the chemiosmotic gradient although it was, in fact, caused by chemical intermediates. In a sense, thus, the famous bacteriorhodopsin experiment was a control experiment for the reconstitution experiments involving respiratory enzymes.

To sum up, our case study from the history of biochemistry supports the general approach to scientific methodology advanced by Deborah Mayo, with the exception of her claim that inquiries into error either are Neyman–Pearson statistical tests, that is, involve the calculation of error probabilities, or are informal analogs of such tests. My analysis of a concrete historical case shows that ruling out error plays a major role in experimental biology, but experimental biologists have developed nonstatistical strategies of controlling for experimental errors at all levels of an experimental inquiry. The main strategy is to vary experimental conditions in a way that rules out disturbing causal influences; this is canonically known as "control experiments."

SUMMARY

My methodological analysis of the oxidative phosphorylation controversy supports the following conclusions. First, the difficulty of comparing the chemical and chemiosmotic theories was due to the fact that the two theories made different predictions, that is, predictions for which there was no corresponding prediction of the other theory. Second, the two theories were not empirically equivalent, nor were they equally well supported by the evidence. The problem of underdetermination as philosophers have perceived it was thus not relevant. Furthermore, I have shown that Duhem's problem was solvable at least in some experimental situations. Third, the Bayesian approach fails to illuminate why some evidence was considered to be better than other evidence. Fourth, the production of evidence is relevant to the question of whether an experimental inquiry constitutes a good test of a hypothesis. Such an inquiry tests a hypothesis only to the extent that it examines the possible errors in inferring to the hypothesis. Such errors may arise at the level of data models, experimental models, or primary hypotheses. Fifth, in experimental biology, controlling for errors does usually not involve the calculation of error probabilities, nor any other statistical methods or informal analogs thereof. Instead, experimental biologists control for possible errors by systematically manipulating the experimental conditions in order to rule out disturbing causal factors and false causal attributions.

5

Experimental Systems: A Life of Their Own?

The philosophy of science used to be mainly occupied with scientific theories and concepts and paid very little attention to experiments. Logical empiricists, for example, placed much emphasis on the logical structure of theories and the conceptual foundations of a few selected sciences, especially theoretical physics. The focus on theories and concepts remained intact, to a large extent, even through the upheavals in the philosophy of science that have been inflicted by Thomas Kuhn and Paul Feyerabend. Anglo-American philosophy of biology, too, has largely remained within the theory-centered tradition almost up to the present; a substantial part of the work done in this area has been on evolutionary *theory* (see Chapter 1).

The neglect of experimentation in twentieth-century philosophy of science seems to be somewhat at odds with the fact that the two major traditions which have dominated the field – logical empiricism in North America and critical rationalism in Europe – are both thoroughly empiricist in orientation. What this means is that both of these traditions firmly hold that theoretical claims are ultimately justified by experience. Why, then, did philosophers devote so little attention to the question of how systematic experience is acquired by scientists in the laboratory or in the field? The reason is not simply negligence, but a particular view of the role of experience in science. The roots of this view lie in the traditional distinction between context of discovery and context of justification (see also the introduction to Chapter 3). What role does experience – experiment or observation – play in this scheme? On the traditional view, the proper role for experience *in the context of justification* is to provide premises for scientific inferences.[1] Popperians, for instance, think that experience enters as "observation statements" that serve as premises in deductive arguments that can be used to falsify theories. Inductivists also think that observation statements function as premises in scientific arguments; however, in their view, these arguments are not deductive but inductive.

Thus, in both critical rationalist and logical empiricist philosophy of science, observation and experiment have the sole purpose of providing premises in arguments that purport to justify scientific theories. Any other roles that experimental and observational practices may have fall into the context of discovery and therefore outside the proper domain of the *epistemology* of science. As a consequence, a large part of the philosophical debates in the twentieth century have centered on the question of what kinds of arguments from empirical premises are legitimate (see the introduction to Chapter 4).

After having explained (not justified!) why the philosophy of science has paid so little attention to experiments, I briefly review how the traditional consensus on the epistemological status of experiments was displaced.

In a book that is equally provocative and influential, Ian Hacking has called for a reexamination of the epistemological significance of experimentation. According to Hacking, experimentation has "a life of its own" (1983, 150, 165). What this means is that a large part of experimental practice is independent of higher-level theories and does not have the purpose of testing such theories. The growth of experimental knowledge thus has its own internal dynamics, which the philosopher or historian who only looks at theories will miss. Furthermore, Hacking argues that answers to some of the old philosophical questions concerning the nature of scientific knowledge can only come from closely examining what experimental scientists do in the laboratory. For Hacking, this concerns, in particular, the question of whether the unobservable entities postulated in scientific theories (such as electrons, genes, or quarks) actually exist, which is part of the problem of scientific realism (see Chapter 9).

Hacking has been joined by many other philosophers and historians of science[2] of quite different persuasions in a movement that Ackermann (1989) has dubbed "the new experimentalism." New Experimentalists are by no means a monolithic block; in fact, a broad variety of positions and approaches to reconstructing scientific knowledge and research have been subsumed under this banner. I now briefly try to isolate what I take to be the main trends in the New Experimentalist movement.

First, there is the idea that experimentation serves purposes in scientific research other than testing high-level theories. For example, an important part of experimentation is *explorative*; that is, it aims at discovering new phenomena or empirical regularities. Explorative experimentation does not require a preexisting theory. Instead, this kind of experimentation provides some of the empirical regularities and phenomena that the scientist then will try to explain with the help of theories. As Friedrich Steinle (1997, 1998) has shown for the example of Ampère's work on electromagnetism, explorative experimentation can be quite systematic; it is by no means a mere trial-and-error procedure. The lack of a guiding general theory does not make explorative experimentation "blind."

A second, related idea of the new experimentalism is that there is experimental knowledge that has its own internal stability which may even survive major transitions in the high-level theories (Carrier 1998, 180). Furthermore, some New Experimentalists reject the view that observation (in the broad sense that includes observation and experimentation) is "theory-laden," a view that came to prominence with Kuhn and Feyerabend (Hacking 1983, 171–185). New Experimentalists do not claim (to my knowledge) that there is such a thing as observation that is independent of *any* theoretical beliefs. Rather, the point is that the theoretical beliefs that inform observation are sometimes independent of the high-level theories that are subject to experimental proof.

Third, New Experimentalists have given new life to the distinction between observation and experimentation. The main difference consists in the fact that, in experimentation, scientists actively intervene with the natural objects or systems they are studying. The question, of course, is whether this difference is in any way epistemologically relevant. The reason that traditional philosophy of science has not made a distinction between observation and experimentation – they were both viewed as possible sources for "observation statements" – is that it was thought that this difference is of no systematic relevance for epistemology (McLaughlin 1993). By contrast, New Experimentalists maintain that this difference matters. For example, Hacking argued that experimental intervention could provide grounds for a certain kind of scientific realism (see Section 9.3). To give another example, which is closer to home, my account of theory testing given in Section 4.7 suggests that the experimentation/observation distinction indeed matters in experimental biology, because scientists often need to actively intervene with a system under study in order to control for errors.

Fourth, some New Experimentalists have challenged the idea that theories are somehow matched to nature on the basis of experimental results. Instead, they opt for a *coevolution* or *co-construction* of theories and experiments. More precisely, theories are used to interpret the results of experiments, to design measurement apparatus, etc. Any friction encountered in this process is removed by adjusting both the theory *and* the devices used to generate data in order to create what has been termed "stable laboratory science" (Hacking 1992; Gooding 1992). This provides a nonrealist explanation for the match between theory and data (Carrier 1998, 178).

Fifth, some philosophers have stressed that much more attention must be given to the details of experimental practice in order to answer some of the old questions concerning scientific inference and theory testing. I have already examined this side of the New Experimentalism in Chapter 4. My conclusion there was that understanding the reasoning of experimental biologists indeed requires that we examine how evidence is produced in the laboratory; it is not sufficient simply to examine the relationship between theory and evidence.

A systematic appraisal of all the different positions and approaches that are associated with the New Experimentalism is beyond the scope of this book. There can be no doubt that we are dealing here with an extremely broad and important development that has fundamentally transformed the history and philosophy of science in recent years and that has offered fresh and rich perspectives on a large number of scientific disciplines (see, for example, Gooding, Pinch, and Schaffer 1989; Pickering 1992; Heidelberger and Steinle 1998). Although most of this work initially focused on experimental physics, experimental practice has also become a major topic in the history of biology. In recent years, historians of biology have produced a wealth of detailed studies of the experimental systems, research tools, and model organisms of various experimental disciplines including physiology, genetics, biochemistry, and molecular biology (see Chapter 6, note 1 for references). This work has departed considerably from older historical works, which have focused mainly on the history of ideas and theoretical innovations.

The goal of this chapter and the following one is to assess what exactly we can learn from studying experimental practice in biology. In addition, I want to examine the relationship of this work to the approach to the epistemology of experimentation in biology that I have pursued in the previous chapters. In the following section, I present an approach to experimentation in biology that accords an important role to experimental systems. In Section 5.2, I reexamine my historical case study from Chapter 4 under this new perspective. In Section 5.3, I discuss the extent to which this new approach is really an alternative to the methodological approach, rather than a complementary way of looking at experimental practice. In particular, I focus on the role of epistemic norms in the development of experimental science. This is followed by some general considerations on the relationship between historical explanations of scientific change and normative standards of scientific methodology (Section 5.4).

5.1   EXPERIMENTAL SYSTEMS AS FUNCTIONAL UNITS OF RESEARCH

The notion of an experimental system is familiar to experimental biologists. In the standard laboratory vernacular it seems to designate a combination of biological materials (e.g., tissues, cells, bacterial strains), instruments, preparation procedures, measurement techniques, and controls that serve to study a particular biological process. As every graduate student knows, the success of a research project crucially depends on whether the experimental system "works" – whatever this means precisely. As a consequence, a considerable part of experimental practice consists in establishing experimental systems that work.[3]

Hans-Jörg Rheinberger (1993, 1996, 1997) has turned this pretheoretical notion of an experimental system into a theoretical concept for explaining the dynamics of experimental research.[4] One of the starting points for his account is a remark found in François Jacob's autobiography, namely that experimental systems are "machines for making the future" (Jacob 1988, 9).[5] What this suggests is that experimental systems have a major influence on the specific directions that an experimental investigation takes. If this is true, obviously, any historical account of some biological discipline that focuses exclusively on theories will fail to understand why the discipline developed the way it did. Instead, the historian of experimental biology must closely study the experimental setups used by biologists in the laboratory and examine their role in the research process.

It should be mentioned at the beginning that Rheinberger does not claim that theories play *no* role in biological research. Instead, "experiment and theory in the life sciences are so intricately interwoven that the function of the experiment as an instance of testing hypotheses appears to be largely marginal" (Hagner and Rheinberger 1998, 363). Thus, according to Rheinberger, experiments and experimental systems have crucial functions in the research process that are not related to testing preexisting theories. In the following, I try to reconstruct Rheinberger's theory of experimental systems.

One of the core postulates of Rheinberger's theory is that experimental systems are "the smallest integral working units of research" (1997, 28). Biological research, on Rheinberger's account, "begins with the choice of a system rather than with the choice of a theoretical framework" (25). Furthermore, he characterizes experimental systems as "systems of manipulation designed to give unknown answers to questions that the experimenters themselves are not yet able clearly to ask" (28). Thus, according to Rheinberger, experimental research in biology does not usually start with a theory or with well-formulated research questions.[6] Experimental systems may precede the problems that they eventually help to solve. What is more, experimental systems do not just help to *answer* questions; they also help to *generate* them. Where this process leads is impossible to predict; experimental systems give rise to unexpected events. Thus, Rheinberger draws our attention to the unpredictable, open-ended nature of the research process – an aspect that seems so fundamental to basic science and that most philosophers of science have completely ignored.

What Rheinberger is advancing is a theory of the dynamics of scientific change. According to his theory, the development of experimental disciplines in biology is not mainly driven by ideas or theories. Instead, the driving force is provided by the intrinsic capacities of experimental systems, which the scientists are constantly trying to explore. It is important to understand that experimental systems contain both the material and cognitive resources required to do experiments. Material resources include the biological tissues or cells that are under study, the preparation tools (e.g., centrifuges), and the measurement instruments (e.g., counters for measuring radioactivity). Cognitive resources (my term) include the practical skills required to operate the apparatus, as well as some theoretical knowledge needed for designing experiments and interpreting the data. That Rheinberger means to include cognitive resources in his experimental systems is evident, as "such systems contain all the conditions necessary for a research process in its entirety" (Hagner and Rheinberger 1998, 359). The developmental dynamics of experimental systems is not bounded, for example, by traditional disciplinary boundaries. However, all this is not intended to mean that experimental systems always develop in isolation. For example, the development of experimental systems sometimes leads to the amalgamation of two (or more) mutually independent systems, a process that Rheinberger refers to as "hybridization" (1997, 135).

An analytic concept that is closely tied to the notion of experimental system is that of "epistemic things." This concept signifies the "material entities or processes – physical structures, chemical reactions, biological functions – that constitute the objects of inquiry" (Rheinberger 1997, 28). Epistemic things can unexpectedly appear and disappear or be reconstituted in new guises as experimental systems are developed. Sufficiently stabilized epistemic things can "turn into the technical repertoire of the experimental arrangement" (1997, 29) and thus become "technical objects."

The main historical episode that Rheinberger's theory of experimental systems is designed to account for is the research on protein synthesis in the 1940s, 1950s, and 1960s that eventually led to the cracking of the genetic code. The experimental system in question is an in vitro system for protein synthesis. In this system, rat liver cells or *E. coli* cells are broken up by mechanical homogenization. The homogenate is then subjected to centrifugation in order to remove cellular debris. In order to observe protein synthesis in such cell-free systems, biochemists add radioactively labeled amino acids and measure the incorporation of this radioactivity into protein. Such in vitro systems were developed and refined over many years of research. In the 1960s, they played a central role in the cracking of the genetic code in the elucidation of the steps of protein synthesis. A brief presentation of Rheinberger's account of this episode will be helpful for grasping his theory of experimental systems.

After Watson and Crick's (1953) discovery of the structure of the DNA molecule, one of the major problems that occupied molecular biologists was the "coding problem," that is, the question of how linear DNA sequences specify the sequence of amino acid building blocks in protein molecules. The critical step toward cracking this "code" was taken by M. W. Nirenberg and J. H. Matthaei in 1961, when they showed that an in vitro system for protein synthesis derived from extracts of *E. coli* cells could be programmed with synthetic RNA molecules such as polyuridylic acid (poly-U).[7] In the in vitro system, poly-U caused the synthesis of an artificial protein containing only the amino acid phenylalanine. In this way, the first genetic code "word" (codon) UUU could be assigned to the amino acid phenylalanine (see also Judson 1979, 470–480). This experimental system proved to be a much more powerful instrument for solving the coding problem than any of the mathematical and cryptographic methods that had been applied to the problem by physicists with the help of the first electronic computers (see Kay 2000).

Focusing in particular on the work done in Paul Zamecnik's laboratory at Massachusetts General Hospital in Boston, Rheinberger describes the development of the protein-synthesizing in vitro system that was later used by Nirenberg and Matthaei. His main conclusions are the following: First, the experimental system came out of cancer research in the 1940s. It was known that cancer cells differ from normal cells, among other things, in the regulation of metabolism. In particular, cancer cells show alterations in the rate of protein synthesis. For this reason, the regulation of protein synthesis seemed to be a promising topic for understanding how cells are transformed into tumor cells, and Zamecnik's early work was directed at this problem.

Second, the experimental system moved across *disciplinary boundaries*, from cancer research into biochemistry, and from biochemistry into molecular biology. This exemplifies one of the senses in which experimental systems have "a life of their own"; they do not respect traditional organizational boundaries. The protein-synthesizing system, Rheinberger argues, had a major impact in redefining the disciplinary landscape of biomedical science, biochemistry, and molecular biology. In particular, the system changed substantially what it meant to be a biochemist or a molecular biologist.

Third, several *technological innovations* were crucial in the development of the in vitro system. One of these innovations consisted in the use of radioactive amino acids as tracers for protein metabolism. The amount of protein synthesized by in vitro systems was exceedingly small, too small to be detected by chemical methods. However, amino acids containing the radioactive isotope ^{14}C could be detected readily by radioactive decay counting techniques.[8] The other major technological innovation consisted in new techniques for cell fractionation that were made possible by new, powerful ultracentrifuges. With these techniques, it was possible to isolate various cellular constituents and separate them by size and molecular weight.

Fourth, the in vitro system eventually served a purpose for which it was not designed. Solving the coding problem was not what Zamecnik and his co-workers set out to do; they were initially interested in the regulation of protein metabolism in cancer cells. Nirenberg and Matthaei, when they started to use Zamecnik's in vitro system, first wanted to use it for synthesizing specific proteins. Furthermore, the use of synthetic RNA molecules initially served a purpose totally unrelated to the cracking of the genetic code; they first used polyadenylic acid as an inhibitor of nucleases (enzymes that digest DNA or RNA). Later, they used these synthetic RNA molecules as *controls* in order to show that the protein synthesis they observed (with natural mRNA) was template-specific. This is how they came across the effect that polyuridylic acid stimulated the incorporation of phenylalanine into protein. Thus, synthetic RNA suddenly moved from the periphery of the experimental system right into its center, and when that happened, the direction of protein synthesis research changed dramatically. Rheinberger concludes that Nirenberg and Matthaei's famous poly-U experiment was a direct consequence of "exploring the experimental space of cell-free protein synthesis according to the cutting-edge standards of the biochemical state of the art" (1997, 212). This example illustrates how experimental systems – construed, as Rheinberger does, as amalgamations of material objects and practical skills – can generate new, unexpected findings pretty much by playing out their intrinsic capacities. Experimental systems can indeed function as "generators of surprises." They can yield answers to questions that were not even asked by the scientists who developed the systems, or perhaps not even by the very scientists who did the experiment that provided important answers.

Fifth, the role of theory. As already emphasized, Rheinberger does not claim that theories play no role in biological research. The cracking of the genetic code in the 1960s was preceded by many years of theoretical work, in which physicists-turned-molecular-biologists invented a large number of possible schemes or mechanisms by which DNA could direct the synthesis of specific proteins. As Kay (2000) shows, some of this theoretical work drew heavily on the mathematical theory of information.[9] However, on Rheinberger's account, this theoretical work developed independent of the work on protein synthesis in Paul Zamecnik's and other laboratories. Initially, this work "was embedded in a research program saturated with considerations of biochemical energy flow and the search for intermediates of metabolic reaction chains. Reasoning in terms of molecular information transfer, therefore, did not and could not restyle the system as a whole and at once" (Rheinberger 1997, 187).

Thinking in terms of information transfer slowly crept into the protein synthesis work. Rheinberger shows how this kind of reasoning became important in Zamecnik's and other laboratories as a way to disentangle the roles of the different RNA molecules that seemed to be involved in protein synthesis in the in vitro system, namely transfer RNA (tRNA) and messenger RNA (mRNA). tRNA was shown to be the long-sought "adaptor" molecule postulated by Francis Crick in 1955, that is, the molecule that basically "reads" the genetic code by binding specific amino acids and transferring them into the protein assembly process according to the "code words" or codons on the mRNA template. mRNA, as already mentioned, was shown to transmit the genetic information from DNA to protein. The theoretical concepts associated with information, that is, "coding," "message," "template," and so forth, were instrumental in understanding protein synthesis, but the in vitro system that produced the epistemic things to which these concepts could be applied came into existence without the help of information-theoretic concepts. These concepts, as Rheinberger puts it, "surreptitiously had crept into the experimental discourse of biochemistry rather than governing it from the beginning" (1997, 221).

In summary, Rheinberger's picture of the research process in experimental biology differs substantially from the traditional philosophical accounts, according to which smart theoreticians think up theories for skillful experimenters to test, or ask questions for experimenters to answer. Biological research, according to this intriguing picture, can only be understood as an ongoing interaction with experimental systems. A large part of this interaction consists in exploring the space of possible manipulations that an experimental system offers. This activity often does not require a well-formulated research question or a specific theory to test. Sometimes, it can only be said in retrospect what question an experiment answered; the question might not even have been asked beforehand.

135

In the next section, I reexamine my case study from Chapter 4 from the experimental systems perspective.

5.2 OX-PHOS REVISITED

At first sight, it might seem that my example of oxidative phosphorylation (presented in Chapter 4) and Rheinberger's example of protein synthesis and the cracking of the genetic code differ fundamentally. In the ox-phos case, we seem to have a clear example of a theoretician (Mitchell) who freely and speculatively invents a theory, which is then put to a series of critical tests by experimenters.[10] In other words, the ox-phos case seems to be much more *theory-driven* than the protein synthesis case. However, it is worth examining whether we could tell an altogether different story about the same case, a story that accentuates the similarities rather than the differences to Rheinberger's story of protein synthesis and the genetic code. This is my objective in the current section. I investigate whether the ox-phos case presents itself differently if we focus our attention on the laboratory context, on the experimental systems and their development, rather than on the competing theories. I begin by examining the context in which Mitchell's theory originated.

The origins of Mitchell's involvement in oxidative phosphorylation research are situated in the postwar years at Cambridge University, where Mitchell obtained his Ph.D. in 1950 for work on the mechanism of action of penicillin.[11] While Slater and most of the other biochemists involved in ox-phos research were more or less straightforwardly trained in the field of intermediary metabolism (Slater 1997), Peter Mitchell's scientific background was more diverse (Mitchell 1981; B. Weber 1991; Prebble and B. Weber 2003). His Ph.D. supervisor was James Danielli, who was interested in the structure and function of biological membranes, especially the transport of water-soluble molecules across phospholipid membranes.[12] Membrane transport physiology was a distinct specialty at that time (see Robinson 1997 for a history of the field). Mitchell's supervisor notwithstanding, the general outlook of the Biochemistry Department at Cambridge during the 1940s, where Mitchell received his scientific training, emphasized traditional enzymology. David Keilin, who had pioneered the work on cytochromes and their role in cellular respiration, was also at Cambridge, at the Molteno Institute. It thus seems clear that Mitchell was situated at the boundaries of several research disciplines, which was probably crucial for his innovative potential. Mitchell (1981, 4) recalls: "I could not but be impressed by the great divergence of outlook, and even mutual antagonism, between the students of membranes and transport on the one hand and the students of metabolic enzymology on the other hand – and I soon determined to try to understand both points of view in the hope that they be brought together."

This interest in bringing together membrane transport physiology and metabolism, which he seems to have acquired early on, shows throughout Mitchell's career. During the 1950s, he worked on the active transport of metabolites across bacterial membranes, where he was interested in a transport mechanism called "group translocation" (Mitchell 1957; Mitchell and Moyle 1958). This designates transport processes across membranes that are thermodynamically driven not by purely osmotic forces, but by the chemical reactivity of the molecules transported. Another theoretical notion was "vectorial chemistry," that is, chemical processes that do not occur simply in a homogenous solution, but in at least two phases separated by a biological membrane. The experimental systems that Mitchell had worked with thus included artificial membranes, which were used to study transport processes, and bacterial in vivo systems for studying the uptake of sugar and similar molecules by bacterial cells.

The chemiosmotic theory, for Mitchell, was an exercise in vectorial chemistry. It combined the concept of group translocation with what was known about oxidative phosphorylation at that time. What Mitchell did, thus, was to *import* certain concepts that had their origins in the field of membrane transport physiology into the field of bioenergetics and oxidative phosphorylation research.[13] Biochemists, of course, were not ready for this. Those among them who were studying the generation and utilization of phosphate-bond energy (such as Lipmann 1941) were mainly interested in elucidating the steps of biochemical reaction pathways and identifying intermediates of individual reaction steps (as exemplified in the work of Hans Krebs; see Section 3.3). This, as already mentioned in Section 4.1, was the background of Slater's chemical theory.

After founding the Glynn Research Laboratory in rural Cornwall in 1965 – it originally consisted only of two people – Mitchell and Moyle had to choose an experimental system that might allow them to demonstrate the operation of the chemiosmotic mechanism. They chose a system in which rat liver or yeast cells were first broken up by mechanical homogenization. Then, mitochondria were isolated with the help of centrifugation. This had to be done, of course, in a way that did not destroy the mitochondria. Such a method had been developed by Albert Claude in the 1940s (Ernster and Schatz 1981, 230). This work is briefly discussed.

For the purposes of the present chapter, it is a happy coincidence that Albert Claude's work also features in Rheinberger's story on early protein synthesis research. Like Zamecnik, Claude started out in cancer research (Rheinberger 1997, 59). Working at the Rockefeller Institute in New York, Claude was using ultracentrifugation to isolate the filterable pathogenic agent that causes sarcomas in fowl (known as "Rous Sarcoma Virus" today). However, he became so intrigued with the possibilities that the ultracentrifuge opened up for studying normal cells that he eventually turned away from cancer research. He developed a method called "differential ultracentrifugation." In this procedure, cells are first broken up by mechanical homogenization. The resulting homogenate is centrifuged at low speed and the sediment (mostly nuclei) is removed. Then the supernatant is centrifuged at a higher speed and again separated into supernatant and sediment. This procedure can be repeated several times. The effect is that the different cellular constituents, that is, the membranes and internal organelles, are separated according to their size and specific density.

In a sediment obtained by centrifugation at $18,000g$, Claude found small particles that were barely visible under a light microscope. He first identified them with the mitochondria. However, in 1943, he concluded that the particles were too small to be mitochondria and renamed them "microsomes." The microsomes had a very interesting career (one of the main themes in Rheinberger's book), because they were later shown to contain the enzyme complexes required for protein synthesis, the ribosomes. During the war, Claude refined the technique of differential ultracentrifugation in order to be able to clearly separate these microsomes from other cellular constituents, including the mitochondria. He showed that the mitochondria are contained in what he called the "large granule fraction," which he obtained by centrifuging mammalian liver homogenates at a centrifugal force of $2000g$ (Claude 1946a, 54), which removes the nuclei and heavy cellular debris and leaves the mitochondria and some other organelles floating. Furthermore, Claude found that mitochondria swell and eventually disintegrate (lyse) when they are washed with distilled water (Claude 1946b, 70). This gave the first indication that mitochondria contain a semipermeable membrane, which functions as an osmotic barrier. In other words, mitochondrial membranes allow the passage of water, but not of molecules dissolved in water. On the basis of Claude's results, Hogeboom, Schneider, and Pallade (1948, 621) developed a technique of isolating mitochondria in an intact state by homogenizing the cells in an isotonic sucrose solution – a technique that has been used ever since in mitochondrial studies (Ernster and Schatz 1981, 230). These newly developed techniques allowed the biochemists to "map" various enzymatic activities to different cellular compartments (de Duve 1975). It was in the course of this work that the enzymes carrying out oxidative phosphorylation were shown to be localized in the mitochondria.

To get back to our main story, when Mitchell began to use mitochondria to measure proton gradients (see Section 4.2), he did not start from scratch. He was bringing an experimental system into his laboratory that already had a long and complex history. The biochemists working under the chemical paradigm were also using this experimental system, but in a different way. Mitchell, with his background in membrane transport physiology, combined the mitochondrial experimental system with concepts he took from there, in particular his concept of vectorial metabolism, that is, metabolism that requires an intact membrane. For this reason, he studied the mitochondria in an intact condition. By contrast, the biochemists looking for chemical intermediates immediately tried to break the mitochondrial system down by solubilizing the membranes with detergents (e.g., Cooper and Lehninger 1956). We could say with Rheinberger (1997, 136) that Mitchell introduced a "bifurcation" into the experimental system; in other words, he started with the same system but took it in a different direction.[14]

Another modification of the mitochondrial experimental system was the production of submitochondrial particles, small vesicles consisting of mitochondrial membrane produced by treating mitochondrial suspensions with ultrasound (Racker et al. 1963). These particles were a compromise between completely breaking down the mitochondria – and obliterating their oxidative phosphorylation function – and working with whole, intact organelles. They revealed a number of things, for example, the functional asymmetry of the membrane (see Section 4.2).

One of the standard approaches in biochemistry at that time was to follow enzymatic activities through a series of purification steps until one ended up with a purified enzyme that catalyzed a particular biochemical reaction step. The problem with this approach in the ox-phos case was that the main enzymatic step that the biochemists were interested in – the respiration-dependent phosphorylation of ADP – was only present in intact mitochondria or submitochondrial particles (which are still huge structures in molecular terms). Thus, the traditional approach of simply tracing enzyme activity through a series of preparative steps was out of the question. However, Racker's laboratory found that an ATP-hydrolyzing activity could be purified from mitochondrial membranes (Pullman et al. 1960). At first, it was not clear whether this enzymatic activity had anything to do with oxidative phosphorylation. In theory, any enzyme that catalyzes a biochemical reaction can also catalyze its reverse reaction; however, it was possible that the ATPase activity purified in Racker's lab was due to some other enzyme. But Racker's laboratory obtained several results that suggested an involvement of their purified ATPase in ox-phos. One indication was that the enzyme was shown to be sensitive to the drug oligomycin (Penefsky et al. 1960), a known inhibitor of oxidative phosphorylation. The mode of action of this drug strongly suggested that it directly poisoned the phosphorylating activity rather than the respiratory enzymes. For example, oligomycin did not block respiration when the latter was uncoupled from phosphorylation by dinitrophenol.

During the 1960s, Racker's laboratory made much progress in characterizing the ATPase biochemically. The enzyme was shown to be a very large, membrane-bound multiprotein complex consisting of a strongly hydrophobic (fat-soluble) part termed F_o and a globular hydrophilic (water-soluble) part termed F_1, which was visible under an electron microscope as knob-like structures on the inner side of the mitochondrial membrane. The F_1 moiety could be detached from the membrane with the help of trypsin (a protein-cleaving enzyme). The soluble F_1 moiety was shown to hydrolyze ATP, but this activity was insensitive to oligomycin. When the F_o portion was added back to the enzyme, it became sensitive to oligomycin (Kagawa and Racker 1966a, 1966b, 1966c).

In spite of the progress made in the biochemical characterization of the ATPase, Racker's lab spent many frustrating years trying to make the ATPase synthesize ATP in vitro. But they were unsuccessful in isolating the ATPase in a fully active state. So far, all the purified enzyme did was to catalyze the hydrolysis of ATP.

During most of the time that Racker's laboratory was trying to purify the ATPase in an active state, they believed that the energy for ATP synthesis came from a chemical intermediate. In 1970, Racker decided to "look at the problem from a different point of view," namely under the assumptions of Mitchell's theory (Racker 1976, 21–22; Racker and Racker 1981, 270; Schatz 1997, 270). The main shift induced by this change in point of view was that, now, attention was given to the functional and structural integrity of the membrane. According to Mitchell, oxidative phosphorylation cannot work without a sealed membrane, a fact that seemed to be borne out by the experimental evidence. Thus, Racker's group realized that in order to reconstitute oxidative phosphorylation, they would have to make artificial membranes.[15] As a consequence, they started to experiment with phospholipid vesicles, trying to create vesicles with functional, purified enzymes embedded in their natural environment – a phospholipid membrane. As we have seen (Section 4.4), this artificial vesicle system proved to be crucial for resolving the ox-phos controversy.

The following conclusions can be drawn from this brief history of experimental systems in ox-phos research. First, the experimental systems that were used by ox-phos researchers were the product of a long and complex history with diverse roots. The methods for preparing mitochondria in an isolated and intact state emerged from a broader research program of exploring the cell with an ultracentrifuge. The method of differential ultracentrifugation initially came out of cancer research, but it was hybridized with experimental systems from classical enzymology in order to "map" the cell's biochemical activities to different cellular compartments. It was in this context that isolated mitochondria acquired the function of "epistemic things" in biochemical research. As Rheinberger's theory predicts, the hybridization of experimental systems created new possibilities and increased the system's resolution for differentiating between various research objects.

Second, the experimental system's resolution was increased by several technological innovations. For example, the adoption of electron microscopy (see Section 9.5) in Racker's laboratory opened the possibility of examining macromolecular structures at the mitochondrial membrane and to correlate structure with biochemical function. This was particularly important for the characterization of the ATPase, but also later for examining the artificial vesicles used for reconstitution experiments.

Third, the refinement of the experimental system and the exploration of its possibilities of manipulation did not depend on a particular theory. The submitochondrial particles, for example, emerged around 1960 from attempts to reconstitute oxidative phosphorylation in vitro. The same is true of the purification and biochemical characterization of the ATPase as well as the respiratory enzymes (the latter was the other main project that Racker's laboratory was involved in during the 1960s). Until 1970, this work proceeded under the assumption that the ATPase was powered by a chemical intermediate, but this did not really matter for the experimental progress in shaping the experimental system toward a high-resolution research instrument. Although it was this work that eventually led to the reconstitution experiments and to the establishment of Mitchell's theory, this work was not done with the *aim* of testing Mitchell's theory. It was part of an enzymological research program that conformed to standard biochemical practice at that time. Thus, the in vitro system eventually served a purpose for which it was not designed.

Fourth, the development of the experimental system, for the most part, did not depend on a *theoretical consensus* on the question of which mechanism was correct. Much of the experimental work (except Mitchell's and Moyle's) proceeded independent of the theoretical controversy.[16]

Fifth, the development of the mitochondrial system showed a tendency toward *simplification* as the system became more refined. Whole mitochondria are extremely intricate biological structures that contain much more biochemical machinery than the enzymes required for oxidative phosphorylation. They carry out the entire Krebs cycle, fatty acid oxidation, and many other metabolic functions. They contain DNA, RNA, and the whole machinery for DNA replication, for protein synthesis, and for importing proteins that are made in the cytoplasm. Furthermore, mitochondrial membranes contain elaborate transport mechanisms for various ions and for taking up and releasing metabolites. In some cells, mitochondria are involved in the regulation of cytoplasmic Ca^{++} concentration, which plays a central role in intracellular signaling.[17] Mitochondria, even in the isolated state, contain a veritable "biochemical bog," to use an expression of Paul Zamecnik's. The complexity of this system severely limits the use of isolated mitochondria as an experimental system in which to disentangle individual biochemical reactions.

Submitochondrial particles are somewhat simpler; however, they still contain a lot of biochemical machinery. Nevertheless, such particles increased the discriminatory capacity of the mitochondrial system, because of their increased manipulability. For example, the possibility of creating "inside out" particles, as we have seen, led to the recognition that the membrane is functionally asymmetric. But the simplest system, clearly, was the artificial vesicle system developed in Racker's laboratory. It contained only the components that Racker and his co-workers included in a particular experiment, for example, ATPase, respiratory enzymes, ADP, and inorganic phosphate (plus, perhaps, contaminants). This system also displayed the greatest manipulability, as individual components could be added or left out according to the experimental design. The artificial vesicle system, in contrast to whole mitochondria and submitochondrial particles, allowed the study of oxidative phosphorylation in the absence of all the other biochemical machinery. It was this strategy that proved to be successful in the end.

Sixth, the reconstitution experiments can be viewed as exemplifying a general experimental strategy of twentieth-century biochemistry. Taking a cell apart and trying to assemble individual processes from purified components in vitro is a powerful way of studying molecular processes, as witnessed, for example, by the in vitro systems for protein synthesis discussed by Rheinberger. This approach was most rigorously pursued by Racker's laboratory throughout the 1960s. It seems that biochemists only considered a mechanism as "solved" once it had been reconstituted in vitro. Indeed, the closure of the ox-phos controversy was mainly a consequence of the successful reconstitution of the process by Racker's laboratory.

142

In summary, the ox-phos case can be viewed as a history of successive branching, hybridization, and other modifications of an experimental system that originated in cancer research in the interwar years. Isolated mitochondria became "epistemic things" because of the experimental possibilities they offered for differentiating the thousands of metabolic processes that occur in living cells. Mitchell's theory was imported from outside this experimental system, and it started to affect its reproduction in 1965. But the further development of the experimental system, in particular its simplification via submitochondrial particles to reconstituted systems, was part of a biochemical research program that proceeded according to the characteristic standards of the discipline which required the isolation of biochemical activities in purified systems. The modifications introduced to the system within the framework of this research program were largely independent of any theory and of the theoretical controversy that went on in the scientific literature.

If looked at in this way, the ox-phos case seems not so dissimilar after all to Rheinberger's story of protein synthesis. Nevertheless, I argue in the next section that methodological considerations have to supplement the experimental systems analysis for a full understanding of the dynamics of research.

5.3 THE ROLE OF EPISTEMIC NORMS

I have shown that my example of research on oxidative phosphorylation in the 1960s and 1970s can be reconstructed both as a series of experimental tests of Mitchell's hypothesis (see Chapter 4) and as a lineage of experimental systems that was largely driven "from behind" by the system's intrinsic capacities, rather than by any particular theory (previous section). Clearly, most historians of science today will regard the experimental systems account as more relevant. This poses a serious challenge to philosophers who believe that scientific research is guided by certain epistemic norms. For this reason, I try to defend the relevance of normative considerations.

To begin, it must be noted that an experimental systems account of the ox-phos case would clearly fail if the claim was that ideas and theories played no role whatsoever in the experimental work. As we have seen, at least some of the experimental work done in the 1960s was directly informed by Mitchell's hypothesis. This is obviously true for Mitchell and Moyle's own research; they designed their experiments explicitly in order to test the chemiosmotic hypothesis. But the same is true for some of the experimental work done by Mitchell's opponents, in particular in E. C. Slater's laboratory in Amsterdam and in Britton Chance's lab in Philadelphia; both Slater and Chance tried to show that the proton transport phenomena found by Mitchell and Moyle, even though they were real, were not part of the mechanism of oxidative phosphorylation (see Section 4.2). Thus, there is no denying that theory mattered. But New Experimentalists are not saying that theories play no role in experimental practice. What they are denying is that learning from experiments *solely* consists in learning about the truth or falsity of theories, and that theory always comes first.

This leaves room for the possibility that learning about the truth or falsity of theories is *one* possible function of experiments, albeit not the only one. To say that experiments sometimes fulfill the function of testing a theory does not exclude experimental work having other functions as well (for example, explorative functions) or the development of experimental systems having its own internal dynamics that is not related to theory testing. If we take methodology to be mainly concerned with issues about theory testing, there is thus no direct contradiction between new experimentalism and a methodological construal of particular episodes in the history of experimental science.

Even though these considerations make room for a reconciliation between the experimental systems approach and more traditional methodology, they do not yet establish that a methodological reconstruction is anything more than a philosophical exercise that the historian of science can dismiss as irrelevant for understanding scientific change. I therefore provide reasons why methodology matters in historical reconstructions of scientific developments. For the following, I mean the term "methodology" to include philosophical attempts to characterize the reasoning behind the generation as well as evaluation of new theories (see Chapter 3).

The first reason why I do not think that an account that focuses exclusively on the development of experimental systems is sufficient is that it cannot explain how *theoretical* controversies are resolved (Weber 2002b). In the ox-phos example, the controversy between supporters and opponents of Mitchell's theory persisted through changes in experimental practice. As we have seen in the previous section, the experimental systems used for studying oxidative phosphorylation underwent a considerable amount of change, from isolated mitochondria through submitochondrial particles to reconstituted systems. During most of this time, there was no consensus concerning the theoretical question of whether a chemical or a chemiosmotic mechanism couples phosphorylation and respiration. Why did this controversy persist into the mid-1970s? And why did a consensus emerge rapidly after Racker's reconstitution experiments? As I have shown in Chapter 4, a methodological account of some of the main experiments can answer these questions. By contrast, a theory of the dynamics of scientific change that relies exclusively on the intrinsic capacities of experimental systems cannot explain the emergence of theoretical consensus. But, surely, the closure of controversies is a significant aspect of the dynamics of scientific change.

At this point, it could be objected that there are alternative explanations for the emergence of theoretical consensus, explanations that do not involve methodological considerations of what is a good test or crucial evidence for a scientific theory. For example, it is possible that consensus is a product of the social dynamics within a scientific community, or even of the broader sociopolitical context within which the research in question is situated.[18] Such a view, of course, would amount to social constructivism, which has had a considerable following in recent science studies. The social constructivist must answer the question of what significant social events took place in the 1970s that led to the emergence of a consensus concerning the mechanism of ox-phos. Furthermore, such an account would have to explain why it was *Mitchell's* theory that was eventually accepted, rather than Slater's. Mitchell, in contrast to E. C. Slater or Britton Chance, lacked institutional power. Furthermore, he did not command the kind of resources in terms of funding, equipment, and collaborators that some of his opponents did.

Allchin (1991, Section 8.2) shows that the scientific community's acceptance of Mitchell's theory was distributed; that is, different scientists accepted it at different times and for different reasons. However, explaining an individual scientist's *beliefs* and explaining the emergence of a community-wide *consensus* are two different matters. The former may be much more influenced by idiosyncratic sociological and psychological factors than the latter.

A social constructivist account of the oxidative phosphorylation controversy has been attempted by Gilbert and Mulkay (1984), who reconstruct the episode in terms of rhetorical strategies used by some of the scientists involved. However, they provide no answers to the questions of why a consensus was eventually formed and why it was Mitchell's theory (rather than that of his more powerful opponent Slater) that was accepted. By contrast, the view that the emerging consensus about chemiosmotic theory was at least in part a consequence of the crucial evidence obtained in Racker's laboratory is plausible. In Chapter 4, I have shown on the basis of methodological considerations why this evidence was crucial.

A different alternative explanation for the emergence of consensus could be sought in the special status that in vitro reconstitution experiments have in biochemistry.[19] On such a view, a consensus was only possible once oxidative phosphorylation was successfully reconstituted by Racker's laboratory. Since Racker and co-workers reconstituted the process in a "Mitchellian" way (i.e., using a system that contained sealed membranes rather than just soluble enzymes), this might explain why the emerging consensus came out on Mitchell's side. However, it is not clear why Mitchell's opponents should have admitted defeat at exactly this point. They had found ways of saving the chemical theory in the past; why not this time? I suggest that after the Racker–Stoeckenius experiment the chemical theory could *not* be saved, because the experiment ruled out the possible error of attributing energy coupling to the membrane when it was, in fact, mediated by a chemical intermediate. But this move brings us immediately back into the methodological game.

Furthermore, I suggest that methodology can explain *why* biochemists value reconstitution so much. Even though it may be intrinsically fascinating to be able to take something apart and put it together again, this can hardly explain the important role that reconstitution experiments play in biochemistry. Instead, I suggest that the great epistemic value of reconstitution experiments is due to the possibilities for controlled experimentation that they offer (e.g., substituting cytochrome oxidase by bacteriorhodopsin). Control experiments, as I have argued in Chapter 4 (Section 4.7), are the main strategy for eliminating errors in experimental biology. Thus, the attempt to explain the emergence of consensus by an appeal to the special status of reconstitution experiments throws us right back into methodological considerations.

A second reason why methodology matters becomes evident if we examine more closely the functionality of experimental systems, as entertained by Rheinberger. The ascription of a function to some entity requires that we state what this function is *for*. A functional claim links some causal activity to some purpose or *telos*, for example, a hammer's potential impact on a nail to the goal of driving a nail into a piece of wood (see Section 2.4). What are the functions of experimental systems for; in other words, what is their *telos*? One possible answer to this question is that the functions of experimental systems are for the reproduction of the system. However, this view is difficult to maintain because experimental systems – unlike biological organisms – are not *self*-reproducing entities. Their reproduction depends closely on human agents and their intentions. Thus, the question arises of why scientists reproduce and modify experimental systems.

The obvious answer to this question is that experimental systems serve the purpose of producing knowledge (and, in some cases, new technology). Experimental systems are maintained and reproduced by scientists if they perform their function of producing knowledge well. A system that fails to yield reproducible results, or one that keeps turning up artifacts instead of robust phenomena or structures, is not functioning well and will eventually be replaced by a system that functions better. But why do scientists value reproducibility? And what is the difference between an artifact and a real biological phenomenon? To ask and seek answers to such questions is already a methodological inquiry (which I pursue in Chapter 9). An experimental system, I suggest, only produces reliable knowledge to the extent that it allows the control of possible errors at different levels of inquiry (see Section 4.7). Thus, methodology is necessary to spell out exactly when an experimental system can be said to "work." The good functioning of an experimental system to a large extent consist in the *reliability* of the knowledge it produces. Reliability is a methodological notion; therefore, the New Experimentalist cannot dismiss methodology as irrelevant.

For my third reason why methodology matters, we need to reflect a little bit on Rheinberger's notion of "epistemic things." If we return to Rheinberger's central case study, ribosomes (RNA-containing intracellular particles that are involved in protein synthesis) did not simply *appear* in the in vitro system all by themselves (as Jesus is said to have appeared to the apostles). Even though they are visible as tiny dark dots in a microscope, interest in ribosomes as research objects depended on *theoretical assumptions*, for example, that these particles play a role in protein synthesis. Epistemic things such as ribosomes do not simply appear or materialize in experimental systems. They are instances of what philosophers of science call *theoretical entities* (see Section 9.1). As such, the concept of a ribosome – whatever its precise description – is part of a theoretical model. Whether this concept has a *referent*, that is, whether these cellular components with the presumed role in protein synthesis actually *exist* had to be established experimentally using reliable methods.

It might be objected that epistemic things do not require any theory. Rheinberger (1997, 28) stresses that, while the research is still in progress, epistemic things "present themselves in a characteristic, irreducible vagueness"[20] and lack a precise definition. I think this is correct. However, even a vaguely formulated theory *is* a theory. As the research progresses, the theory will be refined and its theoretical entities will become more sharply delineated. But in any case, at some point there has to be *evidence* that some theoretical entity really exists; otherwise it will suffer the fate of the high-energy intermediates of oxidative phosphorylation (see Section 4.1). Evidence is a methodological concept.

Richard Burian has made a similar point in response to Rheinberger's notion of "epistemic things." He put it quite elegantly (or, as it were, elephantly): "What must be explored is how one might justify the claim that scientists working with different experimental systems, and thus different *epistemic* things, might nonetheless have good grounds to hold that they had gotten hold of different parts or aspects of the same elephant" (Burian 1995, 130).

According to Burian, scientists sometimes succeed in identifying research objects by working with *different* experimental systems. For example, a great variety of experimental work in molecular biology had as its central research object the DNA molecule. Burian cites, for example, Avery's demonstration that DNA is the transforming principle, Hershey and Chase's experiment showing that DNA is the genetic material of bacteriophages, the crystallographic studies of DNA structure by Franklin and Wilkins, Crick and Watson's double helix model, and Meselson and Stahl's experiments showing the semiconservative replication of DNA. Burian suggests that "[t]he interesting thing is that we have reached a position from which we can understand all of this work in terms of a series of approaches to a common object or class of objects, the DNA molecule or various DNA molecules" (Burian 1995, 130). According to Burian, this asks for "norms and standards according to which some such cross-experimental system identifications are correct, and others incorrect" (1995, 131). Norms and standards are exactly the methodologist's business.

What reasons can scientists have to conclude that the research objects that they study with different experimental systems have a common referent? In other words, how do they establish that the theoretical entities that are being studied with different experimental approaches actually exist, rather than being artifacts created by a particular system? These are ultimately methodological questions, and Rheinberger's account offers us no resources to answer them.[21] I return to these questions in Chapter 9.

A fourth reason for the relevance of methodology becomes evident if we ask how new concepts and theories are introduced in the course of the development of an experimental system. Even if Rheinberger is right that experimental systems develop according to their intrinsic capacities, new concepts and theories are not generated by the experimental system itself. Ideas do not just drop out of a centrifuge. New ideas may arise in response to the behavior of an experimental system, for example, in the face of unexpected anomalies. Therefore, we still can and must ask how these ideas are generated, as I have done in Chapter 3.

These considerations show that an account of some historical episode that focuses on the development of experimental systems and "epistemic things" alone is incomplete. An appeal to epistemic norms is necessary, first, in order to understand how theoretical controversies are resolved, second, in order to spell out what the good functioning of experimental systems consists in, third, in order to show how the existence of theoretical entities is established, and fourth, in order to understand how new concepts and theories originate (perhaps in response to developments in an experimental system).

In the final section, I try to further clarify the relationship between methodology and the historical study of experimental systems.

5.4   LOCAL SETTINGS VERSUS UNIVERSAL STANDARDS

One of the most intriguing consequences of an account of the dynamics of scientific change such as the one developed by Rheinberger is that it accords a major role to historically contingent factors. Experimental systems are always the product of some local laboratory setting. For example, the availability of some new apparatus may have a great impact on the further development of an experimental system; it may alter the direction that research takes in ways that were not anticipated by the people who built or started to use that apparatus. The same is true for other constituents of experimental systems, such as the biological materials used, the preparation techniques, chemical reagents, and so on. According to Rheinberger, the changes that such alterations of experimental systems can introduce are profound; they may change even the research questions that are being asked.

Furthermore, experimental systems are rarely built from scratch; they are usually extensions, hybrids, or other modifications of previously existing systems. Sometimes, modifications of a system occur when a new researcher joins a laboratory, who brings with him or her new techniques, skills, or materials. Thus, instruments, the biological materials available, preparation procedures, previously existing systems, model organisms (see Chapter 6), and human resources all contribute to the local laboratory setting out of which a research process arises. On Rheinberger's account, the course of this research process is causally highly underdetermined by the ideas and intentions of the actors in the laboratory at any particular time, and these ideas and intentions continually change as a consequence of the interaction with the experimental system. In this picture, a scientific discipline changes mainly due to historically contingent factors that are part of local laboratory settings.

I have argued in the previous section that epistemic norms are highly relevant to understanding the development of an experimental discipline. This raises the question of the respective roles played by the local laboratory settings on the one hand and more general, perhaps even universal, methodological norms on the other hand.

It could be suggested that an approach such as Rheinberger's is concerned with the context of discovery, whereas methodology is concerned with the context of justification. However, the experimental systems approach addresses questions that were not usually associated with the context of discovery, for example, questions concerning the genesis of experimental results and of research problems. The received view in traditional philosophy of science was that the design of experiments closely followed empirical predictions derived from theories in accordance with methodological rules (which may sometimes be the case in experimental physics). However, on Rheinberger's account, the reasons for which certain experiments are done can differ substantially from the role that the experimental results play in the further development of the field. Thus, the traditional two-context distinction (see the introduction to Chapter 3) seems inadequate for explaining the relationship between local factors and epistemic norms.

A second question concerns the problem that, although experimental systems always have a history that explains how they came into being, epistemic norms do not seem to have such a history. Where do epistemic norms come from? These two questions are the focus of this final section.

First, we consider the respective roles played by locally contingent factors and more general methodological norms. I suggest that these two kinds of factors influence the research process at *different stages*. It seems to me that the experimental systems approach pursued by Rheinberger and other historians explains how scientists come to choose both the *experimental approaches* they use and the specific *research problems* that they want to follow up.[22] If Rheinberger is right, experimental approaches and research problems are often inextricably linked in experimental biology. Cases where an experimental biologists starts from a theoretical framework and then chooses an experimental system to match the theories to be tested seem rather rare. In this respect, experimental biology does seem to differ from experimental physics, where experiments are more often designed according to a theoretical framework (e.g., the so-called standard model in particle physics). Thus, if we just focus on theories and theoretical ideas, we will not understand why the various disciplines of experimental biology developed in the way they did.

However, I suggest that there are also stages in a research pathway where the need to test theories and, therefore, methodological norms does become relevant. These will typically be stages in which an experimental system has been developed to a point where it actually allows scientists to test a theory or hypothesis that may or may not have preexisted that experimental system. For example, a severe test of Mitchell's theory had to wait for the development of the experimental systems to run its course; it only became possible with Racker's in vitro reconstitution system (which, as we have seen, was largely developed for other reasons). If we consider the example discussed in Section 3.1, Jacob and Monod eventually had to choose between the "repressor model" and the "inducer model," which were both consistent with the PaJaMo experiments. Which kinds of methodological norms governed their choice of the repressor model would have to be investigated, but this represents a typical situation where methodological considerations probably came into play.

Thus, there is no conflict in granting both local laboratory settings and methodological norms a major impact on the research process: they influence the process at different stages in an investigative pathway. These stages need not always be so clearly separated in time as in my case of oxidative phosphorylation.

The second question is more difficult. Where indeed do epistemic norms come from? According to some philosophers, methodological norms can only be established on a priori grounds. For Popper, methodology was based on deductive logic in conjunction with a few "conventions," such as the convention of accepting observation statements as true under certain conditions. Similarly, inductivists such as the Bayesians (see Section 4.5) seek the foundations of methodology in a priori principles such as probability axioms. However, apriorism is by no means the only position available in meta-methodology (i.e., the study of the epistemological status of methodology itself). An interesting alternative is known as "methodological naturalism" and is discussed here briefly.

Various philosophers of science have adopted a naturalistic position with respect to scientific methodology, for example, Larry Laudan's "normative naturalism" (Laudan 1996, Chapters 7 and 9). The central claim of normative naturalism according to Laudan is that "the claims of philosophy [i.e., epistemology, M. W.] are to be adjudicated in the same ways that we adjudicate claims in other walks of life, such as science, common sense, and the law" (Laudan 1996, 155). Thus, "philosophy is neither logically prior to these other forms of inquiry nor superior to them as a mode of knowing" (Laudan 1996, 155). Laudan also rejects the view that "the theory of knowledge is synthetic a priori (as Chisholm would have it), a set of 'useful conventions' (as Popper insisted), 'proto-scientific investigations' (in the Lorenzen sense)," etc. (Laudan 1996, 155). Instead, the normative naturalist construes methodological claims as "theories or hypotheses about inquiry, subject to precisely the same strategies of adjudication that we bring to bear on the assessment of theories within science or common sense" (Laudan 1996, 155). Thus, normative naturalists insist that methodological norms cannot be deduced a priori. They must be studied empirically by philosophers and historians. Philosophers and historians must learn from scientists what good methodological norms are, rather than the other way around. Behind this, there is the assumption that the scientists *themselves* have acquired methodological norms by experience, rather than applying some principles of human reason that are given a priori.

Normative naturalism is also the meta-methodological position preferred by Deborah Mayo, whose methodology I have applied and extended in Chapter 4. According to her, scientists learn from *errors* that they have made in the past and develop strategies to avoid such errors in the future. The canonical form of such a strategy, in Mayo's account, is the argument from error (see Section 4.6). But scientists develop a whole "error repertoire" from "the history of mistakes made in reaching a type of inference" (Laudan 1996, 18).

I think that normative naturalism can also be applied to the approach to experimental reasoning that I have developed in Chapter 4 (Section 4.7). There, I have argued that an important strategy for controlling error in experimental biology is to systematically alter the experimental conditions in order to rule out false causal attributions. On a naturalistic perspective, it can be suggested that such strategies have been *learned* by scientists from errors that they have committed in the past. Thus, no a priori reasoning, neither on the part of the scientists nor on the part of the philosopher of science, needs to be invoked.

It would be interesting to study such learning processes historically in greater detail. To give an example, Ephraim Racker's highly personalized account of the development of bioenergetics (Racker 1976) contains a number of "lessons" that he advises young scientists to take to heart before they commit similar kinds of mistakes again. A nice example is the following:

> *Lesson 5:* Not everything that shines is gold. Not everything that floats after high-speed centrifugation is soluble. We want to make sure that the proteins we study are lean and don't carry fat as life preservers, which prevent them from sinking. If we claim solubilization too hastily, all that is sinking is our scientific reputation. (Racker 1976, 13)

This "lesson" was directed at claims that oxidative phosphorylation has been observed in systems where all the membranes had been dissolved and where the membrane proteins that catalyze ox-phos were solubilized (see Section 4.2). As it turned out, these proteins were not completely solubilized after all; they still contained lipids ("fat as life preservers"). Thus, ox-phos in the absence of membranes was an experimental artifact. This example of a "lesson" that had to be learned by biochemists concerns a very specific type of mistake that arises in biochemical research. However, there are no reasons to believe that scientists have not acquired more general strategies for controlling error in a similar way.

The upshot of this discussion is that we can reject the claim that scientific methodology is given a priori. Methodology is empirical, just like historiography, psychology, or sociology. Methodological analysis is not based on a priori reasoning, but on studying actual scientific practice, historical or contemporary. Furthermore, there is nothing ahistorical or otherworldly about methodological norms; they have not dropped out of Plato's heaven. Methodological norms are just as much a product of history as are the more specific experimental practices that have been studied by historians. As such, they are just as important parts of experimental cultures as are the materials and instruments used for experimentation. This recognition, I suggest, reduces the apparent contrast between historical accounts such as Rheinberger's experimental systems approach and the methodological approach that I have taken in the previous chapter. These two approaches are both necessary in order to gain a full understanding of the workings of experimental biology.

SUMMARY

Experimental systems can be treated as functional units of research, as well as units of historical analysis. They exert a strong influence on the specific directions that research in experimental biology takes. In this chapter, I have reexamined the example of oxidative phosphorylation research from the experimental systems perspective. This analysis suggests that the development of experimental practices in this area of biochemistry was indeed strongly influenced by the availability of certain experimental systems (isolated mitochondria, submitochondrial particles, and artificial vesicle systems) and that the development of these systems was partly independent of the theoretical controversy on the mechanism of oxidative phosphorylation. However, I have argued that only an appeal to epistemic norms – such as the ones discussed in the previous chapter – can explain why the controversy was resolved at a particular time. Furthermore, epistemic norms have to enter in order to explain what the functioning of experimental systems consists in and how the existence of the research objects was established. Thus, both historically contingent factors present in the local laboratory settings and more generally applicable epistemic norms influence the research process, but they may do so at different stages. Finally, I have argued that if we take a naturalistic attitude towards epistemic norms, the latter turn out to be just as much a product of history as experimental systems; they have no a priori status. There is thus no conflict between the experimental systems approach and a methodological approach; rather, the two approaches complement each other.

6

Model Organisms: Of Flies and Elephants

Most of the spectacular advances in experimental biology made in the last 100 years came from research on a remarkably small number of organisms. Here are some examples: During the second decade of the twentieth century, Thomas Hunt Morgan and his students laid the foundations of modern genetics by experimenting with just a single species, the fruit fly *Drosophila melanogaster* (see Section 3.2). The advent of molecular biology, which began in the 1940s, came mainly from research on the bacterium *Escherichia coli* and its bacteriophages (viruses that infect bacteria). Neurophysiologists unraveled the mechanisms of neurotransmission (see Section 2.1) using the giant nerve cells of squid. The rat and its liver played a central role in elucidating the major metabolic pathways such as the urea cycle and oxidative phosphorylation (see Section 3.3 and Chapter 4). The mouse was and still is the favorite lab animal of immunologists, giving rise to a major increase in biomedical scientists' understanding of the mammalian body's defense mechanisms. Baker's yeast, *Saccharomyces cerevisiae*, proved to be a powerful tool for studying the biology of eukaryotic cells, for example, the biogenesis of mitochondria or the regulation of the cell cycle. The nematode worm *Caenorhabditis elegans* was one of the first organisms that allowed biologists to study behavior, embryonic development, and other biological processes at the molecular level (see Section 2.5). In the 1970s, *Drosophila* made a spectacular comeback as a main experimental organism for developmental biology. More recently, it was joined by the see-through zebrafish *Danio rerio*. Most of what is known on the molecular biology of plants came from studies on the mustard weed *Arabidopsis thaliana*. For most of these organisms, the full genomic DNA sequence has now been determined by various sequencing consortia, and this is widely seen as an important complement to the human genome sequence (Kornberg and Krasnow 2000; International Human Genome Sequencing Consortium 2001).

Thus, twentieth-century experimental biologists differ fundamentally from the traditional naturalists who are interested in the *diversity* of lifeforms and whose research might involve hundreds or thousands of different species. Compared to natural environments and natural history collections, genetics and molecular biology laboratories are extremely impoverished in biodiversity. In fact, most laboratories work on only a single species, and a large number of laboratories work on the same species. While this is sometimes lamented as a deficiency (in addition to being a subject for scientific jokes), it is worth examining *why* model organisms played and continue to play such an important role in experimental biology.

There are at least three questions that can be raised concerning the role of experimental organisms in biological research. First, there is the question of why some organisms are especially well suited for studying certain biological processes. Second, we may ask why experimental biologists more often choose to work on an already established laboratory organism, rather than bringing a new organism into the laboratory and initiating research with this new organism. A third question concerns the notion of a *model* organism, that is, to what extent can certain experimental organisms function as models for other organisms, in particular humans?

In recent years, a number of historians of biology have focused on the use of model organisms as an important aspect of biological practice.[1] Most of them view such organisms as being part of a "material culture" of experimental biology. Like Rheinberger's experimental systems discussed in Chapter 5, the research materials and organisms employed by biologists can be units of historical analysis in their own right that can give the historian important insights into the dynamics and organization of laboratory practice. As we shall see (Section 6.2), some of these historians give a number of unexpected answers to the main questions concerning the role of model organisms that I have mentioned. First, they make substantial use of several economic and technological metaphors to describe the role of model organisms; for example, they describe them as "artifacts," "instruments," "craftwork," "tools," or "systems of production." Second, historians have found that it is far from clear that biologists deliberately choose certain organisms for solving specific biological problems. It may very well be the other way around: The choice of certain organisms defines what comes to be viewed as a relevant research problem. This choice is mainly determined by locally contingent factors. Thus, the historical course of experimental biology could have been driven in certain directions for largely contingent reasons. Possibly, our ways of thinking about life today could be quite different, if – more or less by chance – different laboratory organisms had been chosen at certain critical junctures in the past.

155

Obviously, the impact of such historical contingencies is reduced if it is the case that the knowledge obtained from model organisms is of a *general* or even *universal* nature, as some biologists have claimed. But how general is this knowledge really? If we look at the history of twentieth-century biology, we find many examples of biologists who overgeneralized findings that they obtained with a particular organism. For example, the Dutch Botanist Hugo de Vries based his "mutation theory" of speciation on just a single organism, the evening primrose *Oenothera lamarckiana. Oenothera* led de Vries to believe that new species can arise by single mutational events (Weber 1998a, Section 2.3). However, it turned out that *Oenothera* is a highly unusual example, because it contains ring-shaped chromosomes that make it genetically unstable. Another example is Wilhelm Roux's "mosaic theory" of development, which assumed that parts of an embryo specify corresponding parts in the adult animal. Roux based his theory on experiments showing that cutting frog embryos into half yields half frogs. However, Hans Driesch showed that, in sea urchins and other marine invertebrates, fragmented embryos can generate whole animals (Weber 1999b). To my knowledge, Roux and Driesch never resolved their disagreement as to which results were more biologically significant.

These historical blunders notwithstanding, it is a major characteristic of modern experimental biology that it aspires to have some kind of *generality*, or even *universality*. Morgan and his co-workers were not seeking to study the mechanisms of heredity of *Drosophila melanogaster* or the mechanisms of heredity of insects. They thought they were studying the mechanisms of heredity, period. Similarly, Paul Zamecnik and his laboratory – working in a hospital – did not primarily want to know how rats or bacteria of the genus *Escherichia* synthesized proteins; they were initially after the regulation of protein metabolism in human cancer cells, and later after *the* mechanism of protein synthesis (see Section 5.1). Presumably, these biologists assumed – or had reasons to believe – that the knowledge they would gain from experimenting with *Drosophila* flies, rat liver cells, or bacteria would be applicable to a wide range of biological taxa, including the species *Homo sapiens*. Therein lies one of the main differences between disciplines such as genetics, biochemistry, or molecular biology and field studies of the sex life of sea snails. However, it would be naïve to assume, for example, that the mechanism of protein synthesis is *exactly* the same in *E. coli* and in humans (in fact, it is not). Furthermore, there are biological mechanisms for which there must exist considerable differences between different organisms, because of the obvious differences in anatomy and physiology. Such phylogenetically variable mechanisms, too, have been studied with great success in various model organisms, for example, embryonic development in frogs and memory and learning in *Drosophila*.

I try to develop a general framework for answering questions about the generality of biological knowledge obtained from studying particular model organisms toward the end of this chapter. Before I turn to this issue, I present another historical case study. In Section 6.1, I examine how the fruit fly *Drosophila* – already the favorite lab animal of geneticists in the first half of the twentieth century – was turned into a major model organism for molecular studies of development. This case study allows me to address several questions concerning the role of laboratory organisms and some more general points concerning experimental practice in biological research. In Section 6.2, I use the case to critically examine some of the claims recently advanced in relation to model organisms by historians of biology. In Section 6.3, I show that technological and economic metaphors are only of limited value in illuminating the role of experimental organisms. In Section 6.4, I introduce a new analytic concept in order to capture the role of experimental organisms more adequately: the concept of preparative experimentation. In Section 6.5, I take on the challenge posed by those historians who stressed the contingency of experimental organism choice. Finally, Section 6.6 examines the question of what can be learned from the study of model organisms.

6.1   THE MOLECULARIZATION OF *DROSOPHILA*

The early history of modern genetics is closely tied to the use of the tiny fruit fly *Drosophila melanogaster* as an experimental organism. Based on experiments with spontaneous *Drosophila* mutants, Thomas H. Morgan's group at Columbia University in New York produced the first genetic map in 1913. This map was constructed with the help of a genetic process known as "crossing-over" (see Section 3.2). That these so-called linkage maps really represented a physical structure, rather than just being mathematical constructs, was confirmed in the 1930s, when it became possible to map *Drosophila* chromosomes cytologically. Cytological maps were prepared on the basis of dyed giant chromosomes isolated from larval salivary glands (Painter 1933, 1934). These giant chromosomes show characteristic banding patterns that are visible under a light microscope. It was shown that the earlier linkage maps were co-linear to the cytological maps. Thus, the fact that two different experimental techniques provided corresponding results provided evidence that these maps represented real chromosomal structures (Weber 1998b; see also Section 9.2).

The classical genetic methods of linkage mapping and cytological mapping were successfully used by *Drosophila* geneticists for several decades and provided a host of biologically significant findings. For example, it was shown on the basis of such maps that some mutant flies carried chromosomal inversions, deletions, insertions, or translocations. Theodosius Dobzhansky, one of the founders of the synthetic theory of evolution, obtained evidence that such chromosomal rearrangements play a role in the evolution of natural populations (Weber 1998a, Chapter 4). Another interesting phenomenon discovered by classical mapping techniques was gene duplication, which suggested a mechanism by which new genes might arise in evolution. Thus, mapping *Drosophila* chromosomes by the classical methods of linkage mapping and cytological mapping was a powerful technique that rapidly turned genetics into a prolific and prestigious science.[2]

But as powerful as the classical methods of *Drosophila* genetics were, they also had their intrinsic limitations. From the 1920s onwards, geneticists were hoping that they could use these methods to learn something about the internal structure and function of genes. However, this turned out to be difficult. In order to map the internal structure of genes, it was necessary to "split the gene," that is, to observe crossing-over within a single gene. Even though there were results from the so-called complex loci of *Drosophila* that could be interpreted as intragenic recombination, this interpretation was problematic and controversial. The reason was that it was not possible to determine exactly how many genes were present at the complex loci. Therefore, the results obtained could also be due to regular crossing-over between different genes (Carlson 1966, Chapter 17; Weber 1998b). Thus, classical *Drosophila* genetics had reached its limits of resolution (see Section 7.1).

In 1955, Seymour Benzer (1955) published the first genetic fine-structure map of two bacteriophage (bacterial virus) genes. Even though the phage he was working on contains only a single chromosome, Benzer was able to observe crossing-over by double-infecting the *E. coli* host cells with different phage strains. Like the *Drosophila* geneticists before him, Benzer then used the frequency of recombination between different mutations in order to determine the physical separation of the mutations on the chromosome. The advantage of phage compared to *Drosophila* was that Benzer could test millions of individual phage particles for crossing-over. The *Drosophila* geneticists, by contrast, could only handle a couple of hundred flies at a time. Benzer was therefore able to detect extremely rare crossing-over events. This made it possible to produce genetic fine-structure maps of individual genes and to link classical genetic maps to the DNA molecule (Weber 1998b; Holmes 2000b).

Benzer's work is widely viewed as being part of a historical transition that started after World War II and that is called the "molecular revolution" in biology. Even though classical genetic methods played a role in this transition (see Weber 1998b), the main experimental organisms of the new molecular genetics were microorganisms, especially *E. coli* and bacteriophage (Cairns, Stent, and Watson 1992). The main difference from the older *Drosophila* genetics was that it revolved around DNA as the genetic material, whereas the earlier geneticists had assumed, most of the time, that genes are made of protein. The structure of DNA was solved in 1953 by J. D. Watson and F. H. C. Crick. Other major developments in molecular biology include the demonstration by Matthew Meselson and Franklin W. Stahl that DNA replication is semiconservative (i.e., that only one of the two DNA strands is newly synthesized in the DNA copying process; see Holmes 2001), the elucidation of the role of different RNA (ribonucleic acid) molecules in protein synthesis by Paul Zamecnik, Sydney Brenner, and others, the cracking of the genetic code by Marshall Nirenberg and Heinrich Matthaei (see Section 5.1), and the identification of the first molecular mechanism for the regulation of gene expression in the *E. coli lac* region by François Jacob and Jacques Monod (see Section 3.1).

It cannot be said that *Drosophila* vanished into complete obscurity during these exciting developments, but it clearly lost some of the scientific glamor that it had enjoyed in the 1920s and 30s. However, from the late 1970s onwards, *Drosophila* made a spectacular comeback as one of the major multicellular experimental organisms for molecular biology. In particular, it turned out to be an extremely useful model organism for studying the molecular basis of embryonic development. After the nematode worm *Caenorhabditis elegans, Drosophila* even closed the race as being only the second multicellular organism for which the full genomic DNA sequence was determined (Adams et al. 2000). Thus, although *Drosophila* played no role whatsoever in the molecularization of genetics during the 1950s and 1960s, it was turned into a powerful molecular research tool afterwards. The fly has come a very long way indeed since Morgan found, in 1910, the first mutant *white eyes.* In the following, I examine how this molecularization of *Drosophila* was accomplished.

The key to molecularization was a set of techniques that were invented in the 1970s and that came to be known as "recombinant DNA technology" (see Morange 1998, Chapter 16, for a brief history). In 1972, Herbert Boyer and Stanley Cohen, working at Stanford University's biochemistry department, took a piece of viral DNA, which they had cut out with so-called restriction enzymes,[3] and inserted it in vitro into a bacterial plasmid. Plasmids are small, circular DNA molecules containing mainly genes for antibiotic resistance. This plasmid was then transferred into *E. coli* cells (a process which came to be known as "transformation"). Boyer and Cohen used the antibiotic resistance that these plasmids conferred to select for bacteria that had taken up the engineered plasmid DNA. Remarkably, the in vitro-recombined plasmids turned out to be biologically active; that is, the *E. coli* cells replicated the artificially recombined plasmid, including the foreign viral DNA, as if it was a natural bacterial chromosome.

Recombinant DNA technology opened up the possibility of isolating the DNA of specific genes from any organism and of transferring manipulated DNA into other species. This technology forms the heart of what is known as "genetic engineering." What concerns me here is the impact that this technology had on research involving model organisms such as *Drosophila*.

One of the first laboratories to apply recombinant DNA technology to *Drosophila* was David Hogness's group at Stanford University. They developed a method called "colony hybridization," which allows the isolation of any DNA fragments that are complementary to a given RNA molecule. For this purpose, the molecular biologists constructed a so-called genomic library from *Drosophila* DNA. Such a "library" is prepared by fragmenting the DNA into smaller pieces by mechanical disruption (or by cutting it with restriction enzymes). These DNA fragments are then inserted into bacterial plasmids and are transferred into *E. coli* cells. Next, single *E. coli* cells are spread on petri dishes and allowed to grow into small colonies. For colony hybridization, the DNA of these bacterial colonies is then attached to a nitrocellulose filter. In order to isolate a particular DNA fragment, the DNA is hybridized[4] on the filter to a complementary RNA molecule, which is radioactively labeled. After the washing away of excess RNA that did not hybridize, the DNA of interest shows up as a black dot on an X-ray film. Scientists can then go back to the original bacterial colony that produced the hybridization signal. These bacterial cells contain the DNA fragment of interest. Thus, colony hybridization is a selection procedure that allows molecular biologists to pick out a piece of DNA with specific properties from thousands of random fragments. Since bacterial clones are used to amplify the DNA fragments, procedures such as this acquired the name "molecular cloning." The first genes isolated or "cloned" by this method were two genes encoding *Drosophila* ribosomal RNA (Grunstein and Hogness 1975). It turned out that the fly contains thousands of copies of these genes.

The new techniques thus allowed molecular biologists to isolate genes from *Drosophila*. The cloning of the ribosomal RNA genes made use of the fact that the gene products (rRNA) were known and available in purified form.

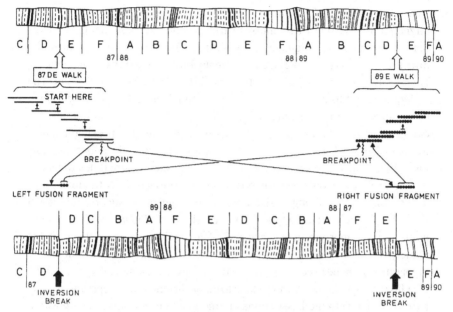

Figure 6.1. Chromosomal walking and jumping. The top of the figure shows a giant polytene chromosome with its characteristic banding pattern; the bottom displays the corresponding region of a mutant chromosome carrying a large inversion. The short black lines represent overlapping cloned DNA fragments generated by the "walk." The two thick black arrows show the inversion breakpoints. The "fusion fragments" represent DNA fragments isolated from the inversion mutant spanning the inversion breakpoints, which are used to "jump" over a large distance on the chromosome. From Bender, Spierer, and Hogness: "Chromosomal Walking and Jumping to Isolate DNA from the Ace and rosy Loci and the Bithorax Complex in *Drosophila melanogaster.*" *Journal of Molecular Biology* 168: 17–33 (1983). Reprinted by permission of Elsevier.

However, for other genes, this was not the case. For this reason, Hogness's laboratory developed an ingenious method to clone *Drosophila* DNA sequences about which nothing is known, except their chromosomal location. This method came to be known as "chromosomal walking."

Chromosomal walking makes full use of the powerful resources of *Drosophila* cytogenetics (see Figure 6.1). A chromosomal walk can start with any fragment of previously cloned *Drosophila* DNA that is located not too far from the region to be cloned. In step 1, the cytological location of the cloned DNA sequence is determined by in situ hybridization to polytene chromosomes (see below). In step 2, the starting DNA fragment is used to screen a genomic library for random fragments that overlap the starting fragment. The location of these fragments on the chromosome can be determined

161

by a technique called in situ hybridization (see below). Then, they can be aligned to the starting fragment by restriction endonuclease mapping[5] in order to determine the fragment that is farthest from the starting point in the direction of the walk. Step 2 is repeated many times and generates overlapping cloned DNA fragments that lie farther and farther from the starting point. The endpoint of a chromosomal walk is the region of interest, that is, the known or assumed location of a gene or gene complex of interest (as determined by classical cytogenetic mapping). In order to save time, it is possible (step 3) to "jump" large distances on the chromosome by using chromosomal rearrangements such as inversions. Rearrangements such as small deletions can also be used to narrow down the position of any cloned DNA fragment.

Chromosomal walking is also possible in other organisms (some human genes have been cloned by this technique). However, in *Drosophila* the technique is particularly efficient because of a special feature of the fruit fly, namely the larval giant chromosomes. Remarkably, such giant chromosome preparations can be used to determine the location of cloned DNA fragments. If radioactively labeled DNA is added to giant chromosome preparations, they hybridize (bind) to the DNA strands in the giant chromosomes at a position where the base sequence is complementary. This position can then be visualized with the help of ordinary X-ray film (see Figure 6.2). In this way, the location of the cloned DNA fragment can be compared with the genetic map. This technique is called "in situ hybridization" and is an especially convenient way of navigating during chromosomal walking.

Hogness's laboratory first used these new techniques in order to clone the *rosy* and *Ace* genes.[6] As Bender, Spierer, and Hogness (1983) report in their paper, while their "walk" was in progress, they learned from E. B. Lewis of inversions with endpoints in the Bithorax complex and in the region in which they were "walking" and used these to "jump" into the Bithorax region. Bithorax is a genetic locus that was first described by the Morgan group in 1915. Mutations at the Bithorax locus cause bizarre phenotypic changes, for example, flies with an extra pair of wings (a most embarrassing condition for a member of the insect order *Diptera*, which means "two-winged"!). Bithorax is classified as a so-called homeotic locus, and mutants at these loci are called "homeotic mutants." Such mutants were described as early as 1894 by the British pioneer of genetics, William Bateson. Mutants at homeotic loci are characterized by transformations of body parts into the form of more posterior or more anterior structures. Hogness's laboratory thus cloned the first homeotic locus rather accidentally. However, because of their spectacular phenotypic effects, homeotic genes were thought to be involved in the control of embryonic development and therefore were of considerable theoretical

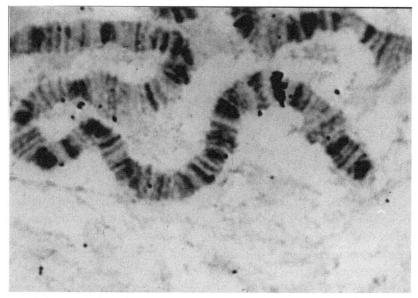

Figure 6.2. In situ hybridization of a cloned DNA fragment containing the *white* gene to *Drosophila* giant chromosomes. The location where the radioactively labeled DNA fragment hybridized shows up as a black band on an X-ray film. Photograph supplied by Walter J. Gehring; reprinted with permission.

interest. On the basis of the phenotypic effects of various homeotic mutations, Lewis (1978) had proposed a genetic model for how the genes residing at the Bithorax locus control the identity of the segments of the fly embryo.

Two other laboratories started cloning work on another homeotic locus, namely Antennapedia. Mutants at this locus are even more bizarre than Bithorax mutants, as they produce flies that have legs instead of antennae attached to their heads. Cloning the *Antennapedia (Antp)* gene was a strenuous effort, which took Walter Gehring's laboratory at the University of Basel three years (Garber, Kuroiwa, and Gehring 1983). Basically, they carried out an enormous "chromosomal walk." Several Drosophila-mutants were useful for this task. For example, the chromosomal inversion *Humeral* allowed them to "jump" over a large distance on the chromosome in order to move closer to the Antennapedia region. The laboratory of Thomas Kaufman at Indiana University used a somewhat different strategy, which also involved much chromosomal walking, to isolate clones spanning the *Antp* region (Scott et al. 1983).

Both Gehring's and Kaufman's laboratories were able to use their chromosomal walk in the *Antp* region to clone another gene that is located in close vicinity to it (it is part of the Antennapedia gene complex): *fushi tarazu* or *ftz*.[7] The Basel and Indiana groups independently found a sequence homology between *Antp*, *ftz*, and *Ultrabithorax* (one of the genes at the Bithorax locus). The homologous sequence turned out to be a highly conserved sequence element of length 180 base pairs (McGinnis et al. 1984c; Weiner, Scott, and Kaufman 1984). It was named the "homeobox." Collaborating with the group of Eddy de Robertis, which was housed at the same department, Gehring and co-workers quickly found that the homeobox is also present in other organisms, for example, the frog *Xenopus laevis* (Carrasco et al. 1984), the mouse (McGinnis et al. 1984b), and other metazoans (McGinnis et al. 1984a). The Basel groups were also able to show that the homeobox genes are active during early development in both fly and frog embryos.

The homeobox had a great impact on molecular studies of development. Using cloned *Drosophila* sequences containing the homeobox as molecular probes for colony hybridization screens, homeobox-containing genes from a great range of organisms that include humans and plants were isolated in a very short time (Gehring 1987). Today, homeoboxes are thought to be functional parts of so-called master control genes in an extremely wide range of organisms. The discovery of homeobox genes is thus widely viewed as a major breakthrough in beginning to understand the genetic control of development at the molecular level. As my account shows, the tiny fruit fly played a crucial role in this discovery.

I return to gene cloning in *Drosophila* in Chapter 7, where I examine the relationship between the classical and molecular gene concepts. In the following sections, I want to explore what we can learn about the role of model organisms from this episode.

6.2   MODEL ORGANISMS AS "SYSTEMS OF PRODUCTION"

Robert Kohler has proposed to analyze the role of experimental organisms in economic terms. He views them as "systems of production," which "give communities of practitioners their distinctive identities and ecologies" (Kohler 1991, 88). On Kohler's account, each experimental organism is part of a small industry of knowledge production that, in the form of publications, career opportunities, scientific prestige, awards, etc., yields returns in the marketplace of knowledge.[8] As in the real economy, the returns depend on supply and demand, which, in turn, may depend on competition. Sometimes, scientists move into new markets in order to evade competition. One way in which this can happen is by *innovation*, in other words, by the invention of a new system of production. One of the historical examples that Kohler details is the move by George W. Beadle and Edward L. Tatum from *Drosophila* into *Neurospora* biochemical genetics in the late 1930s. Since the returns from the *Drosophila* system were diminishing (partly due to international competition, but also due to intrinsic limitations of the *Drosophila* system), Beadle and Tatum took to working on the mold *Neurospora*. The new system turned out to be highly productive, as more than 100 *Neurospora* mutants that were deficient in various biosynthetic pathways were isolated and characterized in a very short time. Thus, the *Neurospora* system was like a factory or production line for mutants in various biosynthetic enzymes. The most widely publicized result of this work was the famous "one gene–one enzyme hypothesis," which stated that each gene makes exactly one enzyme (Beadle and Tatum 1941). What is less well known is that the *Neurospora* system, within just a few years, provided "more knowledge of biosynthetic pathways than two or three generations of biochemists using traditional methods" (Kohler 1991, 122).

Kohler also proposes to view experimental organisms not as *objects* of research, but as *instruments*: "[T]he standard rat and the mutant *Drosophila* are constructed artifacts. 'Standard' animals, produced by generations of inbreeding, are unambiguous examples of engineered instruments" (1991, 89). Although it was initially designed as an instrument for genetics, "*Neurospora* unexpectedly proved more productive of biochemical knowledge. *Drosophila* was abandoned, and the practice of developmental or physiological genetics was transformed. Thus, searching for more convenient and productive forms of known experimental systems, Beadle and his coworkers came to invent new systems of production, the astonishing productivity of which transformed the practice of both genetics and biochemistry" (1991, 89).

This finding of Kohler's is echoed in Rheinberger's theory of experimental systems. Rheinberger has also argued that such systems move across disciplinary boundaries in a way that redefines these disciplines and frequently end up serving purposes for which they were not initially designed (see Section 5.1).

In his admirable study of the rise of *Drosophila* as an experimental organism for genetics, Kohler (1994) adds to the economic perspective an ecological one. In this work, Kohler construes *Drosophila* as a "breeder-reactor" for mutants and for the production of detailed genetic maps. The fruit fly was, as it were, ecologically preadapted to life in the genetics laboratory for various reasons that include its short generation time, high fecundity, low resource requirements, and large number of viable and phenotypically visible mutations. But in order to become a highly productive system for experimental work, *Drosophila* had to be transformed from its wild, natural state into a creature that was domesticated for life in the laboratory: "The transformation of 'natural' creatures begins when they enter into their first experiment, and the more productive they become, the more they come to resemble instruments, embodying layers of accumulated craft knowledge and skills, tinkered into new forms to serve the peculiar purposes of experimental life" (Kohler 1994, 7).

Thus, on Kohler's account, scientists continuously craft their experimental organisms into technological artifacts that are highly fit for the specific tasks of experimental knowledge production. In the case of *Drosophila* genetics, this construction process consisted in inbreeding, mutating, and selecting strains of flies through thousands of generations until a life form was created whose genetic makeup differed considerably from that of wild fruit flies. Kohler (1994, 5–6) views experimental organisms simultaneously as technological artifacts, as biological systems with a distinct natural history that enter into a special kind of symbiotic relationship with the human actors, and as centers of social organization with a distinct "moral economy" that places constraints on the selfish interests of individual scientists.[9]

I now interpret my case study of the molecularization of *Drosophila* from this perspective.

As I have shown in the previous section, several laboratories started to use recombinant DNA technology in order to clone a number of *Drosophila* genes in the late 70s and early 80s. Some of the first genes to be cloned had been studied by geneticists for up to seventy years, for example, *white*, Bithorax, or *achaete-scute*. However, because of the intrinsic limitations of classical *Drosophila* genetics, molecular cloning techniques were a necessary prerequisite for studying these genes at the DNA level.

A striking aspect of the early cloning attempts is the extent to which the molecular biologists relied on *classical* mapping techniques in order to isolate specific DNA fragments containing the genes of interest (see Section 6.1). The geneticists did not know anything about the physiological function of the genes cloned; all they knew was the manifest phenotypic effects of some mutations in these genes and the chromosomal location. The locations were known from classical linkage maps as well as cytological maps. This knowledge was used in Hogness's, Gehring's, Kaufman's, and other laboratories to isolate specific DNA fragments from genomic libraries by chromosomal walking. Thus, for the first time, genetic maps were actually used for the same purpose as city maps, namely for finding one's way around!

Another interesting feature is how closely the classical techniques and the recombinant DNA technology interacted. In the method of in situ hybridization, radioactively labeled DNA was used to directly visualize the location of specific sequences on preparations of polytene chromosomes. An even more spectacular example is provided by those genes that were cloned by directly cutting out DNA fragments from cytological preparations by microdissection (e.g., Scalenghe et al. 1981). In these cases, the traditional preparation technique used for cytological mapping itself became a source of specific DNA fragments (which can be as small as 200 kb) for cloning experiments.

The cloning work required a highly fine-tuned combination of techniques, materials, and suitable fly strains. For example, in order to genetically engineer *Drosophila*, special gene-transfer vectors[10] were designed on the basis of the P-Element, a piece of "selfish DNA" that spreads through fly populations by moving from chromosome to chromosome. Klemenz, U. Weber, and Gehring (1987) designed a vector containing the *white*[+] gene (i.e., the normal wild-type copy of the gene). This feature was highly useful for identifying germ-line transformed flies, that is, flies that had actually inserted the DNA injected into their genome. If *white*[−] flies are used for transformation, the successfully transformed flies can be identified by their wild-type red eye color. Such vectors are remarkable pieces of molecular craftsmanship, for they also contained the sequences necessary to propagate the vector DNA in *E. coli* for amplification, as well as bacterial antibiotic resistance genes for cloning. Germ-line transformation is an important technique that allows molecular geneticists to verify the identity of the genes they have cloned, namely, to determine if the cloned fragments can rescue the mutant phenotype. Furthermore, geneticists can introduce altered versions of genes and regulatory regions in order to examine their functions.

Any well-defined DNA fragment that was cloned constituted a potential resource for further experimental work, because it could serve as a tool for the cloning of other DNA fragments. For example, an already cloned DNA fragment could serve as a starting point of chromosomal walks to isolate sequences that are located in a near chromosomal region. In this way, cloned DNA fragments were used to clone other, totally unrelated sequences. In order to isolate homologous sequences, cloned DNA was used as a hybridization probe in colony hybridization screens. This method could be applied to DNA isolated from any organism, a fact that was crucial for isolating a host of genes from other organisms (including humans) using cloned DNA fragments from *Drosophila*. Thus, some of the products of gene cloning work in *Drosophila* became an important resource for work on other organisms. The most striking example from the early molecular *Drosophila* work is the homeobox, which profoundly transformed molecular studies of development. Recombinant DNA technology transcends species boundaries.

On the basis of these considerations, we can now try to apply Kohler's concepts to the present case study.

I suggest that, with each new gene cloned, *Drosophila* was slowly crafted toward a highly efficient tool for the production of more defined cloned DNA sequences. Kohler's "breeder-reactor" was transformed from a system for the production of mutants and maps to a system for the production of specific cloned DNA fragments. The productivity of the system increased as time went on, and some of its products served as resources for work on other experimental organisms, that is, other systems of production. The early cloning work in *Drosophila* greatly benefited from the enormous material and cognitive resources from classical genetics, which consisted in thousands of well-defined mutant strains, genetic and cytological maps, and skills in manipulating flies in the laboratory.

The enormous productivity of the molecular *Drosophila* system, at least in the early stage of molecularization, can be explained by the development of *hybrid techniques* (Rheinberger 1997, 135) that combined the material and cognitive resources from two different traditions, classical *Drosophila* genetics and recombinant DNA technology. Classical *Drosophila* genetics has been developed by several generations of geneticists and has led to the isolation of thousands of mutant strains and to ever more detailed chromosomal maps showing the location of the mutations. Cytological mapping of polytene chromosomes proved to be an especially powerful technique. Recombinant DNA technology, by contrast, was mainly developed by work using *E. coli*, bacteriophages, and animal viruses as experimental organisms. The basic tools of this technology are a set of purified enzymes that modify DNA in various ways, for example, restriction endonucleases, which recognize specific sequences and hydrolyze DNA phosphodiester bonds at specific locations, or DNA ligase, which can be used to join DNA fragments in vitro.

I suggest that these two parent technologies – classical *Drosophila* genetics and recombinant DNA technology – were used to create a hybrid technology for the identification and isolation of genes. An important part of the hybrid technology was the method of in situ hybridization of radioactively labeled DNA to polytene chromosomes, which was originally developed on chromatin from *Xenopus* oocytes (Gall and Pardue 1969; John, Birnstiel, and Jones 1969; Pardue and Gall 1969). This technique greatly expanded the capacities of the cytological mapping system by allowing researchers to visualize the positions of specific DNA sequences on polytene chromosomes. Thus, one of the most powerful mapping approaches available in *Drosophila* was adapted to the needs of molecular cloning. The technique of chromosomal walking, which allowed the isolation of DNA from specific locations on the chromosome, relied heavily on in situ hybridization. In addition, geneticists made clever use of available inversion and deletion mutants (which had been mapped cytologically) in order to "jump" over large chromosomal regions and in order to determine the exact position of their DNA clones between the breakpoints of known chromosomal rearrangements. This demonstrates how the hybrid system was able to integrate the classical and molecular research traditions in genetics. It enabled the scientists to mobilize the experimental resources that came with these traditions, for example, mutants, maps, DNA-modifying enzymes, and preparation techniques, to create a powerful new system of production.

If classical geneticists continuously modified the organism they were studying, as Kohler and others have shown, the magnitude of the modifications introduced to the experimental organism is even more striking in molecular research. The molecular Drosophilists *genetically engineered* their favorite lab animal until it became a highly productive system for identifying and isolating DNA regions that interested them.

To conclude, the rapid molecular transition in *Drosophila* genetics was the result of a set of classical/molecular hybrid techniques that allowed researchers to mobilize the enormous experimental resources of classical *Drosophila* genetics and to transform them into molecular cloning tools. By "experimental resources" I mean not just the knowledge of mapping techniques, etc. but also *material* resources such as the thousands of genetically defined mutant strains. The hybrid system allowed researchers to mobilize these resources for cloning experiments, and thus to transform them into molecular research tools. In the process, *Drosophila* was transformed from a system for the production of mutants and genetic maps into a system for the production of specific recombinant DNA molecules. This successful molecularization of *Drosophila* was a prerequisite for the spectacular comeback of this experimental organism, which today populates hundreds of cell and developmental biology laboratories around the world.

In the following section, I critically examine what this approach actually explains.

6.3   THE SHORTCOMINGS OF TECHNOLOGICAL AND ECONOMIC METAPHORS

Kohler's approach has opened a rich and interesting perspective on the role of experimental organisms in biological research. The industrial-economic notions of production, resources, returns, technological innovation, crafts, etc., as well as the focus on the "material cultures" of experimental biology, have proven their worth as analytic tools for historical studies of scientific practice, including its social organization, that go beyond the history of ideas and theories. However, in the philosophy of science we must scrutinize the adequacy of such notions for illuminating the nature of science as explanatory concepts. First, I examine the notion of experimental organisms as "instruments," "tools," "systems of production," and "technological artifacts." I argue that these terms cannot be taken literally in this context; they are metaphors. Second, I show that although experimental labor clearly does have an economic aspect, the capacity of economic notions to illuminate the role of experimental organisms in biological research is limited. This paves the road to an examination of the *epistemic* roles of experimental organisms.

According to Kohler, experimental organisms can be viewed as "instruments," that is, as a special kind of technological artifacts. Another technological metaphor is that of a "tool." I begin by arguing that these terms are, indeed, metaphors. In their original usage, all these terms designate objects that were created by human beings in order to fulfill a certain function. Tools are the exemplars of technological artifacts and include an extremely broad range of objects, ranging from the simple stone tools used by our Paleolithic ancestors to skin dead animals to the massive drills that were used to build a tunnel under the Channel. Instruments are also technological artifacts, but their original function was to aid the human senses for purposes of navigation (e.g., compasses or sextants). Scientists, of course, also build instruments in order to make measurements. Scientific instruments range from simple balances to particle accelerators or gravity wave detectors. In what sense can an experimental organism said to be an instrument?

One of the similarities partly underlying the appeal of this metaphor, I suggest, is that laboratory organisms are also used to observe something that our unaided senses could not perceive. However, there are also dissimilarities. In a typical instrument, a *causal input* (e.g., the Earth's magnetic field in the case of the compass) leads to an *observable signal* (the motion of the compass needle). What the user of an instrument is interested in is the process or object that is responsible for the causal input (the orientation of the Earth's magnetic field). This process may or may not be physically attached to the instrument. How does this translate into our examples of experimental organisms? It seems to me that, rather than using an experimental organism to detect some causal input, the experimental biologist is interested in the organism *itself*. To be precise, the experimental biologist wants to understand the biological and physicochemical processes that occur in the organism. The laboratory organism is not the measurement or observation device; it is the object that is being measured or observed. Thus, strictly speaking, an experimental organism is not an instrument. The case is different if an organism is used to measure or observe something else, for example, when dogs are used by miners to warn them about leaking gas. But this is not the typical role of model organisms.

The same is true of the "tool" metaphor. Tools are used to modify an object at which the tool use is directed. However, typical experimental organisms are themselves the objects of experimentation; their function is not to manipulate something else. I say "typical" experimental organisms because it is easy to find exceptions to this claim. In the example of molecular cloning techniques, *E. coli* bacteria are used to amplify DNA fragments. These bacteria may be viewed as proper tools, but not the *Drosophila*. The latter are the *objects* of research.

The instrument- or tool-like character of experimental organisms also arises because scientists spend a lot of work modifying, developing, and standardizing them. Laboratory flies are highly inbred creatures, some of which were deliberately bred to contain certain combinations of genes. In molecular biology, scientists even introduce deliberate changes in specific genes, or they introduce genes from a different, sometimes unrelated organism. Most of the experimental work would be impossible without these modifications of the experimental organisms. However, these modifications do not make these organisms instruments or tools. They are simply *interventions* in a natural object, which are the hallmark of experimentation.

More generally, the whole notion of experimental organisms as technological artifacts is metaphorical. Even a *Drosophila* that has been bred for hundreds of generations, or a genetically engineered fly, is not an artifact – it is still a living creature. Laboratory organisms are no more technological artifacts than, say, basset hounds. Even for those of us who reject vitalism, there is a fundamental difference between a machine and a living organism. This difference, at least from a materialistic perspective, is not that organisms contain substances or elementary forces that are absent in nonliving systems. The main difference is that technological artifacts were built by humans, whereas organisms are a product of evolution. Far from being merely of interest for pedantic philosophers, this difference is of great epistemological significance. Even if most technological artifacts are "black boxes" for most of us, *somebody* knows what components they are made of and how they are connected. As far as I am concerned, the computer I am using to write this book is nearly as mysterious as a fruit fly; however, there exist engineers who understand

171

how it works (I hope). This is not to say that a technological artifact cannot exhibit properties or have uses that the people who built it did not anticipate, but at least they know what parts it is made of and how it is wired up. Not so with fruit flies. If somebody knew exactly what components a fruit fly is made of and how they interact, there would be little need for biological research.

These considerations suggest that notions of experimental organisms as "instruments," "tools," or "technological artifacts" are metaphors of some limited analytic value at best, and misleading at worst. However, there is an important element in these metaphors that we must take seriously. This has to do with the fact that so much laboratory work in experimental biology goes into modifying the experimental organisms and developing the other research materials used. This work is not directly aimed at discovering new phenomena or testing theories. Instead, it provides the *material resources* needed for doing the latter kind of research work. When Beadle and Tatum started to work on *Neurospora*, they first had to isolate and breed hundreds of different strains of this microorganism, and develop methods for genetic analysis (e.g., separating spores for genetic crosses), before it became useful for analyzing biosynthetic pathways. Or, to go back to my case study, Walter Gehring's technicians, students, and postdocs spent three years of extremely hard work at the lab bench just in order to isolate a piece of DNA containing the *Antennapedia* gene, even though they could not really be sure that this would lead to such a significant discovery as the homeobox. What is more, this work was only possible because David Hogness's laboratory had previously worked out the method of chromosomal walking, because molecular biologists had developed recombinant DNA methods, and because several generations of drosophilists had previously accumulated mutants, genetic maps, cytological methods, and the like. I believe that it is this aspect of research in experimental biology that produces the appearance of technological artifacts. Indeed, there might be close similarities between much of the research work going on in an experimental biology laboratory and the research work of, say, an electrical engineering laboratory. Nevertheless, as I have argued, experimental organisms are not instruments, nor tools, nor any kind of technological artifacts. Their origin is biological. Experimental organisms are not *built* by biologists; they are only *modified*. Therefore, we need more adequate analytic concepts to explain the role of experimental organism in biological research. In the next section, I introduce such an analytic concept. But first, I examine the adequacy of the economic notions.

On the view taken by Kohler, the widespread use of model organisms can be explained by the benefits that these organisms bring to the researchers who develop and use them. Indeed, there is a sense in which a scientist's livelihood can come to depend directly on the productivity of an experimental organism. On such a view, certain organisms that are highly productive have a strong tendency to dominate scientific disciplines or subdisciplines because their productivity can sustain a large number of scientists who use them for their research. This is fully equivalent to explaining the spread of some new technology by virtue of the economic returns that this technology yields. Thus, in contrast to the "instrument" and "artifact" metaphors that I have criticized in the previous section, the economic notions of production and returns are not metaphorical; they are bona fide applications of these terms. However, I want to maintain that, in the philosophy of science, we cannot stop at such economic explanations of certain research strategies.

The limit of the economic approach becomes clear as soon as we ask what is being produced by research in basic science. While it may be the case that some biological research aims at producing material goods (e.g., therapeutic proteins or laboratory animals such as the infamous Oncomouse), this is not true of the examples that concern us here. *Drosophila*, *Neurospora*, bacteriophage, and Company were not developed as experimental organisms because of their potential economic value as material goods. Instead, they were developed for advancing basic science. This recognition by no means invalidates the economic approach. Scientists do not just work for the advancement of science. They also work for the advancement of their careers. Science, like any social activity, has specific reward mechanisms. However, this indisputable fact notwithstanding, it remains the case that some research work is valued by scientists and by those who fund them because it contributes to the advancement of knowledge. Kohler's "systems of production" are systems for the production of knowledge. Whatever other benefits experimental organisms may have, for example, for the careers of the scientists involved, these benefits are dependent on the contributions that they make to the advancement of scientific knowledge. If moving into *Neurospora* was beneficial for Beadle and Tatum's careers, for attracting funding and gifted students, and so on, this was because the *Neurospora* system proved to be conducive to producing knowledge. There is much that can be said about the development of experimental organisms from an economic and sociological perspective, as Kohler has done so masterfully; however, it remains the case that these organisms have *epistemic* functions in experimental research, and their success in the laboratory depends on how well they perform these epistemic functions. An examination of the epistemic functions must thus accompany the economic and sociological analysis of "material cultures."

In the following sections, I examine the epistemic roles of experimental organisms in biological research.

6.4 PREPARATIVE EXPERIMENTATION

In the previous section, I have criticized the view of experimental organisms as "instruments" or "tools." However, I have also argued that there are some similarities between the development of technological artifacts and the development of experimental organisms that deserve serious consideration. The main similarity that motivates the technological metaphors, as I have suggested, is the large amount of work that goes into developing the materials that are needed for research – work that is not primarily aimed at discovering new phenomena or testing theories. In order to capture this aspect of experimental practice, I propose the concept of *preparative experimentation*.

As paradigmatic cases of preparative experimentation, I take the production of a collection of well-defined mutant strains of an experimental organism such as *Neurospora, E. coli,* or *Drosophila*; the cloning of specific DNA fragments containing genes or other sequences of interest; and the production of genetically modified organisms for further research.[11] This kind of experimental work is not directly aimed at testing a specific hypothesis, nor do biologists necessarily need a guiding theory for conducting this kind of research. This does not mean that they do not need any theoretical knowledge. Clearly, developing experimental organisms and other research materials requires some knowledge of genetic mechanisms, chemical properties of biomolecules such as DNA or protein, and so on. What I mean by a "guiding theory" is something like Mitchell's chemiosmotic theory (see Section 4.1), that is, a theoretical explanation the confirmation or refutation of which is the direct goal of the experimental work. In fact, as some authors have pointed out, the problems that an experimental organism helps to solve might only arise during or after the work of developing that organism and associated research materials (Lederman and Burian 1993).

The results of preparative experimentation include both research materials (strains, cells, isolated DNA, purified proteins, etc.) and knowledge *about* these material objects. Research materials, for example, a fly strain or a cloned DNA fragment, would be absolutely useless for research without knowledge of at least some of the properties of these materials. Which known mutations does this particular *Drosophila* strain contain? What is the source of this isolated DNA fragment, and how was it selected from the thousands of fragments in the original genomic library? Without this kind of information, any research materials would be worthless. On the other hand, the knowledge about the properties of these materials is sometimes highly fragmentary and will have to be examined by further experimental work.

A great deal of experimental work in biology is preparative. In an extreme case, biologists might be able to do an experiment that supports or refutes an important theoretical claim in a single day, after many years of developing the experimental organisms and research materials that are required to do this experiment including the necessary controls. Alternatively, years of preparative experimentation can be rewarded by an unexpected discovery that was not predicted by any theory, such as the discovery of a highly conserved sequence element in homeotic genes (the "homeobox").

How does preparative experimentation differ from building a scientific instrument? It seems to me that these developmental activities differ both with respect to the intentions of scientists and with respect to the products. When an instrument is built, the scientists involved have to know what kind of causal inputs the instrument is supposed to be sensitive to. A gravity wave detector is designed to respond to gravity waves and to allow discrimination of gravity waves – should they really exist – from disturbing causal inputs (e.g., seismic vibrations). Similarly, a particle accelerator is built in order to detect decay events in certain energy ranges. I am not claiming, of course, that no unexpected things can happen during or after the construction of such a scientific instrument. But it seems to me that there is an epistemologically relevant difference between taking a natural object – a biological organism – and experimentally modifying it in order to analyze its functional organization, and designing a technological artifact – a scientific instrument – in order to detect a specific kind of causal input. It is this difference that I am trying to capture by distinguishing between preparative experimentation and instrument building.

I suggest that the importance of preparative experimentation in biological research is shown by biologists' strong tendency to work on an already established laboratory organism rather than initiating research on a new one. Again, economic or sociological reasons could be invoked to explain this fact. However, I suggest that there are also epistemic reasons. One of the main reasons is the strongly *cumulative* character of the preparative-experimental work done with specific organisms. This can be nicely illustrated with the help of my case study of the molecularization of *Drosophila*. As we have seen (Section 6.1), the scientists who were beginning to study *Drosophila* genes at the molecular level heavily relied on the resources of classical *Drosophila* genetics in order to clone some of the first fly genes. These resources included, first, detailed linkage maps and cytological maps of the four fly chromosomes, second, mutant strains, and third, techniques and skills in breeding flies, preparing chromosomes, dissecting larvae and adults in order to analyze their phenotypes, and so on. All these resources had been accumulated through what I call preparative experimentation by several generations of classical geneticists. A large part of these resources were produced before the molecular revolution in genetics.[12]

Thus, even though the theories and concepts of genetics have undergone considerable change because of the rise of molecular biology, some of the experimental resources that preceded molecular biology could be combined with the resources of recombinant DNA technology in order to turn *Drosophila* into a powerful experimental organism for molecular research. Once the molecular biology of *Drosophila* had acquired momentum, the preparative-experimental resources for this organism increased further due to the increasing availability of materials such as cloned DNA fragments and recombinant vectors for gene transfer and for controlled gene expression. As this kind of resources accumulate, model organisms greatly increase in the experimental possibilities that they offer, and this explains why scientists more often than not choose to work with an already established system rather than starting from scratch.

I now turn to the question of how experimental organisms are selected by scientists.

6.5   "RIGHT CHOICES" OR CO-CONSTRUCTION?

How do experimental biologists choose their experimental organisms? It might be suggested that the best-suited organisms for studying a particular process will be the *simplest* ones that actually contain this process. For example, research on the molecular basis of behavior has been particularly successful in the sea snail *Aplysia* and the nematode worm *C. elegans*. These organisms are among the simplest ones that exhibit some features that might be called behavior, and they even show some simple conditioned reflexes that have been promoted as models for learning and memory. Similarly, the simplicity of the genetic systems of *E. coli* and bacteriophage were certainly important for these organisms' widespread use in molecular biology laboratories. Baker's yeast, *Saccharomyces cerevisiae*, is one of the simplest eukaryotic cells and, partly for this reason, has turned out to be very useful for studying intracellular processes such as protein transport, the biogenesis of mitochondria, and cell cycle regulation. However, simplicity alone cannot explain why certain experimental organisms were as successful as they were. First, it is difficult to spell out exactly what "simple" means. (Is the slime mold *Dyctostelium*, which can form a complex fruiting body from free-living amoeba-like cells, simpler than *C. elegans*?) Second, there are always a great variety of other simple organisms that could have been chosen as well. And third, there are clearly exceptions to the simplicity rule. The mouse is not the simplest animal with an immune system, and *Drosophila* does not have the simplest developmental program. Yet the mouse has been one of the most important model systems for immunologists, and *Drosophila* for embryologists. Therefore, there must be other reasons that some species are better suited than others for studying particular biological phenomena. These reasons cannot lie solely in an organism's physiological, genetic, and morphological makeup, but must also lie in the nature of the specific tasks to be solved, the availability of instruments, techniques, and so forth (Burian 1993a). I just provide a few characteristics that affect the suitability of a model organism:

(i) An organism must be easy to breed in the laboratory.

(ii) Its generation time should be short; otherwise genetic crosses simply take too long for the results to become available.

(iii) If the organism need to be amenable to genetic analysis, there must be viable mutants with noticeable phenotypic effects.

(iv) As I have shown in Section 6.1, certain methods such as linkage mapping can have intrinsic limitations, which may be overcome by using a different organism. As we have seen, linkage mapping had a much higher resolution in bacteriophage compared to *Drosophila* because of the number of individuals that could be handled in a single experiment.

(v) Some organisms have peculiar features that biologists can exploit. The giant chromosomes found in *Drosophila* larval salivary glands (see Section 6.1) are an example of such a peculiarity. Due to the enormous size and the characteristic banding patterns of these chromosomes, cytological mapping of *Drosophila*'s giant chromosomes is a much more powerful technique than comparable methods in other organisms, such as plants or mammals. That *Drosophila* has these chromosomes was a lucky coincidence for geneticists. Another example of scientifically useful peculiarities is the large squid nerve cells, which greatly facilitate neurophysiological measurements.

(vi) Finally, there is the possibility that an organism is by no means ideal for studying a particular biological process, but that any disadvantages are compensated for by the large amount of experimental resources (see Section 6.4) that come with this organism.

There is an interesting challenge to the view that certain organisms are best suited for certain "jobs." A number of historians of biology have argued that biologists do not always deliberately choose certain organisms for certain preexisting research problems. Instead, the research materials and the problems are "co-constructed" (Clarke and Fujimura 1992). This view is based on two kinds of historical findings. First, the choice of experimental organisms as well as other research materials, instruments, and so forth is highly *contingent*. Biologists will use whatever is available to them, given their local situation and practical constraints (compare Section 5.1). Second, the "job" to be done, that is, the set of research problems to be solved, is often *transformed* through the adoption and development of new experimental organisms. The "rightness" of certain choices, then, may be a retrospective rationalization. Thus, the whole idea of scientists choosing the right "tools" for the "job" may be misguided, because experimental organisms can thoroughly transform the "job" to be done.[13]

One of the best examples for this is *Drosophila* itself. As Kohler has shown, the fruit fly was not brought into the laboratory to do genetics. Rather, "it was brought indoors because its habits and seasonal cycles were well-suited to the needs and seasons for academic life" (1994, 20). As far as Morgan is concerned, he was interested in all sorts of organisms and a wide range of biological questions. Initially, *Drosophila* was only one of several organisms that Morgan was experimenting with. However, once the first mutants had been isolated and shown to exhibit Mendelian inheritance patterns, it strongly canalized the focus of Morgan's laboratory. The large number of spontaneous mutants that kept cropping up in the laboratory and the amount of labor that was required to maintain, characterize, and catalog them quickly took up all the time of Morgan and his co-workers. Kohler also argues that genetic mapping took over because of *practical* imperatives, especially the need for "classifying mutants and managing an accumulating mass of experimental data that was rapidly getting out of hand" (1994, 56). I have defended the view that mapping was also of considerable theoretical interest at that time (Weber 1998b); however, practical imperatives certainly played a role. Thus, *Drosophila* was not initially chosen for what it is famous for, namely the construction of the first genetic maps. But once in the laboratory, *Drosophila's* high output of viable mutants dramatically transformed the practice and problems of genetics. Only in retrospect can we say that *Drosophila* was a "good choice" as an experimental organism for genetics.

The contingencies of experimental organism choice may have considerable epistemological implications. For example, it is far from obvious what would have happened if Morgan had not chosen *Drosophila* as an experimental organism to look for mutants.[14] Would genetic mapping have been invented? Or would genetics have taken an altogether different course, leading to different questions, different concepts, and different theories? Because of their counterfactual nature, it is notoriously difficult to answer such questions (which does not mean they are not meaningful).

The view that the choice of experimental organisms is largely a matter of locally contingent factors while, at the same time, having far-reaching consequences for defining relevant research problems could mean that our perception of the biological world is strongly biased toward a small number of organisms that have not been chosen because they allow scientists to discover important biological principles, but simply because they happened to be part of some local "material cultures." What comes to be viewed as an "important" principle could also be determined by these contingent factors. However, this constructivist conclusion only follows if the knowledge obtained from certain experimental organisms cannot be *generalized* to other organisms. I discuss the problem of generalizing from model organisms in Section 6.6. Now, I argue that the impact of locally contingent factors may be more limited than some historians of biology have suggested.

What the historical analyses provided by Kohler and others suggest is that the *initial* choice of experimental organisms can be highly contingent and strongly influenced by locally contingent factors.[15] However, the later *vindication* of this organism as a system that is conducive for producing certain kinds of knowledge may be less dependent on local factors. The proof of the pudding is in the eating, and model organisms can only prove their worth after a while, by enabling fruitful research in many different laboratories, the results of which can again serve as a basis for further research. In such a distributed process, the impact of locally contingent factors may be reduced.

It is now time to move on to the question of what can be learned from laboratory organisms that goes beyond the biology of the species that serves as a model.

6.6 WHAT CAN BE LEARNED FROM STUDYING MODEL ORGANISMS?

How much truth is there in Jacques Monod's legendary saying, "What is true for the coline bacillus is true for an elephant"? To put it less allegorically, to what extent can the findings obtained from studies of a particular organism be generalized to other organisms, for example, to humans? Richard Burian (1993a, 366) has called this question "an especially acute version of the traditional philosophical problem of induction." Kenneth Schaffner, after examining some recent research on the genetic basis of social behavior in *C. elegans*, mating behavior in *Drosophila*, and the pathogenesis of a mouse analog of Alzheimer's disease, concludes that "animal models are ultimately used as a source for extrapolations to humans, though an important subordinate goal is to understand the biology of all organisms" (Schaffner 2001,

224). While the use of mice as an animal model for Alzheimer's disease seems somewhat plausible – given that mice, like humans, are mammals – what can biologists hope to learn about human beings from studying an organism such as the humble nematode worm *C. elegans*? In this section, I try to develop a general framework for answering such questions.

As a starting point, it is helpful to consider the conditions under which we would expect inferences from one organism to another to succeed. For obvious reasons, if all organisms were radically different in terms of their physiological, developmental, and genetic mechanisms, nothing could be learned about other organisms from studying a particular model organism. A model organism must have *something* in common with the organisms it is supposed to be a "model" for. Now, it is known today that all organisms are remarkably similar at the level of certain molecular mechanisms. For example, so far as is known today, all self-reproducing organisms contain DNA as their genetic material.[16] Furthermore, the genetic code is the same in almost all organisms. Known exceptions include only the mitochondrial genetic system and some protozoans. The molecular mechanisms of DNA replication, transcription (RNA synthesis), and translation (protein synthesis) are remarkably similar within the prokaryotes (bacteria) and within the eukaryotes (nucleated cells), whereas they differ between prokaryotes and eukaryotes. The same is true of many metabolic pathways, for example, oxidative phosphorylation (see Chapter 4). Although oxidative phosphorylation comes in different versions in bacteria, it is virtually the same in all the eukaryotes.

But what grounds do biologists actually have for believing that the mechanisms listed above are really the same, or almost the same, in many different organisms? In other words, what makes them think that they have learned some biological principles that are of *general* significance? It could be objected that life on our planet still consists of more than one million species. Only a tiny fraction of these have ever been studied at the molecular level. This seems like a rather thin induction base. In fact, if the generality of molecular biological principles had been inferred solely by *enumerative induction*[17] from the known cases, the grounds for this generality would be very weak indeed. However, I claim that the generality of certain biological principles over a large number of species is *not* inferred by enumerative induction, but by a more sophisticated kind of inductive argument.

This kind of argument proceeds by comparing organisms *phylogenetically*, in other words, by paying attention to their common evolutionary descent. If a mechanism is found to be the same in a set of phylogenetically very distant organisms, this is evidence that it is also the same in a great number of other organisms as well, namely all those that share a common ancestor with the known organisms that are being compared. For example, the genetic code has been shown to be the same in organisms as phylogenetically diverse as mammals, fish, insects, plants, fungi, and bacteria. Now, it is assumed that this standard genetic code arose only once in evolution. This assumption is very likely to be true, because of the extremely large number of possible alternative genetic codes (i.e., possible assignments of nucleotide triplets to amino acids), which makes it highly unlikely that the same code arose more than once in evolution. Given these assumptions, it can be inferred that a large majority of organisms from all the kingdoms of life share the same genetic code. The deviant genetic codes found in a small number of genetic systems are very likely to be *derivative* traits that evolved only in a small number of taxa. For, if changes to the genetic code had occurred more often in evolution, we would not expect to find the same code in such phylogenetically distant organisms as humans and bacteria.

From a logical viewpoint, what we have here is an *argument from parsimony*; that is, we are inductively inferring to a statement from logically weaker premises by making an assumption of *simplicity*. In this case, the simplicity assumption is that the genetic code arose only once. While arguments from parsimony are not generally reliable, this one is: we have strong reasons to believe that the genetic code *did* arise only once. For a penetrating analysis of the problems of parsimony and phylogenetic inference, see Sober (1988).

The case of the genetic code is perhaps the most clear-cut one, because of its remarkable phylogenetic conservation and the extremely low probability that the same genetic code has arisen more than once in evolution. However, in principle, arguments of this kind can be applied to other mechanisms as well.

These considerations suggest that the usefulness of model organisms crucially depends on the extent to which the mechanisms in question are *phylogenetically conserved*. Any extrapolations from model organisms are only reliable to the extent that the mechanisms under study have the same evolutionary origin in the model organisms and in humans. However, we must be clear about what exactly the special kind of inductive argument that I have used on the example of the genetic code establishes. To be specific, what this kind of argument can show is that there is a high probability that a particular organism shares a certain mechanism with other organisms for which it is *known* that they have this mechanism. In other words, if we are asking whether some organism S_1 has a mechanism M, we can use this kind of phylogenetic inference by first checking whether organisms $S_2, S_3, \ldots, S_n$ have M. If we know the phylogenetic relationship of $S_1, S_2, S_3, \ldots, S_n$, we can then estimate the probability that S_1 has M. Alternatively, this kind of inference can be used to estimate how widespread a certain mechanism is in the biological world, that is, in how many taxa it occurs (as in the example of the genetic code). It seems to me that the latter kind of extrapolation is more important than extrapolation to particular species (see below).

Thus, extrapolations or inductive inferences from model organisms to other organisms are possible and can be reasonably sound, provided that they are based on known phylogenetic relationships. Ultimately, the *generality* of molecular biological principles can only be established by such inferences.

It might seem that this conclusion basically answers the question of what biologists can learn from studying model organisms. However, I want to show that the role of model organisms in biological research does not stop here. That is, model organisms have epistemic functions over and above providing a basis for inductive inferences or extrapolations to other organisms (Weber 2001c). It is to these other functions that I turn now.

In order to show that model organisms are not good just for extrapolations, I return to my case study from Section 6.1. As we have seen there, two *Drosophila* laboratories have cloned some fly genes that, due to the drastic effects that mutations in the genes have on the fly's body plan, were implicated in the genetic control of development. It was found that three of these genes, *Antennapedia, fushi tarazu,* and *Ultrabithorax,* share a short, conserved sequence element called the "homeobox." By using DNA probes from *Drosophila,* Walter Gehring's laboratory then discovered that several organisms that are phylogenetically very distant from *Drosophila* also have genes containing a homeobox. This points to an important role of model organisms in modern biological research. Namely, research on model organisms provides important *research materials* for work on other organisms. The cloned DNA fragments containing homeobox genes from *Drosophila* are an example of such research materials. These cloned DNA fragments provided scientists with rapid access to the homeobox genes of other organisms (including humans). Namely, these DNA pieces could be used to "fish" for homologous DNA sequences in genomic libraries prepared from the nuclear DNA of a variety of other organisms (a procedure colloquially known as "zooblots"). It is hard to say whether mammalian homeobox genes could have been discovered without the DNA probes that the *Drosophila* workers had prepared. There are a small number of extremely rare mammalian mutants that are probably due to homeotic genes, for example, the mouse mutant *rachiterata* or people with inborn alterations in digit numbers at their fingers or toes. However, in the pregenomic age, cloning homeotic genes in *Drosophila* was a much easier task than cloning mammalian genes, because – as I have shown – chromosomal walking was greatly facilitated by cytological mapping methods (Section

6.1). Thus, model organisms such as *Drosophila* can provide biologists with valuable research materials that can be used to discover homologous genes in other organisms. This function of model organisms has nothing to do with extrapolation or inductive inference; it is an example of what I have called preparative experimentation (see Section 6.4).

In the example discussed above, a cloned DNA fragment that originated from *Drosophila* research was used to discover genes with similar functions in other organisms. This approach was useful at a time when genomic DNA sequence information was still comparatively scarce. Today, in the age of genomics, model organisms can provide even more direct access to the function of certain genes. Recently, the full genomic DNA sequences of humans, *C. elegans*, *Drosophila melanogaster*, yeast (*S. cerevisiae*), the plant *Arabidopsis thaliana*, and various bacteria have become available. Clearly, the sequencing of the model organisms was not done for pure biological interest in these organisms. Instead, the sequences of model organisms are thought to be of crucial importance for interpreting the human genome sequence. The rationale behind this approach results from the fact that all of these model organisms contain so-called orthologous genes. These are genes that can be traced back to a common ancestor. For example, an initial analysis of the *Drosophila* genome sequence revealed that the fly genome contains orthologs to 177 out of 289 human disease genes (Rubin et al. 2000). Similarly, a comparison of the predicted proteins of yeast and *C. elegans* suggested that "most of the core biological functions are carried out by orthologous proteins [. . .] that occur in comparable numbers" (Chervitz et al. 1998, 2022). Thus, it seems that most organisms share a substantial number of molecular mechanisms that are evolutionarily very ancient. At the molecular level, evolutionary history is "one of new architectures being built from old pieces" (Baltimore 2001, 816).

When the first draft of 94% of the full DNA sequence of the human genome was published in February 2001, biologists expressed surprise that humans contain only about 30,000–40,000 genes (International Human Genome Sequencing Consortium 2001, 860). This is only twice as many as in *C. elegans* or *Drosophila*, suggesting that the additional complexity of human beings is due to more complex interactions between genes, rather than a substantially larger number of genes. Human genes also appear to encode a larger number of proteins than worm or fly genes because of a mechanism known as "alternative splicing."[18] However, there appear to be few protein domains (functional modules of proteins) in humans that are not found in any of the major model organisms.

Thus, in the postgenomic age, biologists are dealing with a universe of genes, proteins, and protein domains many of which are present in a great variety of organisms. A recently formed "Gene Ontology Consortium" writes: "Where once biochemists characterized proteins by their diverse activities and abundances, and geneticists characterized genes by the phenotypes of their mutations, all biologists now acknowledge that there is likely to be a single limited universe of genes and proteins, many of which are conserved in most or all living organisms" (Ashburner et al. 2000, 25).

Most, if not all, human genes have orthologs in other organisms. Therefore, understanding the role of any of these genes in model organisms can shed light on the function of the corresponding human genes. Biologists use the term "functional annotation" for this procedure, because it involves the ascription of biological functions to DNA sequences.

In order to see how the knowledge obtained from studying mechanisms in model organisms is relevant to other organisms, it is helpful to remember that biological mechanisms are often embedded in a *hierarchy* of mechanisms (Craver 2001; Craver and Darden 2001; Weber 2001c). For example, consider the neurological mechanism of long-term potentiation (LTP). This designates long-lasting changes in the readiness of neurons to transmit signals to neurons to which they are connected. LTPs are thought to be involved in learning and memory. The mechanism is part of a hierarchical structure of mechanisms that includes the basic mechanism of neural transmission (action potentials, see Section 2.1) at the next lower level and neural circuits at the next higher level. LTP operates in some very simple neural systems such as those of the sea snail *Aplysia* or *C. elegans*, as well as in the extremely complex central nervous systems of the mouse or humans. This means that a mechanism that has been elucidated in some simple model organism can be transferred into an explanatory structure for a more complex process in another organism. In other words, the basic mechanism of LTP could be worked out in *Aplysia*, where it is embedded in mechanisms such as the conditioned gill-withdrawal reflex. But the same (or almost the same) mechanism is also part of much more complex mechanisms in mammals, namely mechanisms that are involved in learning and memory.

It thus seems that biological mechanisms have a *modular structure*. The mechanisms of neural transmission or LTP are modules that can be part of different mechanisms of varying complexity in different organisms. The mechanisms or modules at the bottom end of the hierarchy are present even in the simplest organisms, whereas the mechanisms at the top end – such as memory and higher cognitive functions – are only found in more complex ones. Biologists have knowledge about a growing store of such modules out of which theoretical explanations of various biological processes can be assembled. Most of these modules have been investigated in model organisms.

The view of biological explanations as sets of modular mechanisms that can be assembled into hierarchies of different complexity can give us some hints as to what kinds of processes can be fruitfully studied with the help of model organisms. Since the mechanisms at the lower levels of the explanatory hierarchies (for example, the basis mechanism of neural transmission) appear to be phylogenetically strongly conserved, it does not matter much in which organism they are studied. For these mechanisms, biologists can choose the organisms that offer the most experimental resources. By contrast, the mechanisms at higher levels of the explanatory hierarchies are generally less conserved; they operate mainly in the more complex organisms. For these mechanisms, the value of model organisms is probably more limited.

To give an example, de Bono and Bargmann (1998) have described a gene from *C. elegans* called *npr-1* that explains some natural variation in the feeding behavior of the worm (discussed in Schaffner 2001, 205–207). Worms that have one variant of the gene feed socially, whereas carriers of a different allele are solitary foragers. Interestingly, there is an orthologous gene that codes for a receptor found in the human central nervous system (neuropeptide Y receptor). However, it is unlikely that that this finding has any significance for explaining human feeding behavior. It is much more likely that this receptor has a more general function in the human nervous system, a function that has nothing *specifically* to do with whether people prefer to eat alone or in large dinner congregations. In other words, in humans the *npr-1* homolog is probably embedded in a neurological mechanism more complex than that in *C. elegans*.

An even more striking, but similar example was provided by Barr and Sternberg (1999). The *lov-1* gene from *C. elegans* is necessary for male mating behavior. The closest human homologues for *lov-1* are *PKD-1* and *PKD-2*, which are kidney disease loci. As such, they are unlikely to be involved in human male mating behavior (I hope I will not be proven wrong on this one). Both the *npr-1* and *lov-1* genes seem to code for receptor molecules that are coupled to cellular signal transduction mechanisms. This suggests that, in terms of the lower-level mechanisms, these proteins perform a similar function in both worms and humans. However, these lower-level mechanisms are probably embedded in quite different higher-level mechanisms in humans, which limits the extent to which the worm is a good model for studying humans.

To conclude, model organisms play different epistemic roles in biological research, only some of which involve extrapolations or inductive inferences from a model organism to a different organism. By placing the knowledge obtained with the help of model organisms in an appropriate comparative–evolutionary framework, extrapolations in the sense of inductive inferences to other organisms are possible and can be reasonably sound. However, such inferences seem to serve mainly for giving us confidence in the *generality* of certain biological principles such as, for example, the genetic code. The main roles of model organisms for illuminating the biology of *specific* organisms – such as humans – are distinct from their role in providing a basis for inductive inferences. Namely, model organisms provide *research materials* (e.g., cloned DNA fragments) and valuable information that are useful for research *directly* on the organism in question. Cloned DNA fragments can be used to isolate genes from other organisms. Comparing DNA and protein sequences with orthologous sequences with known functions, as determined by research on model organisms, can be used for *functional annotation* of genomic DNA sequences, for example, the human genome sequence. Furthermore, specific mechanisms can be studied in model organisms because they are phylogenetically conserved. This is possible because of the modular structure of biological mechanisms, which can be assembled into hierarchies of mechanisms of different complexity. Mechanisms located at the lower levels of this hierarchy tend to be more strongly conserved phylogenetically, and they can be embedded in higher-level mechanisms of varying complexity that are less conserved. This limits the extent to which simple organisms can serve as "models" for more complex ones.

SUMMARY

Experimental organisms such as the laboratory fruit fly *Drosophila melanogaster* play a central role in biological research. In this chapter, I have examined how the fly was transformed from an experimental organism for classical genetic analysis into a model organism for molecular biology, in particular the molecular basis of development. This transition required recombinant DNA technology, which was developed in the 1970s using *E. coli* and bacteriophage. I have shown that, in addition, classical mapping methods and the material resources of traditional *Drosophila* genetics also played an important role. These methods were combined with recombinant DNA technology for the molecular cloning of some *Drosophila* genes that had been implicated in the genetic control of development.

In order to analyze this case, I have first tried to apply the general framework developed by Robert Kohler, who has brought an economic and technological perspective into the historical study of experimental organisms. On such a view, *Drosophila* was transformed from being an instrument for producing genetic maps into an instrument or tool for producing specific cloned DNA fragments. I have then argued that the view of lab organisms as "instruments" or "technological artifacts" is metaphorical and, therefore, of limited value. Instead, I have proposed that there is an important mode of experimentation in biology, namely *preparative* experimentation. This kind of experimentation involves the production of research materials and knowledge about these materials that are required for controlled biological experiments. The great value of laboratory organisms for biological research can be explained by the cumulative character of preparative experimentation. Furthermore, I have argued that economic terms such as "production" or "resources" are not metaphorical in the present context; however, their explanatory scope is limited. Specifically, these concepts fail to illuminate the *epistemic* roles that experimental organisms play.

Finally, I have examined what can be learnt from studying model organisms. The generality of biological knowledge that was obtained by studying model organisms can be established on evolutionary grounds by a special kind of inductive argument. However, model organisms are not just useful for providing a basis for such extrapolations or inductive inferences; they have additional functions in biological research. Namely, they can provide research materials and valuable information (e.g., DNA and protein sequence information) for research directly on other organisms such as humans.

7

Reference and Conceptual Change: Out of Mendel's Garden?

Popular accounts have it that genes were discovered in Mendel's monastery garden and that, since then, scientists have built up an increasing body of knowledge about them (for example, Weiner 1999, 19). However, during the twentieth century, the concept of the gene has changed considerably. Genes are today defined as DNA segments that specify the linear sequences of proteins or RNA molecules. But this was not always so. The concept of the gene dates back to a time when both the material carrier and the biochemical function of genes were unknown. Thus, genes could not have been initially defined in the same way. In spite of this, it is assumed by popular accounts that, throughout the twentieth century, the growing body of genetic knowledge – even though it underwent some dramatic transitions – was *about the same set of entities*, namely the genes that were discovered in Mendel's garden. If this is a caricature of the history of genetics, it is a useful one. For it can serve us as a foil for developing the philosophical problem of reference and conceptual change in genetics.

The starting point of modern philosophical debates on conceptual change lies in the logical empiricist model of scientific progress. According to this model, science advances by older theories being *reduced* to newer theories, typically ones that treat some set of natural phenomena at a more fundamental level (see also Chapter 2). One of the favorite examples for this model of scientific change has been the reduction of classical thermodynamics to statistical mechanics by Ludwig Boltzmann and others. Classical thermodynamics contained concepts such as temperature or entropy, which were defined entirely in macroscopic terms. In other words, in applying these older thermodynamic concepts it makes no difference what matter consists of. More specifically, it makes no difference to thermodynamic theory whether matter consists of atoms or is continuous. (The latter issue was controversial until well into the twentieth century.) By contrast, statistical mechanics assumed that matter consists of discrete particles – atoms and molecules – that are eternally in motion. On this atomistic view, heat came to be viewed as consisting of atomic and molecular motion, and not as a substance of its own (older theories viewed heat as a substance called "caloric"). As a consequence, the thermodynamic concepts were redefined in molecular terms. Namely, temperature was now defined (for some idealized systems like ideal gases) as the mean kinetic energy of the molecules of a system, and entropy was interpreted as a measure of the number of possible microscopic states corresponding to a particular macroscopic state of a system. Ernest Nagel argued that such *redefinitions* of the main theoretical concepts of a science with the help of "bridge principles" are necessary for reduction (see the introduction to Chapter 2).

Nagel's account of reduction and the view of conceptual change it contains were subsequently criticized by Paul Feyerabend (1962). He examined Nagel's example of reduction in thermodynamics more closely and concluded that it did not meet Nagel's own criteria for a successful reduction. Specifically, Feyerabend argued that the second law of thermodynamics, which states that entropy increases in any change of the macroscopic state of a closed system, has not been derived from statistical mechanics. What was derived, according to Feyerabend, was an *analog* of the Second Law, namely a law that states that a closed system of particles is extremely *likely* to move to a state with higher entropy. This is a probabilistic law, whereas the classical Second Law was deterministic. Since Feyerabend held that the meaning of theoretical terms is determined by the theories in which they are embedded, he concluded that the meaning of thermodynamic concepts had changed in the transition from classical to statistical theory in a way that precluded a direct translation of the respective theoretical vocabularies. Feyerabend introduced the term *incommensurability* to express this relationship of scientific theories to their historical successors. The concept of incommensurability, which was introduced independently by Thomas Kuhn, generated a philosophical debate that continues until the present (see Hoyningen-Huene and Sankey 2001 for a comprehensive bibliography and an overview of this debate). The main issues at stake in this debate are the implications of incommensurability for the empirical comparison of theories and, as a consequence, for the rationality of science and scientific realism. I discuss some of these issues in the context of experimental biology in Weber (2002a). Here, my concern is with a particular approach to conceptual change that was developed in response to the challenges raised by incommensurability.

The beginnings of this approach lie in Israel Scheffler's (1967) appeal to the Fregean distinction between sense and reference to face the incommensurability challenge. A term's *reference* or *extension* is the set of objects to which the term applies, whereas the *sense* or *intension* is the descriptive content associated with the concept. Frege (1892) argued that two terms may have different intensions and yet share the same extension. For example, the expressions "evening star" and "morning star" refer to the same object – the planet Venus – but they differ in the way in which they refer to this object. On this traditional view, sense determines reference, but not vice versa. Scheffler raised the possibility that what changes in theoretical transitions in science is the sense of theoretical terms, but not the reference. Thus, a theoretical term may change with respect to the associated conceptual content, but continue to refer to the same entities.

Whether this kind of conceptual change (i.e., changing meaning but not reference) actually exists in the history of science is another question. What is clear is that the examples of conceptual change discussed by Feyerabend and Kuhn were selected as examples where reference did *not* remain constant. In Kuhn's example of the Copernican Revolution, the reference of the term "planet" changed; it excluded the Earth in Ptolemaïc astronomy, whereas it included it in Copernicus's world system. Similarly, it is clear that Feyerabend did not intend his examples of temperature and entropy as cases where the reference of terms remained constant. Of course, it can be argued that – Kuhn and Feyerabend notwithstanding – the Earth was always a planet, but that pre-Copernican astronomers were simply mistaken about what a planet was.[1] But this only emphasizes that there are some vexing philosophical problems associated with reference and conceptual change.

One question that has been discussed extensively is whether historical theories could be said to refer to existing entities even though they were mistaken about the *properties* of those entities. In other words, is it possible that scientists, in some historical situation, referred to a particular set of objects even though the *descriptions* they gave of these objects were false? The traditional accounts of sense and reference (e.g., Gottlob Frege's) preclude this possibility. According to this view, which is called the *descriptivist* theory of reference, a concept's reference is defined as a class of objects that satisfy a certain description. That description gives the concept's sense, and it uniquely determines the reference of the associated term. On this view, the answer to our question must be that it is not possible that theories that give false descriptions refer to existing entities. For example, nineteenth-century theories of the propagation of light cannot be said to have referred to electromagnetic radiation, since they assumed a luminiferous ether, which entered the description associated with the concept of electromagnetic wave. In a similar way, many other theoretical terms from the history of science have suffered reference failure on a descriptivist theory of reference, since their descriptions included false elements. On the descriptivist theory of reference (at least in its classical, Fregean form), the conceptual history of science appears highly discontinuous.

Several authors have sought to remedy this problem by adopting a different theory of reference, namely the so-called *causal theory of reference* initially introduced by Saul Kripke and Hilary Putnam (Kripke 1971, 1980; Putnam 1975a, 1975c). They have argued that a term's reference is not determined by descriptions, but by a certain kind of *causal relationship* between the relevant language community and the objects referred to. On this view, a term's reference is fixed by an initial "baptism," for example, by somebody pointing to an object and applying a name to it. This name is then transmitted to a broader language community, where it continues to refer to the same object. Thus, on the causal theory of reference, a term's reference is not determined by a description (i.e., sense does not determine reference), but by a historical relationship between a linguistic community and the initial introduction of the term into discourse.

The causal theory of reference has a number of consequences, some of which are plausible and some of which are not. One of the more plausible consequences is that we can successfully refer to a set of objects even though our beliefs about these objects are false. For example, I can successfully refer to scorpions even if I mistakenly believe that they are six-legged insects. In other words, I do not need a correct description of scorpions in order to refer to them, since the reference is secured by the history of the language community that I belong to, which has, at some point, fixed the reference of the term "scorpion." Similarly, scientists can successfully refer to a set of objects even though their theories about these objects are false. Thus, although the history of science will look *discontinuous* in terms of reference for the descriptivist, a causal theorist will see more *continuity* of reference (Burian, Richardson, and Van der Steen 1996, 16).

It is widely accepted today that a pure causal theory of reference, according to which descriptions play no role in reference determination, is inadequate for scientific terms. First, some minimal descriptive content must be involved in the reference fixing of *kind terms*, because an individual object may belong to several kinds. Second, additional problems for causal theories of reference are posed by *theoretical* terms in science, because their referents cannot be pointed to ostensively. The referents of theoretical terms such as "electron" are only known through their experimental effects. For these reasons, various authors have adopted mixed theories of reference that postulate both descriptive and causal elements (e.g., Nola 1980; Sankey 1994).

Another development in the theory of reference was the recognition that, for any given term, there may be more than one way in which reference is determined. Philip Kitcher has argued that there can be different modes in which a given term refers (Kitcher 1993, 76–80). According to Kitcher, a term's mode of reference depends partly on the intentions of the speaker. For example, I can use a particular term with the intention of referring to an object that satisfies a particular description, a description that I have in my mind. Thus, for Kitcher, a descriptive theory of reference gives a correct account of how terms refer at least in certain situations. However, in other situations I may want to refer to an object in a different mode, again depending on my intentions. I may have the intention of referring to an object for which I cannot give a correct description, for example, to quarks (because I am not a physicist). In this case, I can still refer to the thing, since the language community I belong to contains experts who know what quarks are. In this case, my usage of the term is "parasitic" on those experts, who have fixed the reference of the term "quark" sometime in the past (no matter how they did this). This mode of reference accords more with the causal theory of reference. Finally, I can refer to an object by pointing to it, which is how terms are initially introduced according to some versions of the causal theory of reference. According to Kitcher, each term has a *reference potential*, which is "a compendium of the ways in which the referents of tokens of the term are fixed for members of the community" (Kitcher 1982, 340). The reference potential of a particular term may contain one or more descriptions that can be used to refer to objects in the descriptive mode. In addition, a term's reference potential may include one or more causal chains of events ranging from an initial "baptism" through transmissions of the term through the language community to the event when the term was acquired by a user. Finally, a term's reference potential is continually in motion as science advances. New theoretical or experimental findings can lead to modifications in a term's reference potential, for example, by introducing new descriptions of the referents or new experimental methods for identifying them.

Kitcher illustrated his account of reference by applying it to the well-known case of the transition from phlogiston to oxygen chemistry. He argued that the terms of phlogiston chemistry, most notably "phlogiston" and "dephlogisticated air," were associated with a heterogeneous reference potential that caused reference failure in some situations, but secured reference in others (Kitcher 1978). On Kitcher's view, some *tokens* of these terms (i.e., individual events when these terms were used by scientists) successfully referred, whereas other tokens failed to refer. For example, there were situations in which the phlogiston chemist Joseph Priestley successfully referred to *oxygen*.

192

One such situation, according to Kitcher, was Priestley's experiment in which he tested the effect of "dephlogisticated air" on his own lungs. (He felt that it had a rather pleasant effect.) In this situation, Kitcher (1993, 100) argues, Priestley's expression token "dephlogisticated air" referred to oxygen.

Kitcher has also applied his theory of reference to the gene concept, arguing that the history of that concept shows a succession of modifications of the term's reference potential (Kitcher 1982). In the course of the advancement of genetic knowledge, faulty ways of referring to genes (i.e., ways that fail to establish a reference) were successively eliminated, while new, correct ways of referring were added to the reference potential.[2] Furthermore, what changed in the history of genetics was not only the reference *potential* associated with the term "gene" (and its predecessors, like "unit-factor") but also the term's reference *itself*. Kitcher suggested that different historic versions of the gene concept had different ways of dividing up chromosomes into genetic units. Thus, there were also *reference shifts* involved. These reference shifts occurred as a consequence of new ways of studying and conceptualizing biological organization, especially during the "molecular revolution" in genetics, when genes came to be viewed as DNA segments.

There is much that I agree with in Kitcher's account. However, Kitcher (like most people) has treated the term "gene" in basically the same way in which he has treated the terms "phlogiston" and "oxygen." Thus, he seems to have assumed that there is nothing specific to the subject matter of genetics that would make it necessary to treat the case of the gene differently. But given what we have learnt in Chapter 2, this assumption may be called into question. For example, we have seen that the natural kinds of biology differ from those of physics and chemistry in that they are variable, not fixed (Section 2.3). In other words, we cannot expect all the referents of a theoretical term to have exactly the same causal dispositions (which we can expect with the genuine natural kinds of physics and chemistry, for example, chemical compounds). Should we not expect this kind of metaphysical difference to have consequences for the way in which theoretical terms in biology refer?

Answering this question requires that we take a closer look at the history of the gene concept. Furthermore, we need to pay careful attention to changing experimental methods of studying genes. Conceptual analysis based on genetic theory alone will not do.

In the next section, I give a historical overview of the conceptual development of genetics. In Section 7.2, I examine the changing modes of reference that accompanied the historical development of genetics in the twentieth century. In Section 7.3, I move one step beyond abstract conceptual analysis and look at some specific examples of genes that have been characterized by both classical and molecular methods. In Section 7.4, I offer some general reflections on some differences between the case of the gene and other examples of conceptual change that have been discussed in connection with the theory of reference, which were mostly examples from the physical sciences.

7.1 THE GENE CONCEPT IN FLUX

Even a cursory look at the history of genetics reveals a bewildering diversity of gene concepts (see Carlson 1966; Falk 1986; Portin 1993; Beurton, Falk, and Rheinberger 2000). In the nineteenth century, there was much speculation about invisible particles that transmit an organism's hereditary properties from parents to offspring via the gametes. Charles Darwin, for example, called these particles "gemmules" and thought that they are given off by various body parts to the reproductive organs in order to somehow represent the organism (Darwin 1868, Vol. II, 374). This is Darwin's "provisional hypothesis of pangenesis." It was adopted and modified, among others, by Hugo de Vries. He presented a theory of "intracellular pangenesis" (de Vries 1889), which was designed to exclude the possibility of the inheritance of acquired characteristics. Darwin, by contrast, explicitly allowed that acquired characteristics, for example, particularly well-developed organs, could be transmitted from parents to offspring with the help of the "gemmules." Darwin's theory of pangenesis was an attempt to explain how an organism's form can be transmitted from parents to offspring while, at the same time, allowing for the heritable variations that were observed in breeding experiments.

Gregor Mendel's hybridization experiments with garden peas, published in 1866 but largely ignored until 1900, have been hailed as a turning point in the history of genetics by several generations of geneticists. Famously, Mendel showed that hybrid progeny of different plant varieties are produced in numerical ratios that approximate the combinatorial series (see Section 3.2). At the beginning of the twentieth century, several biologists including Carl Correns, Erich von Tschermak, Hugo de Vries, and William Bateson called attention to Mendel's paper and repeated his experiments. Bateson attributed to Mendel the view that the characters of his pea plants were determined by paired *Faktoren* or *Elemente* (Mendel's terms) that each individual received from the parents in a pure form, one *Faktor* per gamete. However, Robert Olby has argued that this is a reinterpretation of Mendel's text that is born out of hindsight (Olby 1979). Mendel, in contrast to twentieth-century geneticists, did not view his *Faktoren* as material determinants of binary character states that segregate in the gametes and recombine in different combinations in the offspring. Since Mendel says so little about his *Faktoren*, it is indeed easy to read into his 1866 paper *Versuche über Pflanzenhybriden* a theory that only originated later. If Mendel had a gene concept at all, it was different from the concept formed by the twentieth-century champions of Mendelism.[3]

After the rediscovery of Mendel's work in 1900, there was a notable trend to move away from the highly speculative and metaphysically loaded notions of "gemmules," "pangenes," or "idioblasts" (a term introduced by August Weismann) that were so popular in nineteenth-century biological thinking. The British champion of Mendelism, William Bateson, introduced the term "unit-character" for the Mendelian factors that can take different forms in the gametes and that emerge stably from different unions in genetic crosses.[4] Even though, as Bateson admitted, "what the physical nature of these units may be we cannot yet tell," he suspected – with "some confidence" – that "the operations of some units are in an essential way carried out by the formation of definite substances acting as ferments" (Bateson 1909, 266). In addition, Bateson suggested that recessive "allelomorphs" (known as "alleles" today) consist in the absence of some substance from the gametes, whereas dominant factors consist in the presence of such a substance (1909, 76).

Around the same time, the Danish botanist Wilhelm Johannsen introduced the term "gene," which he adapted from de Vries's "pangene." Johannsen explicitly wanted the concept of the gene to be free of any hypothesis concerning its material constitution, as is evident, for example, from the following passage from the second edition of his influential textbook *Elemente der exakten Erblichkeitslehre* (1913, 143–144):

> *Das Wort Gen ist also frei von jeder Hypothese.* Es drückt nur die Tatsache aus, dass Eigenschaften des Organismus durch besondere, jedenfalls trennbare und somit gewissermaßen selbständige "Zustände," "Faktoren," "Einheiten" oder "Elemente" in der Konstitution der Gameten und Zygoten – kurz, durch das was wir eben *Gene* nennen wollen – bedingt sind. [*The word "gene," then, is free of any hypothesis.* It merely expresses the fact that the organism's properties are determined by special, separable and therefore autonomous "states," "factors," "units," or "elements" in the constitution of the gametes and zygotes – in short, by that which we wish to call *genes*. (Author's translation; emphasis in original)]

Thus, Johannsen suggested to use the term "gene" for *whatever it is* that determines an organism's properties and is passed down through the gametes. A similar, minimal "definition" (of course, this is not a definition in the strict sense) was preferred by one of the American pioneers of genetics, Edward M. East (1912). The fact that these geneticists chose to keep the concept of the gene free of any assumptions concerning the physical nature of its referents does not mean that they did not believe in the *existence* of genes. Some accounts of their gene concept as "calculating units" or "symbolic notation" (e.g., Portin 1993, 175; Falk 1986, 144) suggests an instrumentalist interpretation. However, Johannsen's and East's writings can also be given a realist reading.

In the 1910s, the gene concept was given a new meaning by T. H. Morgan and his students (see also Waters 1994, Schwartz 2000). In 1915, they published a small book entitled *The Mechanism of Mendelian Heredity* in which they used their results from crossing experiments with different *Drosophila* mutants to lay down the basic theoretical and experimental principles of the new genetics. Of particular interest for the present study is their rejection of Bateson's concept of unit characters. Morgan and his colleagues used the case of the *Drosophila* eye color mutants to make their case. Up to 1915, the Morgan group had identified twenty-five different mutations that affect the eye color of the fruit fly. In general, this means that simple Mendelian behavior will not be observed with the eye-color trait. However, it remains the case that a mutation in any one of these factors can produce a *difference* with respect to the normal, red eye color. But this is all that is needed:

> It is this one different factor that we regard as the "unit factor" for this particular effect. However, since it is only this one factor and not all 25 which causes the difference between this particular eye color and the normal, we get simple Mendelian segregation in respect to this difference. In this case we may say that a particular factor (p) is the cause of pink [eye color, M. W.], for we use cause here in a sense in which science always uses this expression, namely, to mean that a particular system differs from another system only in one special factor. (Morgan et al. 1915, 209)

What Morgan and his students were proposing here was that the Mendelian units should not be identified with determinants for particular traits, as Bateson had done when he introduced the concept of unit character. Instead, a Mendelian factor is something that can cause a *difference* in some trait. Thus, no matter how many different factors influence a particular trait, if all these other factors are kept constant, this difference will behave in the Mendelian way. Thus, genes are merely *difference makers* (Sterelny and Griffiths 1999, 87); they are not able to produce specific traits all by themselves.

Morgan and his co-workers then proceed to show that the converse is also true; namely, a single factor can affect more than one trait. One of the examples they cite is the gene *rudimentary*, which affects the size and shape of the fly wings. But the same gene also affects other traits, such as the phenotype of the legs or the number of eggs laid. The same is true for most of the *Drosophila* mutations that had been identified in Morgan's laboratory. Thus, in contrast to Bateson's concept of unit character, there is *no one–one relationship* between genes and traits. The relationship is *many–many*; that is, any trait can be affected by many genes, and any one gene can affect many traits.

Another major development to come out of research on *Drosophila* was that genes were shown to be associated with chromosomes. Even though Morgan was hesitant to fully admit this claim to the theory of genetics in his preface to the 1915 book, C. B. Bridges did a series of elegant experiments that provided strong evidence for a physical association of genes with the chromosomes (Bridges 1914; 1916). Furthermore, another student of Morgan's, A. H. Sturtevant, produced the first genetic map in 1913, which suggested that genes are arranged linearly on the chromosome (Sturtevant 1913). This genetic map was based on the crossover frequencies between different genetic loci (see Section 3.2). W. E. Castle questioned the linear interpretation of the genetic map and suggested a three-dimensional arrangement of genes within the chromosome. However, this so-called rat-trap model of the chromosome was rather short-lived (see Section 9.2).

As already mentioned, William Bateson thought that there are only two alleles for each unit-factor: a dominant allele that is produced by the presence of some germinal substance and a recessive allele that corresponds to the absence of this substance. This "presence-and-absence hypothesis" was rejected by many examples of multiple alleles, i.e., different mutations of the same gene. One of the best examples was the Drosophila *white* gene, for which Morgan's group isolated several alleles. This raises the question of how the geneticists *knew* that these alleles were really variants of the same gene, rather than different genes affecting the same trait. Remember that the latter case was possible according to the many–many relationship of genes and traits.

Morgan et al. (1915, 155ff.) discuss the case of *white* and *eosin*, both sex-linked mutations that affect eye color in *Drosophila*. If a white-eyed male is crossed to a red-eyed female, the Mendelian ratio of 3 red to 1 white is observed in the F_2 generation. If an eosin-eyed male is crossed with a red-eyed female, the same ration of 3 red to 1 eosin is observed. If a white-eyed male is crossed with an eosin-eyed female, the result is 3 eosins and 1 white. No red-eyed flies (the wild type) are produced in the latter cross. Furthermore, eosin and white map to exactly the same position; that is, they show the same distance to other loci as determined by crossover frequencies. Morgan et al. discussed two possible explanations for these results. First, white and eosin, together with the wild type (red), could constitute three different alleles of the same gene, that is, a case of *multiple alleles*. Alternatively, the alleles eosin and white could belong to different genes that are *completely linked*; that is, there is no crossing-over between them. This would also explain why they mapped to the same chromosomal position. But Morgan et al. considered the first explanation, the one with multiple alleles, to be correct. The reason was the behavior of the two alleles when they were *both* present in an individual. Such heterozygotes always showed a mutant phenotype. Morgan and his co-workers reasoned that, had these alleles been variants of two different genes, the offspring of crosses of eosin and white mutants should have produced flies with normal, red eyes. The reason is that "each mutant type contains besides its mutant factor the normal (dominant) allelomorph [allele, M. W.] of the other type" (1915, 168). By contrast, if the two alleles are variants of the same gene, then "neither of them brings in the normal allelomorph of the other; hence the wild type is not reconstituted" (ibid.).

The test that Morgan and his students applied here to decide whether two alleles affect the same or two different genes was later termed *complementation*. To test for complementation, two mutant alleles *a* and *b* must be observed in the heterozygous condition, which can be produced by crossing two homozygous mutants: *aa* × *bb*. Now, if the alleles are variants of two different genes, this cross will produce only offspring that show the normal, nonmutant phenotype. If we write the nonmutant versions of these alleles as "+," such a cross can be described as *aa*++ × ++*bb*. This cross will produce only offspring with the genotype *a*++*b*. If the alleles are recessive,[5] this genotype will show the nonmutant phenotype. The reason is that every individual inherits one normal copy of each of the two genes. Geneticists say that the two mutant alleles *complement*. If, by contrast, the two alleles are variants of the same gene, then the cross *aa* × *bb* yields only the genotype *ab*, which shows the mutant phenotype. In this case, no normal copies of the gene are present that could complement the mutant allele.

Complementation has been the main operational criterion for identifying genes since the days of the Morgan school. The complementation test picks out genes as *units of function*. Historically, there has been much debate as to whether such functional units – as revealed by complementation tests – correspond to genetic units identified by other criteria, especially recombination (crossing-over) and mutation (Portin 1993, 181–183). In the initial period of *Drosophila* genetics, it was not possible to observe recombination between alleles of the same gene, suggesting that genes might also be units of recombination. Furthermore, it was possible that the gene is the smallest unit that can undergo mutation, thus making genes units of mutation. However, Portin (1993, 181) points out that the belief that these units necessarily coincide was never generally accepted in the genetics community.

Drosophila, to be sure, did not do the geneticists the favor of producing only mutants for which each of these three criteria coincided. Some genes (e.g., *white*) initially seemed to be units both of function and of recombination. Whether they were also units of mutation could not be tested back then. But beginning in the late 1920s, an increasing number of mutants were discovered that did not fit into the basic scheme of units of function = units of recombination = units of mutation. For example, a series of alleles were isolated from the complex locus *achaete–scute*, which affects the number of bristles on the fly body (Carlson 1966, Chapter 17; Weber 1998b). Some of these alleles complemented, but they did not recombine. To make matters worse, some combinations of alleles complemented *partially*, that is, only with respect to some aspects of the phenotype. One possible explanation for this phenomenon was that the *achaete–scute* alleles belonged to a single gene, but that there was complementation between different parts of that gene (intragenic complementation). This "subgene hypothesis" was controversial. Alternatively, it was possible that the *achaete–scute* alleles belonged to different genes that were so close together that crossing-over between them was too rare to be observed. Perhaps these genes, for some reason, just failed to complement properly. It was not possible, in the 1930s, to distinguish between these possibilities, and the problem was never really resolved for the complex genetic loci of *Drosophila*. The later finding that recombination did, after all, occur between different *achaete–scute* alleles did nothing to improve the situation, since it could not be decided whether this recombination occurred within a single gene or between different genes. The same was true of other claims of intragenic recombination.[6]

The rapid pace by which new genetic phenomena were discovered in the 1920s and 1930s posed an increasing number of difficulties for the classical concept of the gene. Another such difficulty came from the finding of the so-called position effect. This effect, which was first described in a *Drosophila* mutant named *Bar* discovered in 1917, designates a dependence of a gene's phenotypic effect on its chromosomal environment. Richard Goldschmidt used this effect (among other findings) in order to mount a sustained attack on the gene concept (Carlson 1966, Chapter 15). Specifically, he argued that classical genes do not exist. On his view, chromosomes cannot be neatly divided into genes. Instead, chromosomes form whole "reaction systems" that exert their physiological action as wholes. Mutations, according to Goldschmidt, are not localized changes in specific genes; they are chromosomal rearrangements of various size. Later, Goldschmidt replaced the whole chromosome as a unit of physiological action and mutation by a hierarchy of genetic structures (Dietrich 2000).

Goldschmidt's positive views were never accepted by the genetics community. However, in his doubts about the adequacy of the classical gene concept he was joined by several of his colleagues in the 1950s. For example, Pontecorvo (1952) proposed to define genes solely as units of physiological function, and not simultaneously as units of function, mutation, and recombination. Especially the discovery of so-called pseudoalleles (alleles that behave differently according to whether they are located on different or the same chromosome), as well as the complex loci such as *achaete–scute* led to problems in the application of the classical criteria for identifying genes. An increasing number of *Drosophila* mutants was discovered for which it was difficult to decide how many genes were affected by these mutations. Thus, it seems that the classical gene concept ceased to be unambiguously applicable to an increasing number of genetic loci.

In 1953, Watson and Crick published their double helix model of the structure of DNA (1953). Even though there was already some evidence linking the genetic material to DNA before this event (e.g., Avery, McLeod, and McCarthy 1944), these findings had not yet had much of an impact on the practice of genetics. Classical genetic mapping worked just as well under the assumption that genes are made of protein (the predominant view in the 1920s and 1930s; see Olby 1994, Chapter 7), or some other substance. However, the excitement generated by Watson's and Crick's discovery started to transform thinking about genes immediately.

One geneticist whose work was strongly influenced by the Watson–Crick model was Seymour Benzer (Holmes 2000b, 124, 128). In 1954, Benzer invented a technique of genetically mapping mutations in bacteriophage T4 that had a much higher resolution than all the previously existing techniques. The technique Benzer used resembled the method of linkage mapping developed by *Drosophila* geneticists (see Section 3.2), but it was able to detect recombination frequencies that were at least three orders of magnitude lower. Benzer's trick was to first infect *E. coli* strain *B* with different strains of mutant phage. Strain *B* is a "permissive host"; in other words, it allows mutant phage to reproduce. Then Benzer recovered phage particles grown on the *B* strain and used them to infect a different *E. coli* strain, namely K12(λ). The latter strain is a "nonpermissive host"; that is, it only allows wild-type phage to grow. If, now, a crossing-over event has occurred between two mutations, the resulting wild-type phage will grow on K12(λ) and thus become visible as a plaque on the bacterial lawn. This method allowed Benzer to detect extremely rare crossover events. Using this technique, Benzer (1955) produced a fine-structure map of the *r*II region of bacteriophage T4 (*r* stands for "rapid lysis," which was the detectable phenotypic effect of mutations in that region).

Benzer then used a complementation test to identify functional units on the fine-structure map. Benzer crossed mutant phage strains with mutations at different positions on the genetic map and checked whether they complemented each other. When two mutations allowed the phage to grow on the nonpermissive host, this indicated that these mutations are located in different functional units. By contrast, when two mutations did not allow phage growth on the nonpermissive host, this showed that the mutations affect the same functional region on the chromosome. The reason is that, in the first case, each of the two chromosomes can supply the function that the other one is lacking. Benzer's data indicated the existence of two such complementation units.

Probably because of the existing doubts concerning the classical gene concept mentioned above, Benzer first hesitated to identify his functional units as genes. Later, he devised the term "cistron." However, the most exciting interpretation of Benzer's results was that his complementation units corresponded to two separate genes. Indeed, Benzer suggested toward the end of his 1955 paper that each unit "might control the production of a specific polypeptide chain" (353). Since Beadle and Tatum's one gene–one enzyme hypothesis was widely accepted by that time, it seems that Benzer did identify his "units" with genes. Under the assumption that the fine-structure map was a representation of the DNA molecule, it would follow that genes are linear segments on the DNA molecule. In 1955, there was no direct evidence for this assumption, but such evidence was obtained later (see Weber 1998b for details).

Benzer's work also resolved the long-standing problem of the relationship between the unit of function, the unit of recombination, and the unit of mutation discussed earlier. According to Benzer, the unit of function was the "cistron," that is, the units picked out by a complementation test. The unit of mutation or "muton" Benzer assumed to be the single DNA nucleotide, even though he had no direct evidence for this. Finally, Benzer's data suggested that the unit of recombination or "recon" consisted of approximately twelve nucleotides.[7]

Toward the end of the 1950s, the question of how genes direct the synthesis of proteins – known as the "coding problem" – came into sharper focus. As we have seen in Section 5.1, this problem was eventually solved with the help of protein synthesizing in vitro systems. But in addition, the laboratory of Charles Yanofsky provided an important piece of evidence in the emerging picture of the gene as a segment of DNA that specifies the sequence of amino acids in proteins. Yanofsky and co-workers were working on the genetic locus *trp* that directed the synthesis of the amino acid tryptophan in *E. coli*. They prepared a fine-structure map of *trp* similar to Benzer's map. Furthermore, they determined the amino acid sequence of a protein that is part of the enzyme that was synthesized by the *trp* gene, tryptophan synthetase. Several mutant versions of this protein were sequenced. Then, by comparing the genetic fine structure map with the amino acid sequences of various mutants, Yanofsky and co-workers were able to show that the genetic map was *colinear* with the amino acid sequence (Yanofsky et al. 1964). This work established a direct relationship between the genetic map of an individual gene, which was conceived as a linear segment of DNA, and the protein product specified by this gene (Weber 1998b).

The classical method of linkage mapping thus played an important role in the molecularization of genetics. Due to the work of Benzer, Yanofsky, and others, the gene was transformed from a hypothetical entity with an unknown material basis into a molecular structure (a segment of DNA) and a physiological function (determining the amino acid sequence of proteins).

In the years that followed, there were a host of new findings that complicated the view of genes as uninterrupted stretches of DNA. In particular, the colinearity of genes with amino acid sequences turned out to be not strictly valid in eukaryotes. It was found in the 1970s that eukaryotic genes are interspersed with so-called introns. Introns have no coding properties and are "spliced" out after a gene is transcribed into an RNA molecule. The resulting messenger RNA (mRNA) contains only "exons," regions coding for protein.

Another source of complication stems from the different mechanisms that regulate gene expression. The first such mechanism was elucidated in the 1960s by François Jacob and Jacques Monod and came to be known as the "operon model" (see Section 3.1). Jacob and Monod showed that the activity of the *E. coli* gene *lac*, which is responsible for the production of the enzyme β-galactosidase, depends on several genetic regions called *y*, *o*, *z*, and *i*. Although these regions were identified as separate "genes" initially (Jacob and Monod 1961, 334), they turned out to be part of a single functional unit or "operon." An operon includes a "structural gene," which determines the amino acid sequence in the requisite protein. But an operon also includes "regulatory genes," for example, an operator region. The operator region can inactivate the gene when a certain protein (the repressor) is bound to it. The repressor protein is made by a different gene called *i*. Today, these "regulatory genes" are not recognized as separate genes anymore (except for the gene that codes for the repressor protein); they are simply called "regulatory regions" or "control regions."

As one might expect, gene regulation in eukaryotes turned out to be much more complex than in bacteria. The activity of such genes is regulated by various control regions (e.g., "enhancers" and "silencers"), which may be located at a considerable distance from the structural gene. Furthermore, there are control regions that affect more than one gene, and there are genes that are affected by more than one control region.

In recent years, an increasing number of genetic gadgets have been discovered which show genes to be even more complex entities than was thought previously. For example, the time-honored "one gene–one polypeptide" hypothesis has been refuted. There are genes that can give rise to several different polypeptides, depending on how their introns are removed ("alternative splicing").

Thus, contemporary molecular biology has shown genes to be much more complex than just simple, uninterrupted segments of DNA that code for protein.[8] What seems to survive is an abstract, mixed-functional characterization of genes as nucleotide sequences that determine the linear sequences of one or several polypeptide or RNA products in a specific cellular environment (Waters 1994, 2000). "Mixed-functional" means that genes are individuated by *both* the stuff they are made of (DNA or RNA) and their functional role in the cell (determination of protein and RNA sequences). As a consequence, the term *molecular* gene concept must not be taken to mean that the gene has been defined in purely physicochemical terms. Even though a particular molecular structure (DNA or RNA) has become part of the very meaning of the gene concept, a specific *biological* function (determining the linear sequence of a gene product in a specific cellular environment) is also part of the concept. The gene concept, for the time being, remains irreducibly biological.[9]

This historical overview was mainly concerned with conceptual content, that is, with the changing ways in which geneticists construed the structure and function of genes. In the next section, I turn to the question of reference.

7.2 CHANGING MODES OF REFERENCE

From the foregoing historical overview, there can be no doubt that the meaning of the gene concept changed several times during the twentieth century. Major transitions include the shift from Darwin's gemmules to de Vries's pangenes, from Mendel's elements to Bateson's unit-characters, from there to the classical gene, from the classical gene to Benzer's cistron (called the "Neoclassical" concept by Portin 1993), from the cistron to the protein-coding genes of molecular biology, and from the simple protein-coding genes to the more complex entities recognized as genes by contemporary geneticists. In each of these transitions, genes were reconceptualized with new properties, while some properties that genes possessed on the older view disappeared. To alleviate the confusion, I have tried to summarize the different gene concepts discussed in Table 7.1. Before I turn to the issue of reference, I should like to make four remarks about this table.

First, the table (as well as the account given in Section 7.1) only provides a few frozen historical cross-sections from the complex development of the gene concept; there have been transitional conceptions that combine the features of the different concepts that I have listed in other ways. The table should also not be read as a linear time chart. Several concepts have historically overlapped, e.g., the unit character and the classical concepts. Furthermore, there was not a unanimous consensus concerning an adequate gene concept throughout its history. In fact, there was considerable controversy almost throughout.

Second, Table 7.1 shows that there is no single feature that all the gene concepts I have listed possess. Instead, the different historical versions of the gene concept seem to be linked at most by a family resemblance. Furthermore, it seems that the first concept on the table (pangenes) has not much in common with the concepts toward the end of the list (the neoclassical and molecular concepts). Between these points, there are some features that persisted for awhile, only to disappear again. One of the most stable features is the Mendelian behavior; however, it become optional at some point, namely as soon as the gene concept was also applied to microorganisms (Mendelian behavior is only observed in diploid, sexually reproducing organisms).

Third, we expect the features that I have listed under the rubric "functional role" to be an important part of the meaning of the gene concept, because they state the place that genes were assigned in the causal nexus of the organism. The gene concept was not introduced in order to distinguish a particular cellular structure (in contrast to, e.g., the chromosomes concept). It was introduced to single out a certain biological *function*. In other words, what distinguishes genes is the particular *causal role* that they play in an organism. However, what this causal role was thought to be changed several times in the history of genetics. Whereas geneticists at the beginning of the century sought the causal role of genes in the determination of traits, the Morgan school saw genes essentially as "difference makers," that is, as the substrate for mutations that cause phenotypic differences in the sense explained in Section 7.1. But with the molecular gene concept, this causal role changed again, since that role is now seen in the synthesis of RNA and protein molecules.

Fourth, apart from the functional role, it is far from obvious which of the features that I have tabulated are really part of the *meaning* of the gene concept, and which ones are *empirical claims* about genes. If it is part of the meaning of the classical gene concept that genes are located on chromosomes, then

Table 7.1. *Historical Overview of Different Gene Concepts*

Time	Concept	Mendel's Laws	Gene-Trait Relation	Functional Role	Material Basis	Structure	Identity Criteria
Late nineteenth century	pangene	no	?	pangenesis	particulate	open	?
1900–1919	unit-character	yes	one–one	trait determination	open	open	trait
1915–1950s	classical	yes	many–many	phenotypic difference maker	chromosomes	subgenes (?)	complementation
1950s	neoclassical	optional	many–many	cistron	DNA	linear	complementation
1960s	molecular	optional	many–many	protein coding	DNA	colinear w/protein	gene product
1970s–present	contemporary	optional	many–many	protein/RNA coding	DNA/RNA	intron/exon (optional)	gene product(s)

205

the statement "genes are located on chromosomes" is an *analytic* statement. If, by contrast, being located on chromosomes is not part of the meaning of the classical concept, then the latter statement is *synthetic*. Initially, it seems that the association with chromosomes was an empirical claim about genes. Thus, this was not part of the meaning of the classical concept.[10] On the other hand, being a nucleotide sequence seems to be part of the very meaning of the contemporary gene concept (as Waters 2000 argues); thus, the statement "genes are nucleotide sequences" became analytic at some point. The problem is that it may not be possible to decide precisely which statements about genes are analytic and which ones are empirical. In fact, some philosophers have argued that the analytic/synthetic distinction cannot be meaningfully drawn in general (e.g., Quine 1953). Even if we do not accept this as a general claim about all kinds of different statements, we may have here an example where the boundary between analytic and synthetic, that is, between conceptual and empirical, is not sharp. Furthermore, this boundary seems to be subject to historical change (Burian, Richardson, and Van der Steen 1996, 10). At any rate, I will be concerned not so much with the meaning of the gene concept, but with its *reference*. Within the semantic framework that I am using here, problems in applying the analytic/synthetic distinction pertain mainly to considerations of meaning.

As this discussion shows, it is not obvious how the different versions of the gene concept hang together. Why are these different concepts, which do not all share a single meaning element, all called *gene* concepts? Is this done merely for rhetorical reasons, for example, out of reverence for a great scientific tradition? I suggest that such a conclusion can be avoided with the help of the referential approach, if it can be shown that different gene concepts are united by an extensional overlap between different stages of the term's historical development.[11] The question, then, is whether such referential continuity can actually be established. (The following discussion owes much to Burian 1985 and Burian et al. 1996.)

To begin with, we may ask how the question of reference stability presents itself under a purely descriptivist account of reference. On this account, the question can be reformulated as follows: Do the different descriptions associated with different gene concepts pick out the same set of entities? If we compare the more recent gene concepts with their nineteenth-century predecessors, the answer must be no. For example, according to current genetic theory, there are no such things as pangenes. The description of pangenes, that is, particulate, circulating entities that carry the form of an organism within them, does not apply to anything from the ontology of modern genetics. The reference of the term "pangene" (or "gemmules") is the same as that of the term "phlogiston": the null set. The descriptivist will diagnose a clear case of reference failure. The same is true of the unit-character concept. Once it was shown by Morgan and his students that the relation between genes and characters is many–many, the term "unit-character" had to be abandoned. There are, in the more advanced genetic theory, no such entities that determine exactly one trait, nor are there traits that are affected by only one gene. The term "unit-character," too, suffered reference failure. If the classical gene concept that succeeded the unit-character concept is taken as a unit of function, a unit of recombination, and a unit of mutation, then – according to the descriptivist account – it has suffered reference failure, too. Thus, as expected, the classical descriptivist theory of reference represents the history of the gene concept as a history of reference failures. In other words, all but (perhaps) the most recent concepts did not refer to anything; their extensions were empty. Let us now have a look at how the causal theory of reference fares.

The causal theory (in its original form) asks us to go back in history, to the time when a particular term was introduced, and then examine how it was transmitted to the scientific community and to subsequent generations. Assuming that the causal theory can handle cases where the *words* that express a given concept are substituted (e.g., "gene" for "factor"), it might be suggested that the reference of the gene concept was fixed when Mendel postulated the existence of differentiating factors to explain the results of his crossing experiments. This "baptism" linked the terms "differentiating element" and "factor" to *whatever it was* that caused the character differences in Mendel's pea plants. This term was then, after being ignored for a while, causally transmitted into the twentieth century, when de Vries, Correns, and Tschermak (independently) discovered Mendel's work and successfully repeated his experiments. It was then transmitted to Bateson, Johannsen, East, and Morgan and to generations of geneticists. The fact that the term's reference, in the causal theory, is independent of the actual descriptions that these geneticists gave of genes allows the possibility that this reference remained the same throughout the rapid theoretical changes that took place in twentieth-century genetics. Thus, in this view it is irrelevant that, for example, Bateson mistakenly assumed a one–one relation between genes and traits. Because the reference of the concept is independent of any specific description, he referred to the same class of objects as Morgan and subsequent geneticists. As this example shows, the descriptivist and causal theories give different answers to the question of whether the reference of the gene concept was stable.

The latter point becomes even more evident when we ask about the fate of the pangene concept. The causal theorist could argue that Darwin, de Vries, and others introduced a term that referred to some class of entities present in the gametes that are responsible for the transmission of an organism's properties from parents to offspring. Since this reference is independent of any description of pangenes, it could be argued that it was, as it were, inherited by the later versions of the gene concept. In this view, Darwin's term "gemmules" referred to what we call genes today, even though Darwin was dead wrong about the properties that these entities have. Thus, while a descriptivist will say that the term "gemmule" suffered complete reference failure, the causal theorist can argue that this term really referred to the same things as more modern gene concepts.

I think the latter example shows that some of the answers given by the original causal theory are extremely implausible. This theory can be used to show that Mendel's term "elements" and Darwin's term "gemmules" referred to the same objects, since they both have historical connections to the twentieth-century concepts of the gene. But surely, this is absurd. In such a view, theoretical terms can never fail to refer. It is all too easy to show, with the help of the causal theory, that Darwin's term "gemmules" really referred to genes. It could even be shown in this manner that Aristotle's term "*psyche*" (soul) really referred to DNA. Intuition prompts us to say that, surely, Mendel's factors must be more closely related to modern genes than Darwin's pangenes or Aristotle's souls.

The latter problem may be mainly a consequence of the idea of an initial baptism. It might be possible to fix this problem within a purely causal theory. However, such theories have a host of other problems (see the introduction). Therefore, pure causal theories have long been abandoned in favor of accounts that allow certain descriptions to be involved in the fixing of reference. In Kitcher's theory of reference potential, each scientific term is associated with a whole compendium of different ways of referring, for example, different descriptions and different historical connections to events where the term was applied previously. The salient question now is if we can find some modes of reference that could have induced a certain degree of referential stability at least for certain time periods in the history of genetics. I examine one particular candidate for such a mode of reference.

Robert Nola has suggested that fixing the reference of theoretical terms always involves theoretical beliefs. These theoretical beliefs may include, for example, a description of the causal role of the intended referents in producing the experimental effects (Nola 1980). Following Nola's suggestion, it could be proposed that the reference of "gene" (or one of its predecessor terms like "unit-factor") was fixed by such causal role descriptions at various occasions in the early twentieth century.[12] For example, Bateson or Johannsen could have done this by pointing out some Mendelian crossing experiments in conjunction with the following reference-determining description:

Reference of "gene" (1): Whatever (a) is transmitted in a pure state in the gametes, (b) segregates in numerical ratios that approximate Mendel's laws, and (c) determines the phenotypic character differences observed.

It is important to understand that *this is not a definition*. A definition would give the concept's *sense* by linking it with the senses of some defining terms. By contrast, (1) is supposed to referentially connect the term "gene" to a class of entities, namely the class of entities that play the causal role described on the right-hand side. On this account, "gene" only refers if there actually exists a class of entities that plays the causal role stated in the description. Whether this is actually the case was far from clear at the beginning of the twentieth century, as witnessed by the theoretical debates at that time.[13] However, if some tokens of the terms "gene" or "factor" were used on some historical occasions with the intention of referring to a set of entities that satisfy (1), then these expression tokens could have referred.

If we construe such a description as part of the reference potential of the gene concept, then we see why earlier conceptions such as "pangenes" or "gemmules" failed to refer; their reference potential did not contain a description of Mendel's laws. By contrast, on some occasions, Bateson and Johannsen both could have produced expression tokens with the intention of referring to entities that satisfy (1), in which case their reference might have been successful. The causal powers that early geneticists attributed to the gene, namely the ability to be transmitted in a pure state across generations, the disposition to exhibit Mendelian inheritance, and the ability to determine character differences, are properties that genes actually possess. For this reference, it was irrelevant that Bateson, for example, also held false beliefs about genes, such as the belief that there is a one–one relation between genes and traits. The minimal description of a gene's causal role was sufficient to connect the concept to a class of existing entities. However, in Bateson's case, there must also have been occasions where his terms failed to refer, namely occasions where he had the intention of referring to a set of entities that exhibit a one–one relation between characters and factors. There are no such factors, and this element of the reference potential of Bateson's term "unit-character" was thus defective.

If we now turn to Morgan and his associates, the reference potential of "gene" acquired new elements: some defective, others not. As we have seen, Morgan and his associates rejected the notion of "unit-character" and recognized that a many–many relation exists between genes and characters. Thus, they removed a defective element from the reference potential of Bateson's concept. A correct element that they added to the reference potential was the description of a gene as something that lies on the chromosomes. Furthermore, Morgan delineated Mendel's laws to the effect that the second law (independent assortment) only obtains if the genes are located on different chromosomes (see Section 3.2). A defective element they introduced was the idea that genes are units within which recombination cannot occur (units of recombination). This element was later removed by Benzer, when he showed definitively that recombination can occur within single genes. At any rate, it does not seem that this criterion played a major role in identifying and mapping genes; the main criterion for delineating genes was the complementation test. Thus, we could attribute to the Morgan school the following reference-fixing description:

Reference of "gene" (2): Whatever (a) is located on a chromosome, (b) segregates according to Mendel's first law, (c) assorts independent of other genes according to Mendel's second law if these other genes are located on a different chromosome, (d) recombines by crossing-over, (e) complements alleles of other genes, and (f) undergoes mutations that cause phenotypic differences.[14]

These new elements of the term's reference potential constitute an improvement over earlier concepts such as Bateson's, because they reduce the occasions on which expression tokens of "gene" failed to refer.

A somewhat difficult question is whether (1) and (2) establish a reference to the same *natural kind*. As we have seen in Section 2.3, natural kinds are thought to be classes of things that share some causal dispositions. The kinds picked out by (1) and (2) seem to satisfy this condition. However, I have also endorsed the (reductionistic) thesis that only physical and chemical kinds (e.g., molecules with the same structure) are genuine natural kinds, because only they have unchanging causal dispositions. This implies that the kinds picked out by descriptions (1) and (2) must have variable causal dispositions, since they are clearly not molecular kinds. And indeed, this appears to be the case. However, as Ellis (2001, 2002) has argued, some biological kinds are sufficiently robust in order to qualify as natural kinds. Thus, I shall continue refer to such kinds as natural kinds. I point out the consequences of allowing different kinds of natural kinds later (Section 7.4).

An interesting aspect of these classes is that (1) is more inclusive than (2), since (1) is also satisfied by *groups* of genes, or large chromosomal fragments (even whole chromosomes). By contrast, (2) contains criteria like (c), (d), and (e), which discriminate between groups of genes and single genes. Thus, if my analysis is correct, there was already some reference shift from the gene concept of the 1900s to the classical gene concept formulated by Morgan in 1915. However, this reference shift is not very dramatic; it is of such a kind that the reference established by (2) is *included* (as a proper subset) in that of (1).

After having examined the reference potential of the classical gene concepts, we can now ask what happened to this reference when the gene concept was molecularized.

The classical gene concept was initially applied to *eukaryotic* organisms such as insects, mammals, or plants. By contrast, the molecular gene concept was initially closely linked to *prokaryotic* organisms, such as bacteria and bacteriophage. The finding that bacteria and phage also contain genes was once greeted as an exciting discovery. It is usually attributed to Salvador Luria and Max Delbrück, who showed that bacteriophage undergoes random mutations (Luria and Delbrück 1943). One of the main differences between eukaryotic and prokaryotic genes is that the former undergo a *regular* process of sexual recombination, which is the basis of Mendel's laws. By contrast, prokaryotic genes recombine *irregularly* (by the primitive forms of sexuality exhibited by bacteria). Clearly, if the reference potential of the classical gene concept contains a description of Mendel's laws, then bacterial genes are outside the extension of that concept. Thus, the reference of the gene concept was considerably expanded due to the advent of bacterial genetics in the 1940s, which shows us another way in which the introduction of new laboratory organisms can impact the development of experimental biology (see Chapter 6).

What elements constitute the reference potential of the term "gene" as it was applied in bacteria? The fact that Luria and Delbrück's famous experiment that showed the operation of a process of random mutation was considered to be strong evidence for the existence of genes in bacteria suggests that the reference potential of the relevant gene concept contained the following descriptive element:

Reference of "gene" (3):   Whatever undergoes random mutations that cause phenotypic differences.

If tokens of the term "gene" were applied on some historic occasions with the intention of referring to a class of entities satisfying (3), this expanded the reference of the gene concept, since (3) also applies to groups of genes (or large chromosomal fragments). However, Seymour Benzer's work reintroduced an element of the reference potential that was already extant in the Morgan group's use of the term, namely the complementation criterion:

Reference of "gene" (4): Whatever maps as a complementation unit.

This contracts the reference of "gene" by excluding groups of genes or large chromosomal fragments. The gene concept associated with (4) has been labeled the "Neo-classical" concept (see Table 7.1).

Beginning in the 1960s, the causal role attributed to genes changed substantially. This role was no longer primarily sought in the determination of character differences, but in what Philip Kitcher has termed "immediate function" (Kitcher 1982, 354). The immediate function of molecular genes is the determination of the linear sequence of building blocks in protein. Thus, the reference of "gene" was now laid down as the following class:

Reference of "gene" (5): The class of DNA sequences that determine the linear sequence of amino acids in a protein.

Here, something interesting happens. While the reference-fixing descriptions (1) – (4) always used the word "whatever...," (5) directly identifies a class of entities as the reference of the term "gene." The operator "whatever..." (which we could express more formally as "the set whose elements...") is not needed anymore. Thus, the gene is denoted by (1) – (4) as a purely hypothetical entity that is solely identified by its causal role. By contrast, (5) identifies genes by a certain material constitution (DNA sequences) in addition to their causal role (the determination of amino acid sequences), thus rendering genes much more concrete.

While it is clear that the modes of reference of the gene concept have changed dramatically in the transition from classical to molecular genetics, it is far from obvious in which way the reference *itself* has also changed. If we take the mode of reference (4), which picks out the set of all cistrons, then the extension might be approximately constant. So far as I know, the complementation test picks out roughly those DNA sequences that determine the amino acid sequence of a functional protein (as was shown by Yanofsky and co-workers; see the previous section). However, applying the cistron concept sometimes poses difficulties because of the existence of intragenic complementation (Crick and Orgel 1964). Furthermore, there could be difficulties with identifying cistrons in eukaryotic organisms (due to introns and regulatory regions).

212

If we go back to the earlier modes of reference, especially (2), it is far from clear how the extensions are related. One problem in answering this question is that (2) contains Mendel's laws, which do not apply to bacterial genes. We could try to avoid this problem by modifying (2) in the following way:

Reference of "gene" (2*): Whatever (a) is located on a chromosome, (b) re-combines by crossing-over, (c) complements al-leles of other genes, (d) undergoes mutations that cause phenotypic differences.

(2*) would include bacterial genes, but it is still not clear whether this exten-sion exactly corresponds to (5). Furthermore, (2*) is historically implausible, since the Mendelian inheritance patterns were clearly an important element of the classical gene concept.

Another problem is created by the genetic control elements that were first found by Jacob and Monod in the *lac* region in *E. coli*. It is known today that many of the gene mutations found by classical geneticists are not alter-ations in the protein-coding region of a gene, but changes in the regulatory regions. In fact, the phenotypic effects of such regulatory mutations can be much more spectacular than mutations that effect the amino acid sequence of proteins. Particularly striking examples are some of the homeotic mutants in *Drosophila* (already mentioned in Section 6.1), which alter the fly's body plan. Since regulatory regions can be quite remote from the protein-coding structural gene, particularly in eukaryotes, it could be concluded that the classical gene concept refers to a much larger chromosomal segment than the molecular gene concept. However, there are biologists who prefer to use the term "gene" in a way that includes the "entire functional unit, encom-passing coding DNA sequences, noncoding regulatory DNA sequences, and introns" (Alberts et al. 1994, G-10). The problem with this view is that it makes genes extremely complex entities (Epp 1997). There are different kinds of reg-ulatory elements, and they sometimes interact with several genes in complex ways. Furthermore, some regulatory elements are quite unspecific; they will influence the expression of any gene in its vicinity. There is something to be said for defining the gene to include just the protein- or RNA-coding regions, which makes it a much simpler entity.

In general, scientists are not interested in such "semantic" issues. (In sci-entific slang, "semantics" is a derogatory term for unsubstantial, pedantic disputes over words.) Since it does not seem to matter for scientific practice whether regulatory elements are included with genes or not, we must seriously entertain the possibility that scientific concepts such as "gene" come with a certain amount of semantic flexibility. Kenneth Waters has argued cogently that the exact application of the gene concept to specific problem situations depends on the investigative context (Waters 1994, 2000). Thus, depending on the kind of process under investigation, for example, the specific stage of gene expression, the term "gene" might include regulatory regions, while it might exclude them in other contexts. This solution would explain the apparent lack of consensus on the question of what exactly a gene includes. Clearly, it would bode ill for philosophers to try to regulate the use of scientific terms on the basis of considerations of reference. For this reason, I abstain from prescribing that the term "gene" should include regulatory regions just in order to save the referential continuity of the term. Our philosophical theory of reference should be able to handle this kind of semantic flexibility, which seems to pose no problems for scientific practice, because scientists know the exact meaning of their peers' utterances from the investigative context.

To conclude, both the reference potential and the extension of the term "gene" have changed continually as the gene concept has developed. At the beginning of the twentieth century, basically anything that exhibited Mendelian inheritance was called a "gene" or a "factor." This would include even groups of genes or large chromosomal segments. The gene concept employed by the Morgan school acquired modes of reference that contracted the extension of the concept, since it offered refined differentiating capacities. The extension was expanded again with the discovery of bacterial genes. Bacterial geneticists eventually developed a functional test for delimiting genes, which introduced a new mode of reference. This mode of reference was associated with the term "cistron." With the advent of molecular biology, genes were reconceptualized as DNA sequences that determine the linear sequences of protein molecules. As a consequence, genes were no longer exclusively identified by their causal role; the molecular gene concept picks out genes by their material constitution as well as their causal role. Whereas the molecular gene concept in its simplest form roughly corresponds to the concept of cistron with respect to reference, the concept of cistron sometimes poses difficulties in its application, especially in higher organisms. Finally, the reference of the molecular gene concept includes regulatory regions in some investigative contexts, while it excludes them in other contexts.

These considerations suggest that, on a large historical scale, the reference of the term "gene" has not been very stable – quite on the contrary, it seems to have been floating at most times. Whenever new investigative techniques (e.g., linkage mapping) or new laboratory organisms (e.g., bacteria and phage) were brought to the study of genes, the concept acquired new modes of reference, causing considerable shifts in its extension. The picture that emerges thus differs considerably from the well-known examples of conceptual change in the physical sciences (astronomy, chemical revolution, thermodynamics, mechanics), which show long periods of referential stability punctuated by short periods of referential shift.

In the next section, I examine a problem that has been completely ignored in the philosophical and historical literature on the gene concept, namely the question of how classical and molecular genes are related in one particular organism.

7.3 CLASSICAL AND MOLECULAR GENES IN *DROSOPHILA*

So far, philosophers have approached the problem of reference and the gene concept by analyzing different historical versions of the gene concept at an abstract level (e.g., Kitcher 1982; Burian 1985; Burian et al. 1996). In this section, I take a different approach. I want to move one step beyond abstract conceptual analysis and look at the problem from an *empirical* point of view, namely by examining the historical fate of a few specific *Drosophila* genes. I make use of the fact that *Drosophila* geneticists have characterized some *classical* genes that have been known for up to 70 years (!) at the *molecular* level. I suggest that one of the problems that beset existing attempts to referentially relate classical and molecular genes is that people have tried to compare *classical* genes from *Drosophila* with *molecular* genes from *E. coli* and bacteriophage. Clearly, the latter organism is where molecular genes have first been described; however, *Drosophila* genetics, too, has been molecularized. Therefore, I return to my case study of the molecularization of *Drosophila* presented in Chapter 6.

As I have shown in Section 6.1, *Drosophila* genes could only be analyzed at the molecular level once recombinant DNA technology became available in the 1970s. In contrast to Benzer's bacteriophage system, the *Drosophila* mapping system lacked the resolution required for extending linkage mapping to the molecular level. But once scientists had learned how to isolate or "clone" specific DNA fragments, it became possible to look for DNA fragments containing *Drosophila* genes in so-called genomic libraries. As we have seen in my case study, there were two basic strategies to isolate genes from *Drosophila*. One strategy identified DNA fragments containing specific genes by using the gene products of these genes. This approach was used in cases where the gene products were known and could be isolated biochemically, for example, the ribosomal RNA or genes. The other strategy, which employed techniques such as cytological mapping and chromosomal walking (see Section 6.1), was used to isolate genes for which the gene products were unknown or could not be isolated biochemically, for example, *white*, *Antennapedia*, or the Bithorax genes. It is the latter strategy that I want to examine more closely here.

The salient question to ask, I suggest, is the following: How did the molecular *Drosophila* geneticists *know* that a particular DNA fragment that they have cloned contains, for example, the *white* gene? The *white* gene was first identified by classical genetic methods; in fact, a *white* mutant was the very first mutant to be identified by Morgan in 1910. Subsequently, classical geneticists isolated over a hundred alleles of the *white* gene, which all mapped to the same position on the X-chromosome. The position of the gene was determined by cytological mapping on salivary gland chromosomes. Around 1980, the laboratories of Gerald Rubin and Walter Gehring used two different methods to clone DNA fragments containing this gene. Rubin's lab used a *white* mutant that was known to be the result of a transposable element insertion. Thus, they could use a DNA probe containing this transposable element (named *copia*) to detect a DNA fragment containing the *white* gene in a genomic library (Bingham, Levis, and Rubin 1981). Gehring's lab, by contrast, isolated the white gene by using Hogness's technique of chromosomal walking, using the already cloned heat shock gene as a starting point for their "walk" (Goldberg, Paro, and Gehring 1982). In the chromosomal walking technique, DNA fragments picked from the genomic library are located on the chromosome with the help of in situ hybridization, that is, by checking where a radioactively labeled DNA fragment binds on cytological preparations of giant chromosomes (see Figure 6.2). Thus, as I have argued in Chapter 6, classical genetic mapping methods played a crucial role in the isolation of some of the first molecular genes of *Drosophila*. In addition, the molecular geneticists employed another technique in order to make sure they had isolated the right gene. Namely, they inserted the cloned DNA fragment into a *white* mutant with the help of P-element mediated germ-line transformation. This transformation "rescued" the phenotype, showing that the cloned DNA fragments contained all the sequences necessary for normal function of the *white* gene. This is a functional test that resembles the classical complementation test discussed in Section 7.1.

In a next step, Rubin's laboratory determined the full DNA sequence of the cloned DNA fragment (O'Hare et al. 1984). In this DNA sequence, they were able to identify a *molecular* gene, namely a 2.6-kb (kilobase)-long transcription unit containing five exons and encoding a hydrophobic and therefore a probably membrane-bound protein.[15] In addition, it was possible to provide a molecular analysis of some of the known mutant alleles of *white*. For instance, a group of *white*⁻ alleles that clustered at one end of the locus and that did not appear to obliterate wild-type function completely had been hypothesized to be "regulatory mutations" by classical geneticists. Indeed, molecular analysis showed these mutations to reside in the 5' upstream region of *white*. To give another example, the molecular analysis of *white-ivory* confirmed a result that had already been obtained on the basis of linkage mapping, namely that this mutation is an intragenic tandem duplication. Remarkably, it was shown that some alleles contained insertions into the gene's introns, suggesting that alterations within an intron may affect gene expression.

I suggest that this example has interesting implications for the relationship of classical and molecular genetics. What the researchers initially cloned was a chromosomal fragment that was identified as the *white* locus using *classical* cytogenetic methods. On the basis of complementation analysis, this locus was assumed to contain one gene, the boundaries of which had been mapped cytogenetically. Molecular analysis then showed that this chromosomal fragment indeed contained a generic *molecular* gene, that is, a protein-coding DNA sequence in five exons and an upstream regulatory region. Thus, in the case of *white*, the classical and molecular gene concepts pick out approximately the same chromosomal region – assuming that the regulatory sequences are considered to be part of a molecular gene.

Another interesting example is provided by the *Drosophila* gene *rudimentary*, which had also first been described by Morgan in 1910. The alleles at this locus exhibit a complex complementation pattern, with some alleles exhibiting intragenic complementation and other showing no complementation at all. The locus encodes the enzymatic activities for the first three steps in pyrimidine biosynthesis. It had been shown that all three activities are encoded by a single gene in *Drosophila* and in hamsters, whereas up to three different genes are needed in other organisms. Different genetic studies of this locus have found different numbers of complementation groups, that is, classes of alleles that behave identically in a complementation test (Scott 1987). It was expected, from the trifunctional nature of the enzyme encoded, that three complementation units (i.e., inferred functional parts of the gene) would be found, with each enzymatic activity corresponding to a complementation unit. However, at least seven units were found by complementation analysis. This was attributed to protein–protein interactions between the subunits in the multimeric protein because the complementation observed appeared to be intragenic. For our purposes, it is interesting to note that classical genetic analysis indicated the presence of a *single* gene (in *Drosophila*), in spite of the complex behavior of the locus in test crosses of different alleles. When the locus was cloned (using a previously available probe of hamster DNA), a single transcription unit was indeed found (Freund et al. 1986;

Segraves et al. 1984). Thus, we seem to have here another example in which classical genetic methods correctly predicted a protein-coding molecular gene.

This locus Bithorax was studied intensely before molecular cloning methods became available in *Drosophila*, most famously by Edward B. Lewis (this work earned him a Nobel Prize, which he shared with Eric Wieschaus and Christiane Nüsslein-Volhard). Interest in this locus stemmed from the fact that it is the site of homeotic mutations, which suggested a crucial role in early embryonic development. The genetics of this locus turned out to be exhilaratingly complex, especially because of the existence of several mutations that enhance or silence the effect of other mutations at the same locus. It was thus assumed that the locus contains a whole gene complex rather than a single gene. Lewis (1978), on the basis of his famous theoretical model of the determination of segment identity, postulated "at least eight genes." However, an alternative model of the genetic organization of the Bithorax complex was presented by Sánchez-Herrero et al. (1985). They isolated Bithorax mutants from a mutagenesis experiment and crossed these with strains carrying large Bithorax deletions. Crosses that failed to produce viable larvae were used to identify lethal alleles. These lethal alleles were then examined for complementation. Three complementation groups were found and named *Ultrabithorax (Ubx)*, *abdominal-A (abd-A)*, and *Abdominal-B (Abd-B)*. Figure 7.1 shows the two alternative models of the genetic organization of the Bithorax locus. This simplified model of the Bithorax locus was not immediately accepted (Lawrence 1992, 214). However, the molecular analysis eventually revealed three protein-coding genes corresponding to *Ubx*, *abd-A*, and *Abd-B*. Thus, the classical complementation criterion for genes successfully predicted the molecular protein-coding genes, while Lewis's analysis, which was based solely on the phenotypic effects of different Bithorax mutants, predicted too many genes.

Like Bithorax, the homeotic locus Antennapedia was assumed to contain a gene complex. However, in contrast to Bithorax, a single classical gene named *Antennapedia* had also been identified on the basis of genetic analysis. In fact, the molecular analysis found a single protein-coding gene (but two nested transcription units), which is located in the genetic map region bracketed between known inversion breakpoints defined by different mutant *Antp* alleles. The Antennapedia complex contains a number of additional genes, for example, the segmentation gene *fushi tarazu (ftz)*, which mapped between the *Antp* and the *Deformed* locus. *ftz* was also cloned in both Gehring's and Kaufman's laboratories (Kuroiwa, Hafen, and Gehring 1984; Weiner, Scott, and Kaufman 1984) and was shown to contain a single protein-coding gene.

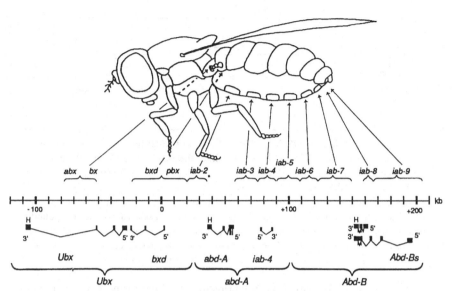

Figure 7.1. Two alternative models of the genetic organization of the homeotic Bithorax locus. The thin black line represents a physical map of the chromosomal DNA. The "genes" above the line are based on the phenotypic effects of various mutations on different parts of the fly body. The three genes shown below the map are based on complementation analysis of lethal recessive alleles isolated from a chemical mutagenesis experiment. *Ubx*, *abdA*, and *AbdB* were shown to contain one molecular protein-coding gene each. Thus, classical complementation analysis and the molecular analysis correspond with respect to the number of genes present. Reprinted with permission from E. B. Lewis: "Clusters of Master Control Genes Regulate the Development of Higher Organisms". *Journal of the American Medical Association* 267: 1524–31. © 1992 American Medical Association.

Thus, in the case of the Antennapedia complex, molecular studies confirmed most of the expectations from classical genetic analysis.

As already mentioned in Section 7.1, one of the first genetic loci to reveal the full extent of the possible genetic complexity of *Drosophila* was *achaete-scute*, as was shown by H. J. Muller, N. P. Dubinin, and others in the 1930s. It is a classical example of a step-allelic series,[16] and first-generation *Drosophila* geneticists had hoped to learn something about the internal structure of genes by complementation analysis of the *achaete-scute* series of mutations. However, these efforts met with little success (Weber 1998b). More recently, mapping of various *achaete-scute* deletion mutants (Garcia-Bellido 1979) led to the subdivision of the locus into four regions: *achaete, scute α*, *lethal of scute*, and *scute β*. A fifth region, named *scute γ*, was mapped by Dambly-Chaudière and Ghysen (1987). These maps were based on *classical* methods. The locus was cloned in 1985 and revealed a complex arrangement

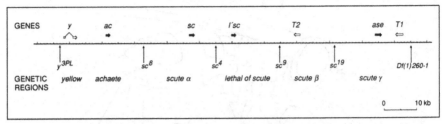

Figure 7.2. The *achaete-scute* locus (ASC). The horizontal line represents a physical map of the chromosomal DNA. Below the line, the genetic regions as identified by classical deletion mapping (i.e., the classical genes) are indicated. The tiny arrows above the line represent the transcribed regions as revealed by the molecular analysis. Filled arrows indicate the four protein-coding genes found at the locus, while the role of the other transcripts identified is unclear. This is a case where classical and molecular genetic analysis did not agree on the number of genes present. From Campuzano and Modolell: "Patterning of the Drosophila Nervous System: The achaete-scute Gene Complex". *Trends in Genetics* 8: 202–208 (1992). Reprinted by permission of Elsevier.

of nine transcription units, the relationship of which to the genes was not clear at first (Campuzano et al. 1985). However, it was eventually possible to identify one transcription unit for each of the *achaete, scute* α, and *lethal of scute* regions (Ghysen and Dambly-Chaudière 1988). A fourth transcript was located to the opposite end of the region and named *asense. scute* β, which is separated from *scute* α by *lethal of scute*, is now thought to be a regulatory element affecting the expression of the *scute* protein. Finally, Alonso and Cabrera (1988) found, on the basis of a full DNA sequence analysis of the cloned *achaete-scute* locus, four homologous protein-coding genes (all encoding transcription factors), corresponding to *achaete, scute, lethal of scute*, and *asense*. It is thought today that the complete function ascribed to the *achaete-scute* region on the basis of genetic experiments are contained in these four transcription factor genes.[17] However, in this case the classical genetic and the molecular analysis did not correspond. The complexity of the locus stems from the interspersion of coding sequences and regulatory sites, such that different regulatory sites act on the same gene, and the regulatory sites act on different genes. This example also shows the complex nature of eukaryotic genes, which comprise not just protein-coding regions, but often include regulatory sequences that involve a far larger segment of the chromosome (Figure 7.2).

These examples should suffice for the purposes of my analysis. I have tabulated the molecular analysis of a few additional *Drosophila* genes in Table 7.2.

Table 7.2. *Some of the First Drosophila Genes to Be Cloned*

Year of Cloning	Locus	Year of First Description	Cloning Method	Genetic Analysis	Molecular Analysis	Gene Product(s)
1975	rRNA	1975	colony hybridization with rRNA	none (found entirely by molecular methods)	multicopy genes (tandem repeats)	18S and 28S rRNA
1978	heat shock	ca. 1970	colony hybridization with mRNA	none (found entirely by molecular methods)	multicopy gene, no introns	70 kD heat shock protein
1981	*white*	1910	transposon tagging/ chromosomal walking	>100 recessive alleles, single complementation group	single protein-coding gene, many transposon insertions	hydrophobic protein (membrane transporter)
1983	*rosy*	1938	chromosomal walking	complex locus, intragenic complementation	single protein-coding gene	xanthine dehydrogenase
1983	Bithorax complex	1915	jumping from rosy/Ace chromosomal walk	homeotic locus, 3 complementation groups	three protein-coding genes	DNA-binding proteins (transcription factors with homeodomains)
1983	*Antennapedia*	1949	chromosomal walking	homeotic gene	2 nested transcription units containing single protein-coding gene	DNA-binding protein (transcription factor with homeodomain)

(continued)

Table 7.2 (*continued*)

Year of Cloning	Locus	Year of First Description	Cloning Method	Genetic Analysis	Molecular Analysis	Gene Product(s)
1983	*Notch*	1918	chromosomal jumping using a Notch inversion mutant	haplo-insufficient dominant alleles	single protein-coding gene, many transposon insertions	related to epidermal growth factor
1983	*Gart*	1981	DNA fragments complementing yeast ade8 mutation	none (found by molecular methods)	alternative processing of transcript	2 proteins with different enzymatic activities (purine biosynthesis)
1984	*fushi tarazu*	1981	from Antp chromosomal walks	segmentation gene	single protein-coding gene	DNA-binding protein (transcription factor with homeodomain)
1984	*rudimentary*	1910	cDNA of the homologous hamster gene CAD as probe	complex locus, intragenic complementation	single protein-coding gene	multifunctional enzyme (pyrimidine biosynthesis)
1985	*achaete-scute*	1916	chromosomal walking	complex locus, step-alleles	four homologous protein-coding genes	transcription factors (involved in neurogenesis)
1986	*PCP*	1986	Along with *Gart*	none (found by molecular methods)	Nested with in intron of Gart gene	Pupal cuticle protein

222

What the preceding analysis shows is that the classical gene concept was not simply abandoned with the development of molecular biology. As I have shown, *Drosophila* geneticists started to clone a number of genes in the 1980s, when recombinant DNA technology had become available. This involved the task of isolating specific DNA fragments from a so-called genomic library, which contains millions of random fragments of nuclear DNA. Some fragments were isolated with the help of RNA molecules, for example, the ribosomal RNA and heat shock genes. However, a number of DNA fragments were identified on the basis of the known map location of *classical* genes on the chromosomes. Thus, the operational criteria for identifying these genes were the criteria of the classical gene concept. DNA fragments so identified were then sequenced and examined for the presence of molecular genes. In most cases, this approach was successful. What this suggests is that the classical gene concept, along with some techniques adapted from classical *Drosophila* genetics, was used to *identify* molecular genes. Thus, even though the classical gene concept had long been abandoned at the theoretical level, it continues to function in experimental practice up to the present.

I think the last point is highly relevant to the old question of the relationship between classical and molecular genetics. In the great debates on reduction in genetics, it was usually assumed that there are two different sciences that enter into some kind of explanatory or reductive relation with each other (e.g., Kitcher 1984). Russell Vance (1996) was the first to challenge this assumption and to point out that these so-called two sciences are actually much more closely intertwined. What my account of the molecular cloning of the first *Drosophila* genes shows is that there are not just explanatory or reductive relations between classical and molecular genetics. There is, in addition, a *pragmatic* relation between the classical and molecular gene concepts such that the former can be used to find instances of the latter, even though the two concepts are not strictly coextensional.

7.4 BIOLOGICAL VARIABILITY, ESSENTIALISM, AND FLOATING REFERENCE

We are now in the position to determine what kind of conceptual change the case of the gene instantiates. From the investigations in Section 7.2 it is clear that we do not have a case of variable meaning with fixed reference. Both meaning and reference have changed considerably as new experimental approaches and theoretical ideas have developed. This by itself would not be news to philosophers, as strict reference stability (i.e., fixity of reference under conceptual change) has been given up by most writers on physical concepts (see the introduction). Furthermore, it is also clear that the case of the gene does not exhibit a "punctuated equilibrium" pattern (i.e., long phases of referential stability intermitted by short periods of revolutionary change). What we seem to have, instead, is some kind of *freely floating reference*.

By this term, I mean the following. As the practice of genetics continuously generated new ways of detecting, localizing (mapping), and describing genes, some DNA segments moved in, others out of the term's extension. This kind of conceptual change differs substantially from the typical cases that have been studied in the physical sciences, such as phlogiston, mass, and temperature. The latter terms shifted in reference during scientific revolutions, but were fairly stable at most times. The reference of the term "gene" was never really stable, and perhaps is not even stable today. Remarkably, this floating of the term's reference seems not to have diminished its theoretical importance or practical usefulness.

In this final section, I relate some metaphysical considerations that will provide an explanation for this mode of referential change. This explanation has to do with the nature of *natural kinds* in biology (see also Sections 2.3 and 9.1).

I suggest that geneticists were not tracking a single natural kind of entities that they called "genes." Instead, different historic versions of the concept referred to *different* natural kinds, which were not coextensive. As I have argued, at least the classical and neoclassical gene concepts denoted perfectly fine natural kinds, just like the molecular gene concept (if natural kinds are understood as classes of objects that share some causal dispositions). Yet these kinds arc all different, perhaps in ways that could even generate incommensurable concepts in the sense of Feyerabend and Kuhn. But this appears not to have caused major theoretical problems; geneticists simply switched freely between different modes of reference and frequently changed the set of things they were referring to when using the term "gene."

Oxygen is always the same stuff, and mass always the same property – they do not come in different flavors.[18] But genes are different: Pick a random pair of genes from the biosphere and the probability that they are identical is exceedingly small. Every individual organism contains a set of different genes, sometimes two or more physically different copies of the same gene. Different individuals of the same species may differ considerably with respect to their genes. Finally, members of different species have different genes. *Genetic variation* within and between species is a fundamental feature of biological organisms – and an important one, for that matter, because biological systems owe their very existence to variation. It is exactly at this point that

224

we must recognize a fundamental difference between physical and biological systems.

The great evolutionary biologist Ernst Mayr has been arguing against *essentialism* in biology for several decades (e.g., Mayr 1982, 45–47, 55). In the physical sciences, there exist natural kinds of entities that share *unchanging essences*, for example, a set of causal dispositions (see also Section 2.3). For example, the essence of oxygen is that its atomic nuclei contain exactly sixteen protons. All oxygen in the entire universe, and only oxygen, has this essential property; and it had this property ever since oxygen was formed some time after the big bang. By contrast, Mayr has argued, biological systems lack such essential properties. Biological organisms *and their genes* are highly variable and therefore do not have a fixed set of essential properties. It has long been known to taxonomists that the high variability of biological species has consequences for their classification.

What seems to have been overlooked to date is that the high amount of variability present in the genetic systems of biological organisms might have consequences for the formation of biological concepts other than the species concept. I claim that the case of the gene is a prime example for this. My analysis in the previous sections suggests that the variable nature of the genetic material allowed this material to be conceptualized in different ways, depending on the available methods, the organisms under study, and the theoretical interests. On this analysis, earlier versions of the gene concept referred to natural kinds; in other words, they referred *successfully*.[19] However, the natural kinds denoted, for example, by the classical gene concept of the Morgan school or the neoclassical concept of microbial geneticists, are different from the natural kinds denoted by more recent, molecularized versions of the gene concept. It seems to me that the situation is thus entirely different from the well-known cases of conceptual change in the physical sciences. If we take the example of the Chemical Revolution, the term "phlogiston" did not refer to a natural kind, although the term "dephlogisticated air" might have referred to a natural kind (oxygen) in some situations. Similarly, the Newtonian concept of mass does not denote a natural property according to relativistic mechanics; only the concept of relativistic mass does. Thus, in these cases, reference is *all or nothing*; these theoretical terms (or particular tokens of these terms) either referred or failed to refer.[20] By contrast, on my account of the gene concept, different historical versions of this concept *did* refer, but to *different natural kinds* that are not coextensive.

My analysis also reveals that the extensions of different versions of the gene concept were not completely disjoint. The best evidence for this is the finding that the molecular cloning of various classical genes of *Drosophila*

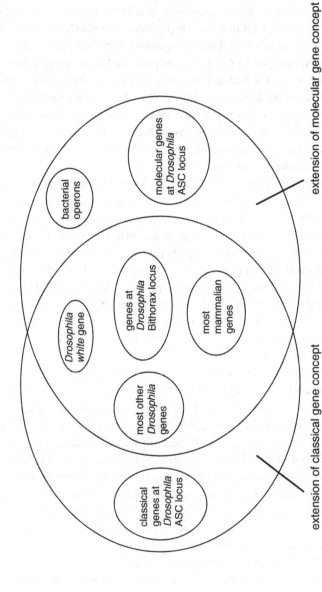

extension of classical gene concept   extension of molecular gene concept

Figure 7.3. Referential relationship between the classical and molecular gene concepts. A relation of partial referential overlap is possible because genes come in different subtypes (depicted as small sets).

showed that the chromosomal regions picked out by the classical gene concept in most of the cases contained molecular protein-coding genes and associated regulatory regions. These regulatory regions are sometimes viewed as lying outside the gene; however, this is controversial even among molecular biologists. We could say that if the regulatory regions are considered to be part of a molecular gene, then some molecular genes identified in *Drosophila* corresponded roughly to classical genes. In fact, as I have argued, the classical gene concept was *used* to identify the DNA fragments containing these molecular genes in the first place. For this, apparently, a full coextensionality of the classical and molecular gene concepts was not necessary; a partial overlap of the respective extensions was sufficient. This overlap was such that several *subtypes* of the natural kind picked out by the classical gene concept also lie in the extension of the molecular gene concept (see Figure 7.3). For example, the subtype denoted by the expression "the *D. melanogaster white* gene" corresponds in both classical and molecular genetics. In other cases, there is no such correspondence, for example, in the case of the *achaete-scute* (ASC) locus. I suggest that such a relationship of partial referential overlap as is depicted in Figure 7.3 is only possible because genes come in different subtypes that can be classified into natural kinds in different ways, depending on the investigative methods and theoretical interests extant in a particular stage of the development of genetics.

To conclude, twentieth-century genetics exhibits a special kind of conceptual change that does not exist in the physical sciences, at least not in the better known examples that have been examined in connection with the theory of reference so far.[21] This kind of conceptual change is characterized by what I call *floating reference*, meaning that changes in experimental methods and in theory continuously altered both the concept's modes of reference and its extension. Apparently, this in no way affected the concept's fruitfulness for research – perhaps it was even vital.[22] Furthermore, it is the nonessentialistic nature of biology's subject matter that makes this complex kind of conceptual change possible.

SUMMARY

The meaning of the term "gene" has changed several times in the history of twentieth-century genetics. If we distinguish between a term's sense and its reference, it is possible that the term's sense has changed, but not its reference. According to Philip Kitcher's theory of reference, each scientific term is connected with a reference potential, which is a compendium of different ways in which a scientific community determines a term's reference. A term's reference potential can be heterogeneous, that is, contain both correct and incorrect modes of reference. In this chapter, I have examined both the reference potential and the reference connected to different historical versions of the gene concept. The versions examined include Bateson's concept of "unit-character," Johannsen's introduction of the term "gene," the Morgan school's so-called classical gene concept, the "Neo-classical" concept of microbial genetics in the 1950s, the molecular concept of the 1960s, and the contemporary concept that originated in the 1970s.

My analysis shows that these concepts all differ with respect both to the associated modes of reference and to reference (extension) itself. However, all twentieth-century versions contain modes of reference capable of connecting the term "gene" to natural kinds, if we take natural kinds to be classes of objects that share a set of causal dispositions. Furthermore, I have examined a number of cases where classical genes of the fruit fly *Drosophila* – some of which were already known to Morgan and his associates – have been cloned molecularly and analyzed at the DNA level. In these cases, the classical gene concept was actually *used* to identify molecular genes, suggesting a certain amount of referential continuity. In spite of this, the gene concept exhibits what I proposed to call *floating reference*, a phenomenon not previously found in the cases from the physical sciences that have been studied in connection with the theory of reference so far. I have explained the possibility of floating reference in terms of the nonessentialistic character of biological concepts, which allows biologists to lay down different natural classifications, depending on the investigative methods available as well as on theoretical interests. Thus, the ontology of living organisms has consequences for concept formation that have not been noted previously.

8

Developmental Biology and the Genetic Program: Explaining Ontogeny

One of the most pervasive ideas in modern biology is that there is a genetic program written into the nucleotide sequence of genomic DNA (see Mayr 1982, 55–56). When a fertilized egg cell initiates the process of embryonic development, what it does is read or interpret the instructions encoded in this program to generate the adult form or phenotype. Thus, the notion of genetic program seems to provide some kind of explanatory framework for developmental biology today. As many people have noticed, this notion bears a striking resemblance to Aristotle's biology.[1] But given that Aristotle's metaphysics differs fundamentally from that of modern science, the question arises of whether such a view can be adequate. Why should ontogenetic processes in biology be explained any differently than other biological processes, for example, the transmission of nerve impulses?

As we have seen in Chapter 2, typical reductionistic explanations in experimental biology essentially consist in applications of physicochemical principles to biological systems composed of interacting molecules and macromolecular aggregates. In addition, experimental biologists give functional explanations, which I have analyzed in the tradition of causal role functions (as opposed to etiological functions) in Section 2.4. The question that arises now is whether the explanations of ontogenetic processes given by modern developmental biologists also exemplify this basic explanatory approach.

At first sight, this might appear to be the case. After all, some of the pattern-forming processes in a developing embryo have been explained at the molecular level, that is, in terms of patterns of gene activity in the embryonic cells of various animals. The best understood model system for developmental biology is the fruit fly *Drosophila*, which we have already encountered in Chapters 6 and 7. Some of the protein–DNA interactions that govern the early development of the fly embryo have even been determined at atomic resolution (see, e.g., Billeter et al. 1993). Does the impressive level of molecular detail exhibited by current developmental explanations not demonstrate that the reductionist research program of modern experimental biology has finally conquered developmental biology?

To some extent, this is certainly the case (Rosenberg 1997a, 1997b; cf. Laubichler and Wagner 2001). However, as I have already indicated, there are tensions between modern developmental biology and reductionistic explanations such as the ones I have analyzed in Chapter 2. One source of this tension is the idea mentioned at the beginning, the idea of a genetic program.

There are two important kinds of objections to the idea that DNA contains or constitutes a genetic program for embryonic development. The first kind of objection complains that this idea is inherently *intentional*. What this means is that biologists sometimes talk as if DNA sequences had some kind of content. The way this is usually put is that DNA contains *information* about the primary structure of proteins and RNA molecules. The latter are then said to express the genetic program. But informational content is not something that sits comfortably with a purely chemical structure. Normally, we attribute to molecules physical properties such as spatial structure, molecular weight, or absorption spectra, and chemical properties such as reactivity with certain classes of other compounds. Informational content or *meaning* is something that we normally attribute to concepts, thoughts, or statements. These are the kinds of things that have *semantic* (i.e., meaning-related) properties, whereas molecules only have physical and chemical properties. Thus, this line of argument concludes, any talk about DNA containing a program, code, or genetic information must be metaphorical.[2]

The second kind of objection claims that the received view in developmental biology attributes to DNA a privileged role in ontogeny that it does not have. This objection begins by noting that various kinds of resources are necessary for building an organism. A fertilized egg cell receives much more from its parents that just DNA, and these other resources are just as important for making a new organism. Furthermore, this objection continues, an organism's development is influenced by so many environmental factors that to say that it is programmed by a DNA molecule is at best misleading, and at worst an ideological construction that serves the purpose of legitimizing genetic determinism and/or risky genetic technologies.

The second criticism is usually offered in order to support a certain kind of alternative to the standard view in developmental biology. This alternative is known as *developmental systems theory* or DST.[3] DST is a set of connected claims about development, evolution, the nature-versus-nurture problem, heredity, and epigenetic inheritance that are aimed toward breaking away from what some call the "hegemony of the gene." Proponents of this view think that too much weight is being placed on genes and DNA in explaining organic development (and behavior) in contemporary biology, whereas epigenetic, environmental, and cultural influences on the phenotype are unduly neglected. Thus, the DS theorists' concern is not merely with the possibly metaphorical status of notions such as information, coding, instruction, specification, programming, and so on in the context of biology. The DS theorists' critique cuts deeper into the heart of current biological thinking.

Most DS theorists also oppose reductionism. Therefore, their reasons for challenging some standard notions surrounding genes, DNA, and development are quite different from the concerns that inform the first objection, which come from the kind of reductionism defended in Chapter 2 of this work.

Of course, DS theorists are aware that much biological research today involves macromolecular structures other than DNA (e.g., proteins, lipid membranes, and polysaccharides) and that most biologists are aware of the multitude of environmental factors that can affect development. However, they argue that the respective roles of these various factors in an organism's development are inappropriately described in standard presentations of developmental and behavioral biology. Thus, the new perspective proffered by DS theorists draws on the same stock of knowledge about cells and biomolecules and their interactions as the standard view it counters, but it proposes to conceptualize these interactions in a different way.

In particular, DST wants to steer clear of the idea that DNA and genes are some sort of "master" molecules and move toward a more "democratic" view of development, where the life of individual organisms is viewed as a cyclical process that emerges out of a complex developmental system. Although genes and DNA form an important developmental resource in such systems, their causal role is said to be completely on a par with those of other factors, which may or may not lie within the bounds of an individual organism's skin.

My goal in this chapter is to examine, in the light of these two objections, how current biology explains pattern formation in embryonic development. I begin by clarifying a number of conceptual issues raised by the second objection, the developmental systems view (Section 8.1). In Section 8.2, I examine an example of a developmental explanation in the best-understood developmental system, namely the fruit fly *Drosophila*. This examination shows that, interestingly, developmental biologists attribute information not just to DNA but also to concentration gradients of certain molecules. So-called morphogenetic gradients are said to contain positional information for the cells along the gradient. In the *Drosophila* system I examine, this gradient is established by the mother. At first sight, this appears to support the case of DST; however, I show in Section 8.3 that this appearance is deceptive. Furthermore, I show that explanations of this type can be rendered as functional explanations in a way that is compatible with the reductionism that I have defended in Chapter 2. Finally, in Section 8.4, I examine whether DNA and genes play some kind of privileged role in current explanations of biological ontogeny.

8.1   A CRITIQUE OF DEVELOPMENTAL SYSTEMS THEORY

DST has never been a very tightly knit theory, and it means different things to different people.[4] But fortunately, Kim Sterelny (a sympathetic critic) and Paul Griffiths (a proponent) have provided a very clear textbook account that can serve as a basis for our discussion (Sterelny and Griffiths 1999, Chapter 5). It is this version of DST that I shall critique.

Sterelny and Griffiths present DST in five steps that are supposed to take the reader from the standard view of the role of genes in development to what at least one of them (Griffiths) thinks is the correct view.[5] It is best to quote these five steps in full (Sterelny and Griffiths 1999, 95):

Step One. Organisms inherit a great deal more than their nuclear DNA. The epigenetic inheritance of nongenetic structures within the cell is a hot topic in current biology. Organisms also behave in ways that structure the broader environmental context of their successors. For instance, many birds inherit their songs through the interaction of their developing, species-specific neural structures with the adult songs to which they are exposed. So an organism inherits an entire *developmental matrix*, not just a genome.

Step Two. The orthodox view of development is that all traits develop through the interaction of genes with many other factors. So genes are neither the only things that are inherited nor the only things that help to build an organism. There is more to evolution than changes in gene frequencies. But genes might still be "privileged causes" of development, which control, direct, or act as an organizing center for everything else. If gene selectionism is to get off the ground, it must demonstrate that genes play some such privileged role.

Step Three. The notion of genetic information and its relatives cannot be made good in a way that singles out genes as privileged causes of development. Every reconstruction of the notion that genes contain information about outcomes of development turns out to apply equally well to other causes of development.

Step Four. A range of further attempts to draw a distinction between the role of genes in development and the roles of other developmental factors fail. These attempts are either mistaken or overstated (for example, the idea that genes are copied "more directly").

Step Five. Developmental systems theorists conclude that for their biological importance, genes do not form a special class of "master molecules" different in kind from any other developmental factor. Rather than replicators passing from one generation to the next and then building interactors, the entire developmental process reconstructs itself from one generation to the next via numerous interdependent causal pathways.

I now turn to the critical evaluation of these steps.

Step One. The claim that organisms "inherit a great deal more than their nuclear DNA" might seem uncontroversial at first sight. Of course, DNA alone cannot produce an organism; it takes a whole egg cell with a complex internal organization to start a new life. The cell already needs to be fitted with a lot of biochemical machinery for its genome to become active, for example, the complex protein-synthesizing machinery that is necessary for expressing genes. This machinery must be passed on by the mother. Furthermore, cell biologists have reasons to think that many of the structures needed to build a new organism cannot be synthesized *de novo* (from scratch) by the cell. The cell membrane that encloses the egg cell and the internal membranes that it needs to carry out its biochemical functions provide examples. You can only make a membrane from a preexisting membrane. Another example is the cylindrical structures called centrioles, which are required for cell division. It appears that they, too, must be passed on by a mother cell. Thus, it is certainly true that developmental processes require more than just DNA, and most of these things must be passed on from the mother to the egg cell. This much is absolutely uncontroversial. Where the problems begin is with the authors' use of the concept of *inheritance*.

For example, in Step One, Sterelny and Griffiths talk about the "epigenetic inheritance of nongenetic structures." This requires clarification. Epigenetic inheritance, as it is commonly used in biology, is *not* identical with the passing on of material structures other than DNA from the mother to an egg cell, or from a somatic cell to its daughter cells. In order to qualify as epigenetic inheritance, some *phenotypic difference* at the level of populations of organisms or populations of somatic cells must be transmitted across generations. This may occur in the form of *trait* inheritance or *somatic cell* inheritance. A trait is epigenetically inherited exactly if there exist population-level phenotypic differences with respect to this trait that are accounted for by differences in some structure that is not a sequence of nucleotides in DNA.

An example is provided by a mechanism called genomic imprinting. Some mammalian genes are chemically modified by the addition of methyl groups to the DNA bases at certain specific positions (DNA methylation). These genomic imprints mark a chromosome according to whether it is derived from the father or from the mother. The chemical marks are passed on when a chromosome divides meiotically (i.e., in the process that generates egg or sperm cells). Thus, even the autosomal (non-sex-linked) chromosomes from the two parents appear to be chemically and biologically distinct (Reik and Constancia 1999; Surani 2001).

At the level of somatic cells, epigenetic inheritance may occur during embryonic development. In some cell lineages, modifications in gene activity take place that are transmitted to the daughter cells in mitotic cell divisions. For example, certain genes that are not used or that would disturb the proper functioning of a cell in certain tissues can be silenced in such a way that all the descendants of the cell inherit the inactivated state of the genes.[6] Recent evidence suggests that such stable modifications of gene expression patterns are determined by chemical modifications in the chromatin. The chromatin contains, in addition to the genomic DNA, protein complexes that tightly control the way in which DNA is packed up in the cell nucleus. An especially important class of such proteins is the histones. Originally, it was thought that histones play a purely structural role in packaging DNA. However, it is becoming clear now that histones play an active role in controlling gene activity (see, e.g., Lyko and Paro 1999; Beisel et al. 2002; Rea et al. 2000; Strahl and Allis 2000). The histones bound to a certain DNA region can be chemically modified so that the gene residing in this region is silenced or permanently active. The chemical modifications involved in this include methylation,[7] acetylation (addition of an acetic acid residue), and posphorylation (addition of a phosphate group). When the DNA is copied prior to mitotic cell division, the chemical modification states are preserved by the chromatin. Thus, these modifications are passed on to the daughter cells.

Histone modification seems to be more common in the phylogenetic tree than other epigenetic mechanisms. For example, *Drosophila* is not known to have genomic imprinting mechanisms, but histone-mediated epigenetic mechanisms of gene regulation are thought to play an important role in cell differentiation in this organism.

It must be emphasized that the passing on of materials other than DNA from mother to offspring, even if necessary, is not a sufficient condition for the occurrence of epigenetic inheritance in either of the two senses discussed above (trait inheritance and somatic cell inheritance). There is no guarantee that some extrachromosomal structure that is passed on is capable of transmitting a mark to future generations. By a "mark" I mean some structural difference that accounts for a phenotypic difference. When cells divide, they pass on many extrachromosomal structures (membranes, organelles, centrioles, etc.). But only in the case of DNA and the epigenetic mechanisms discussed above (so far as we know today) will a physical *difference* in these structures be found in the next generation of cells or whole organisms.

Thus, the concept of epigenetic inheritance needs to be carefully distinguished from the passing on of cellular constituents other than DNA. Only if the passing on of non-nucleic-acid structures from mother to offspring occurs in such a way that *differences* in these structures account for corresponding differences in the offspring do we have an instance of epigenetic inheritance. Epigenetic inheritance is a special kind of parent–offspring (or cell–daughter cell) covariation that may or may not occur when some extra–chromosomal structure is passed on to the next generation. How frequently this occurs in nature is another matter. So far, I am only making a conceptual point. By glossing over natural distinctions, DST runs the risk of muddling the issues rather than clarifying them.

The notion of epigenetic inheritance having thus been clarified, it is time to venture a comparison of genetic and epigenetic inheritance mechanisms. The evolutionary theorists John Maynard Smith and Eörs Szathmáry have claimed that, in contrast to the known epigenetic mechanisms, DNA is an *unlimited hereditary replicator* (Maynard Smith and Szathmáry 1995, 42). What this means is that a DNA molecule has an unlimited number of alternative states given by the possible combinations of the building blocks A, G, C, and T. Furthermore, when a DNA molecule replicates, it transmits its state to subsequent generations. By contrast, Maynard Smith and Szathmáry argue, other replicating structures have only a limited number of alternative states that they can transmit to future generations.

There is a difficulty with this suggestion. For the term "unlimited" cannot be equated with mathematical infinity. A genome cannot be infinitely long; therefore, the number of possible DNA sequences for a given genome length is finite. But this means that the number of different states is not really unlimited; it is just extremely large. What is more, recent evidence suggests that the number of alternative states that can be transmitted by some of the epigenetic mechanisms is also very large. This appears to be the case, for example, with the chromatin modifications (e.g., histone methylation) discussed above. Referring to these mechanisms, Strahl and Allis (2000, 45) have gone so far as to conjecture that "every amino acid in histone tails has specific meaning." This would mean that the number of alternate states that can be transmitted is enormously large.

While it might be the case that the number of possible permutations of A, C, G, and T in a DNA molecule is still larger, we must ask what significance the difference between two extremely large numbers has, if neither of them is mathematical infinity.[8]

However, the number of alternative states that can be transmitted by a hereditary replicator in principle may not be the crucial issue anyway. The crucial issue is the *functional relevance* of the states of the replicator. It is here that the most significant difference between genetic and epigenetic inheritance mechanisms probably lies. The known genetic and epigenetic inheritance mechanisms differ clearly with respect to the extent to which their alternative states can have functionally relevant effects. DNA sequences have their functional effects through an all-purpose mechanism that is totally blind to the identity of the input states, namely transcription (RNA synthesis) and translation (protein synthesis). The protein synthesis machinery treats all sequences basically the same.[9] By contrast, the mechanisms that functionally respond to methylated or acetylated histones recognize only some very specific patterns, probably with the help of specific enzymes. Thus, only some specific input states are recognized by the epigenetic systems (the same is true of the genomic imprinting mechanisms).

As a result, the extent of phenotypic differences that are made possible by DNA sequence variation is probably far greater than that for the known epigenetic mechanisms (Weber 2001c, 247). In order to make this claim plausible, consider what it would take to turn, say, a crocodile egg into a dinosaur. I am not claiming that such "Jurassic Park" scenarios (see Crichton 1990) are technically feasible. But for all we know about developmental biology, *if* something like Jurassic Park (i.e., making dinosaurs from crocodile eggs) is possible, then this will involve substituting the organism's DNA. Epigenetic modifications alone will almost certainly not be sufficient, even if they might also be necessary for this science-fiction scenario to work.

We could even go further and claim that there exists no *environment* in which a crocodile zygote with a crocodile genome would develop into a dinosaur. As far as is known today, this kind of phenotypic difference can only be caused by differences in genomic DNA sequence. To my knowledge, DS theorists have never explicitly denied this. But accepting that the difference between a crocodile and a dinosaur is, to a first approximation, a difference in DNA sequence is tantamount to affirming the primacy of DNA in inheritance.

Closer to home, the genetic difference between a gorilla and a human may be astonishingly small, as these two primate species share a majority of their genes. But their phenotypic difference is, to a first approximation, nonetheless due to a difference in DNA sequence: probably differences in the genes' complex regulatory regions. It is more than doubtful that there exists an environment or an epigenetic modification that could turn a gorilla zygote into a human being, or vice versa. Even though environmental factors can sometimes have rather drastic effects on an organism's phenotype, the scope of the changes made possible by differences in DNA sequence is far greater. If the genotype (i.e., the DNA sequence) is held constant and the environment is allowed to vary, there are strict limits to the range of phenotypic variants that will result. By contrast, if the environment is held constant and the genomic DNA sequence is allowed to vary, almost everything is possible, even changes to an animal's fundamental *Bauplan*.

Consider, for example, the homeotic mutations in *Drosophila* (see Sections 6.1 and 8.3). In theory, it is conceivable that some externally added chemical induces the same developmental aberration as a homeotic gene mutation. There are chemicals that are known to strongly interfere with the development of limbs, for example, thalidomide. However, such chemicals seem to have quite specific effects and they seem to act only at certain points of the developmental pathway. In contrast, changes in DNA can affect any aspect of the phenotype.

Thus, DS theorists are exaggerating the relative magnitudes of effects that can be traced back to environmental or epigenetic factors relative to the magnitude of the possible changes that are due to differences in DNA sequence.

My rebuttal of DST's claims concerning the primacy of DNA in inheritance should not be misunderstood as a commitment to genetic determinism about human traits. It may very well be the case that some of the traits that interest us most in humans are nongenetic: for example, certain behavioral and psychological dispositions. My claim that DNA has the greatest *potential* to cause heritable phenotypic variation does not entail that all phenotypic variation is, in fact, genetic (the potential may not be realized). It even allows that there exist traits that depend more strongly on nongenetic (e.g., environmental) factors than on genetic ones. What I am claiming is that if all the traits of an organism are taken into consideration, including its fundamental *Bauplan*, then altering its DNA sequence has a greater potential for phenotypic change than altering either the environment or epigenetic factors. Only by changing its DNA could you turn a gorilla into a human, or a crocodile into a dinosaur.

At this point, DS theorists are likely to object that the standard concept of inheritance is biased. They propose to "redefine inheritance so that every element of the developmental matrix that is replicated in each generation and which plays a role in the production of the evolved life cycle of the organism counts as something that is inherited" (Sterelny and Griffiths 1999, 97). For example, they count the transmission of nest-specific songs in songbirds as an example of inheritance. The nestlings of some songbirds, apparently, acquire their song and some other habits after they are born, by interacting with their parents. A nestling transferred to a different nest early enough would acquire the song of the new nest rather than that of its genetic parents. Sterelny and Griffiths describe this as a structuring of the environment by the parents such that the offspring will replicate the song. Thus, it seems as if something is inherited here over and above the birds' genes.

I have no quarrels with admitting this to the category of inheritance. However, this kind of inheritance must be conceptually distinguished from *both* the passing on of extrachromosomal structures from one generation to the next and from epigenetic inheritance. The bird songs are a case of trait inheritance that is neither genetic nor epigenetic. We have here a classic case of *cultural transmission*. It arises from an interaction between the parents and the developing nervous system of the offspring, which, apparently, is flexible enough to allow some nongenetic phenotypic plasticity. In contrast to both epigenetic and genetic inheritance, the song phenotype is fixed *after birth*. DS theorists simply skate over these biologically significant distinctions and end up in a conceptual muddle. The sense of uneasiness and the many "yes, but ... " responses that some of us experience when reading texts from DS theorists probably arise from this fact.

Another of the authors' concerns is to show that something other than genes can be a *replicator* in the sense of evolutionary theory (see below). There is a sense in which bird songs can be said to replicate; perhaps they may even be subject to a process of cultural evolution.[10] However, this does not in any way affect the standard view of the role of genes in development. The latter view is not committed to all inheritance being genetic; in other words, it allows genetic and epigenetic inheritance to peacefully coexist with the cultural transmission of traits (see the remarks concerning genetic determinism above). But this means that the example of bird songs provides no grounds for accepting any of the claims of DST or for rejecting them. It is totally irrelevant to its central claims.

So what is the advantage of redefining the concept of inheritance, as DS theorists suggest? The main consequence of such a redefinition appears to be that we lose the distinction between genetic, epigenetic, and cultural inheritance. Of course, DS theorists will consider this to be a plus for their theory, as they want to move beyond the traditional nature-versus-nurture problem. But what is really gained by this move? Where the standard view in genetics sees three different kinds of transmission processes, DS theorists only see one kind of process: the passing on of what they call the "developmental matrix." I fail to see why such a loss of differentiating capacity should constitute an advance. In fact, an unsympathetic spirit might view it as a regression to a less advanced stage of biology, e.g., the nineteenth century. Our thirst for unity must not blind us to the diversity of different processes that occur in nature, or else the aim of science will be defied.

My suspicion is that all that is gained by flattening the distinction between different kinds of inheritance is to immunize DST from the empirical evidence, which currently speaks against it.

Step Two. Here, the authors bring in a set of issues that we have not yet touched upon, namely evolution and gene selectionism. Ever since G. C. Williams's influential book *Adaptation and Natural Selection* (Williams 1966), biologists and philosophers have argued about a provocative thesis, namely that genes are the units of selection. What this means is that Darwinian natural selection is best described as a process of differential replication of single genes. The literature on this subject is voluminous, and there is no point in discussing it here.[11] But there are some lingering confusions with respect to this problem that keep being propagated by DS theorists. The source of these confusions is the idea that any copying process found in biology is an instance of replication in the evolutionarily relevant sense. Whenever DS theorists see a structure that appears to be copied in biological growth, they talk as if it were an instance of replication. I shall argue that this is a mistake. For the only evolutionarily relevant sense of replication is *hereditary* replication, that is, replication of a structure such that a change in this structure can be transmitted to future generations.

To illustrate the difference between mere copying and hereditary replication, I shall briefly discuss an example from cell biology. Cell organelles like mitochondria, chloroplasts, and peroxisomes are not formed de novo by cells. They can only be made by growth and fission from preexisting organelles. Thus, in a sense, a cell can make copies of some of its organelles. Does this mean that the organelles replicate? As mitochondria and chloroplasts contain DNA, which complicates the issue, let us look only at the peroxisomes,[12] a DNA-free organelle. The salient question is whether some structural alteration to a cell's peroxisomes can have any evolutionary consequences. To the best of our current knowledge, this is not the case. Even though there are mutants in peroxisome biogenesis (e.g., in yeast), these mutations reside in nuclear genes. A striking class of mutations affect peroxisome biogenesis in such a way that the cell contains only so-called ghosts. These are empty peroxisomal membranes. But by transformation with a functional copy of the mutant gene, a cell can be rescued with respect to the peroxisome phenotype. Thus, a cell can make functional peroxisomes from ghosts if it is given the necessary nuclear genes. But no physical or chemical alteration of peroxisomes is known that would be transmitted to the next generation. Only nuclear gene mutations are.

The upshot of this example is that, at least in this case, the nuclear genes are the only hereditary replicators.[13] For something to be a hereditary replicator, it must be able to mutate in a way that has evolutionary consequences. In other words, it must be possible that some structural modification is transmitted to future generations. I have described this as transmitting a mark to future generations. If this is not possible, then the structure cannot give rise to cumulative evolutionary change and, therefore, it cannot be a unit of selection. Peroxisomes, even if they are being copied by the cell in some sense, cannot be modified in such a way that this modification will still be present in future generations. Therefore, peroxisomes do not replicate in an evolutionarily relevant sense.

Of course, it cannot be excluded, on the basis of our current knowledge, that there are biochemical structures other than nucleic acids (DNA or RNA) that are true replicators.[14] But DS theorists (and others including Sterelny, Smith, and Dickison 1996) are far too quick in granting the status of replicator (or unit of selection) to biological structures that are not capable of sustaining cumulative evolutionary change.

Finally, Sterelny and Griffiths's claim that gene selectionism requires that genes play some sort of organizing role in development is plainly false. All that is required for gene selectionism to "get off the ground" is that genes are true replicators (in the sense explicated above), and that no more inclusive genetic structures are true replicators. The old gene selectionists (Williams and Dawkins) make no assumptions concerning the role of genes in development.[15]

Step Three. The concept of genetic information has turned out to be problematic (see note 2 for references). In the way biologists sometimes use the concept, it seems to presuppose some kind of *intentionality*. In other words, DNA sequences are often talked about as if they had some sort of meaning or content, much like a sentence or statement. But this is illegitimate, some critics say, because a cell does not interpret its DNA in the way in which the human mind interprets sentences. There is no understanding involved. Nobody would want to claim that a rock tumbling down a mountain interprets the slope and gravitational field. The rock's trajectory is merely a causal consequence of the latter. So why do biologists have this strong tendency to say that DNA sequences are read or interpreted by the cell, instead of just stating its causal relations to other cellular constituents?

It must first be noted that there is also a weak information concept in current biology. This concept is not problematic. In this weak sense, for example, DNA can be said to contain information about protein molecules. But this claim, if interpreted correctly, does not state a causal relation between DNA and protein. Instead, it states at least a three-place relation between biological structures and a cognitive agent (the scientist) who uses his or her knowledge about biological structures and some background knowledge (in particular the genetic code) to draw inferences about other biological structures. The whole discipline of bioinformatics depends on such a notion of information in molecular structures. But according to this weak sense, we can just as easily say that protein contains information about DNA (even though a reduced amount of information, due to the redundancy of the genetic code). In fact, the weak information concept applies to any kind of system where there is reliable covariation between a set of signals and a set of alternative states of the system. In this sense, soil pH values could also be said to contain information about plant growth, since pH affects the growth of most plant species.

I shall refer to this weak concept as the *pragmatic information concept*. Some authors refer to it as the "causal" information concept (e.g., Sterelny and Griffiths 1999, 101). This term is misleading because it suggests that this kind of information somehow resides in the causal relations of things.[16] But this is not the case: the information concept discussed here is sensitive to the investigator's interests and state of knowledge. Information in the weak sense can be attributed to any factors that can be used to predict some events. What these factors are depends on what we want to predict and what we know about the system in question.

The pragmatic information concept fails to pick out a special role for genes or DNA. Anything that causally affects an organism's phenotype can be said, in a certain sense, to contain information for this phenotype. This much is absolutely uncontroversial. The salient question is whether there is a legitimate stronger information concept that accords a special role to the so-called genetic material, such that only nucleic acids are bearers of genetic information, whereas all other cellular constituents are viewed as mere stuff or raw materials for development that wait to be instructed by the master molecule. The whole idea of a DNA-based genetic program depends on this notion. The bone of contention here is the so-called *intentional* or *semantic* information concept, according to which DNA carries some sort of meaning (like a statement).

There are authors who think that there is a good, intentional information concept, and that it does pick out nucleic acids as the sole or at least main carrier of genetic information. I discuss such an attempt in Section 8.3. For the time being, it suffices to note that the DS theorists' critique is correct insofar as the pragmatic concept of information is concerned. The more controversial questions are whether there is a legitimate intentional or semantic information concept in biology, and – if there is – whether this concept picks out DNA as unique. Discussion of this problem is postponed to Section 8.3.

Step Four. DS theorists claim that all the attempts to attribute a special role to genes as opposed to other developmental resources have failed. These (allegedly) failed attempts include the following claims (see Sterelny, Smith, and Dickison 1996, 381): (1) The causal influence of genes is more direct than that of other developmental resources. (2) There is a causal asymmetry between genes and other factors like, for example, proteins. (3) Genes have a greater potential to cause phenotypic variation than other factors, for example, intracellular chemicals or environmental factors. (4) Genes are copied with greater fidelity than other developmental resources. (5) Genes are causally more important than other factors in the production of phenotypes.

Some of these attempts are indeed quite feeble. (1) and (5), as they stand, lack clarity and are probably difficult to spell out. (4) is probably true, as DNA replication is equipped with elaborate proofreading mechanisms and is less error-prone than transcription or protein synthesis.[17] However, this is hardly why biologists attribute to genes a special role in development. The question only becomes interesting if we raise the question of *why* cells spend more energy to ensure that DNA is copied with high fidelity than to ensure high-fidelity transcription. This suggests that genes and DNA play a special role after all. At any rate, the most serious contenders are (2) and (3), the thesis that there is some causal asymmetry between genes and proteins and that the range of possible phenotypic variation is greater if DNA sequences are allowed to vary. I discuss (2) in Section 8.4.

Concerning (3), I have already shown in my critique of Step One that DS theorists are probably wrong in thinking that the range of potential phenotypic variation that is attributable to environmental changes or to epigenetic mechanisms is anywhere near the range of potential changes that could arise through differences in DNA sequence. Only the latter can change any aspect of the phenotype, including an organism's fundamental body plan. The phenotypic effects of the known epigenetic inheritance systems seem rather limited in comparison. If there are environmental factors that can cause a comparable range of phenotypic differences, for example, to an organism's body plan, these changes will not be heritable. Thus, DNA clearly plays a unique role in inheritance. The DST theorists' assertions to the contrary are simply not supported by the empirical evidence at present; they are based on pure make-believe.

242

Step Five. DS theorists finally arrive at their thesis of *causal parity*, which denies genes any special role in development and inheritance. According to them, it is the entire developmental system that replicates itself from one reproductive cycle to the next; genes are just one of many causal pathways that connect two subsequent cycles.[18]

An important feature of the causal parity thesis that should be noted is its *conceptual independence* from DST's claims about genetic information. For the thesis could be false even if everything DS theorists said about genetic information were correct. In other words, genes could be causally privileged in a way that cannot be adequately expressed using the concept of information. Conversely, DS theorists could be wrong about the concept of genetic information even if it were true that the causal role of genes is on a par with other cellular constituents. For it is not conceptually necessary that genes are privileged as carriers of information on the basis of their *causal* role. In theory, there could be explications of the concept of genetic information that do not make reference to a specific causal role. In Section 8.2, we shall see that biologists attribute information to developmental systems components other than DNA, without claiming that the causal role of these components is equal to that of DNA.

Thus, the two claims that the concept of information does not pick out a special role for genes and DNA and that the causal role of the latter is on a par with that of other cellular components should be assessed independently. This is what I shall attempt. The concept of information in developmental biology will be the subject of Section 8.3, and the causal parity thesis is treated in Section 8.4. But to make this debate less abstract, it is helpful to examine a concrete example from current developmental biology. This example is about the as yet best understood developmental system: *Drosophila*.

8.2   PATTERN FORMATION IN *DROSOPHILA*

According to contemporary developmental biology, the structure of an organism is not preformed in the fertilized egg; it is generated anew in each generation by a set of processes called *pattern formation*. One of the best understood examples is the genesis of the segmented body pattern of arthropods, especially *Drosophila*. The embryonic development of the fruit fly is widely considered to exemplify a set of important principles in developmental biology: the formation of repeated units with the subsequent differentiation of these units into different body parts. Most animal groups, including mammals, exhibit such repeated units (e.g., vertebrae or digits). For our present investigation, the generation of the segmented body plan of *Drosophila* is interesting because most of the components of the developmental system that accomplish this feat have been characterized at the molecular level.[19]

I begin by introducing some of the crucial components of *Drosophila*'s developmental system. For the sake of simplicity, I am only concerned with those components involved in the formation of the anterio–posterior (from head to tail) pattern of the insect body.[20] The components may be grouped into five classes:

1. An initial intracellular *morphogen gradient* in the fertilized egg cell, formed by the protein product of the gene *bicoid*, a so-called egg-polarity gene. Morphogens are molecules that vary in concentration along some axis and that convey *positional information* to the embryonic cells.[21] The *bicoid* gradient is generated by the synthesis of protein from an mRNA species that is derived from the mother (by transcription of maternal genes) and that is deposited in the egg cell at the anterior end by maternal nurse cells.

2. A set of six *gap genes*, for example, *hunchback*, *Krüppel*, and *knirps*.[22] These genes interpret the morphogen gradient defined by the *bicoid* protein by regulating the expression of other members of the class and of the pair-rule genes (see below) in a spatially localized manner.

3. A set of eight *pair-rule genes*, for example, *even-skipped*, *fushi tarazu*, or *hairy*. Interactions of the protein products of the gap- and pair-rule genes with the regulatory regions of other genes from these two classes generate a pattern of gene expression along the anterio–posterior axis that delimits fourteen so-called parasegments – the fist appearance of the typical segmented pattern of the insect body. The first three of these parasegments will later form the head, the next three will form the thorax, and the remaining eight will form abdominal segments. The gene *even-skipped* delimits odd-numbered parasegments, whereas *fushi tarazu* delimits even-numbered parasegments.[23]

4. A set of at least ten *segment polarity genes*, for example, *engrailed*, *hedgehog*, or *gooseberry*. These genes determine the anterior and posterior boundaries of the parasegments.

5. Two complexes of *homeotic selector genes*. The Bithorax complex contains three such genes (*Ultrabithorax*, *abdominal-A*, and *abdominal-B*; see Figure 7.1), and the Antennapedia complex contains five selector genes (for example, *Deformed*, *Sex combs reduced*, and *Antennapedia*). The Bithorax complex determines the identities of parasegments 5–13, whereas the Antennapedia complex determines the identity of parasegments 1–5 (see Chapter 6 for the discovery and subsequent molecular cloning of these genes).

These five classes of components act more or less sequentially, that is, (1) before (2), (2) before (3), and so on. Furthermore, the components of each class act on the components of the next class. Together, these interactions generate spatial patterns of gene activity that commit the cells in the so-called blastoderm stage to their fate just a few hours after fertilization. Because of the action of the egg-polarity, gap-, pair-rule, segment polarity, and homeotic selector genes, the cells of the developing embryo activate some of their genes and inactivate others. This pattern of gene activity establishes the body plan of the fruit fly along the anterio–posterior axis.

At the beginning of the developmental process, the only anterio–posterior pattern that the embryo contains is the *bicoid* protein concentration gradient.[24] Development in *Drosophila* begins with a series of divisions of the cell nucleus without accompanying divisions of the cell membrane. Therefore, the embryo first forms a so-called syncytium, a many-nucleated cell with a couple of thousand cell nuclei. Crucial steps in cell differentiation already occur in this syncytial stage, when the cell nuclei and their genomic DNA are freely accessible to cytoplasmic factors. Most of the protein products of the egg polarity, gap-, pair-rule, segment polarity, and homeotic selector genes are transcription factors. These are proteins that bind to DNA at specific sites (called regulatory regions) and enhance or suppress the activity of the gene nearby. A gene that is active is said to be expressed or transcribed; that is, RNA copies are made from this gene, which are used by the cell to synthesize specific proteins. Thus, this early phase of *Drosophila* development mainly consists in the synthesis of transcription factors from the egg polarity, gap-, pair-rule, segment polarity, and homeotic genes. The interactions of these transcription factors with the regulatory regions of other as well as their own genes[25] generate a spatial pattern of gene expression that determines the fate of each of the embryonic cells.

The first transcription factor to act is the *bicoid* protein. As this protein forms a gradient along the anterio–posterior axis, its concentration drops below a certain threshold about halfway through the embryo (see Figure 8.1). When the *bicoid* concentration is above this threshold, the gap gene *hunchback* is expressed; below the threshold, *hunchback* is repressed. The *hunchback* protein is itself a transcription factor. It affects the activity of another gap gene, *Krüppel*. The latter is only active when the concentration of *hunchback* protein lies between two thresholds. As a result, *Krüppel* is expressed in a single band. In a similar fashion, *hunchback* activates and represses the activity of the gap genes *knirps* and *giant*. The gap genes further interact with the regulatory regions of various other gap genes and the

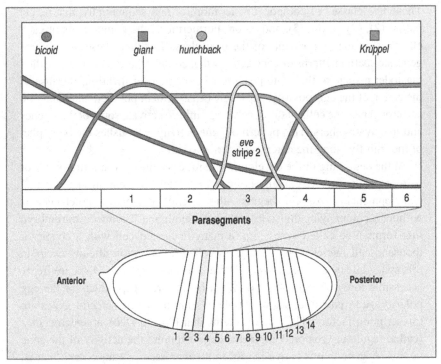

Figure 8.1. Pattern formation in the *Drosophila* embryo by pair-rule and gap genes. This diagram shows the specification of the second stripe of expression of the pair-rule gene *even-skipped (eve)*. Pair-rule genes are expressed in seven transverse stripes. The concentration gradients of the morphogens *bicoid* and *hunchback* activate the *eve* gene in a broad domain. The other gap genes *giant* and *Krüppel* repress *eve* in parts of this domain, allowing its expression only in a narrow stripe in the third parasegment. The other stripes of *eve* expression are specified in a similar fashion by the action of additional genes. From Wolpert et al. (2002). Reprinted by permission of Oxford University Press.

pair-rule genes to create spatial domains of gene expression. These domains define the parasegments of the insect embryo (see Figure 8.1).

As a result of these events, the initial *bicoid* morphogen gradient is transformed into a complex concentration profile of the protein products of gap- and pair-rule genes (mostly transcription factors) along the anterio–posterior axis of the embryo (a part of this profile is shown in Figure 8.1). These proteins in turn selectively activate segment polarity genes that determine the orientation and boundaries of the fourteen parasegments. Finally, the homeotic selector genes give each of the parasegments a unique identity. This means

246

that the cells in these segments are programmed to differentiate in a specific manner by the homeotic selector genes. Thus, even though all the cells contain the same complement of genes, they differ in the expression pattern of these genes, which is determined by their position in the blastoderm embryo. Later, when the larva metamorphoses to form the finished fly, these cells will remember their position in the blastoderm stage. This information, as it were, is stored in the expression pattern of homeotic selector genes in each cell.[26] As Alberts et al. (1994, 1094) put it: "The control regions of the homeotic selector genes act as memory chips for positional information."

Thus, in a nutshell, the morphogen *bicoid* provides positional information, while all the other developmental genes described here function in the interpretation of this information.

The simplified account of the formation of the segmented pattern of an insect body that I have given should suffice to address the conceptual questions at hand. An attempt to analyze the concept of positional information will be given in the following section. For the time being, the following important conclusion must be drawn from the present case: Even though it is true, in a sense, that some morphogen gradients must be passed on from the mother to the egg cell, it would be a mistake to view such gradients as *hereditary replicators* (which DS theorists are prone to do), because alterations in these structures are not passed on to future generations.

There are, of course, genetic mutations in the *bicoid* gene itself that are passed on. They belong to the class of so-called maternal effect mutants. In this class of mutants, the phenotype of a larva depends solely on the genotype of the mother and is independent of its own genotype. One could say in such cases, perhaps, that the phenotypic trait is transmitted epigenetically, but only for one generation (*bicoid* larvae are not viable). However, it remains the case that the relevant *replicator* here is DNA-based, because this is how the mutation can be passed on to subsequent generations.

As I have shown in Section 8.1, it is not enough to be merely *copied* by the cell for something to be a hereditary replicator. The replication process must be able to preserve one out of a number of possible alternative states when the cell divides or the organism reproduces. This is not the case for the kind of morphogen gradients discussed here; even though they can be manipulated into alternative states that have phenotypic effects, these states are not passed on to subsequent generations. Thus, we have here a clear difference between DNA, which is a hereditary replicator, and morphogen gradients, which are not hereditary replicators. To lump them together as "developmental resources" in the manner of DST is just not helpful.

8.3   THE CONCEPT OF INFORMATION IN DEVELOPMENTAL BIOLOGY

As was already mentioned in Section 8.1, philosophers and some biologists have objected to the use of notions such as genetic information, instruction, specification, representation, end coding in genetics because they seem to presuppose intentionality, that is, some kind of meaning or content like the thoughts of a conscious being. Developmental systems theorists reject these notions on different grounds, namely because they grant genes and DNA a privileged role in development which, in their view, they do not have.

Although this problem has been a matter of debate for some years now, the concept of positional information has (to my knowledge) not yet been subjected to this kind of scrutiny. For our critique of DST, this notion is interesting because it seems to suggest that embryonic development depends on other kinds of information, that is, information that is not stored in the organism's genome. But how fundamental is the concept of positional information, and what does it really mean? Is positional information a species of genetic information? Or is it a metaphor that serves the purpose of communicating science to students and the general public?

Clearly, for Wolpert (1969), who introduced the concept, positional information seems to have been more than a metaphor to popularize scientific findings. A reading of this classic text reveals that he clearly intended this concept to play an important theoretical role. Wolpert defined positional information as the "specification of position" by "a mechanism whereby the cells in a developing system may have their position specified with respect to one or more points in the system" (1969, 1). Wolpert contrasts the specification of positional information with what he calls "molecular differentiation," which refers to the "control of synthesis of specific macromolecules" (ibid.) Thus, in Wolpert's terminology, the expression of genes by the synthesis of protein and RNA molecules is called molecular differentiation. Interestingly, Wolpert thought, in 1969, that "[t]he specification of positional information in general precedes and is independent of molecular differentiation" (ibid.). Furthermore, he speculated, "there may be a universal mechanism whereby the translation of genetic information into spatial patterns of differentiation is achieved" (ibid.). Thus, for Wolpert, positional information was a fundamental explanatory concept that supplements genetic information and that is also needed to explain pattern formation in development.

I take it that this idea sits well with DST, because it grants developmental resources other than DNA or genes an informational or organizational role. Note that morphogens can be either proteins that are directly genetically encoded (such as *bicoid*) or small molecules that are the product of secondary

248

metabolism (such as retinoic acid, which was shown to function as a morphogen in the development of avian limbs). Either way, morphogens require a preexisting structure to form their characteristic concentration gradient. In the case of *bicoid*, maternal mRNA molecules that are attached to the egg cell's cytoskeleton at its anterior end provide this structure (of course, the structure of this RNA is determined by the mother's nuclear DNA). Thus, positional information cannot be provided by zygotic DNA alone, which might spell good news for DST.

The critical question for substantiating this idea is what exactly the concept of positional information means. Surely, an embryonic cell does not interpret a morphogen gradient in the manner in which we interpret a text or a sign. Such a concept of interpretation requires mental capacities that cells clearly do not have. But what else could these notions mean?

In a more recent review, Wolpert admitted that the concept of positional information "places a great burden on the process of interpretation" (Wolpert 1989: 3). In other words, how can cells make use of the information that their position in a morphogen gradient contains?

The *bicoid* story provides an answer to this question: A cell's response to the concentration of a morphogen seems to be mostly a matter of appropriate thresholds in the activation of gene expression by cytoplasmic transcription factors. Although it might have been conceivable that each concentration value along a morphogen gradient specifies a different behavior of the cells, this is not how such gradients are interpreted in the case of *bicoid*. Instead, the segmented pattern of *Drosophila* is generated by an interaction of several gradient-forming transcription factors and by the activation and repression of genes when the concentrations of these transcription factors are above certain thresholds (see Figure 8.1). These thresholds are thought to be determined by finely tuned binding affinities of the transcription factors to the promoter regions of their target genes. Thus, it is the strength of certain DNA–protein interactions that ultimately explains pattern formation.

This suggests that at least the mechanism of action of the *bicoid* morphogen can also be explained without the concept of positional information. The formation of the concentration profiles of the different proteins shown in Figure 8.1 is nothing but a complex, spatially heterogeneous biochemical reaction. The spatial heterogeneity is initially introduced into this reaction by the *bicoid* gradient and is transformed, by a series of molecular interactions, into a more complex pattern that generates the fourteen parasegments of the *Drosophila* embryo. Chemical reactions are known to be capable of forming very complex spatial patterns by a few simple molecular interactions.[27]

The response of cells to morphogen concentrations can thus be described without using the notions of reading, translation, or interpretation. Another way of saying that cells can translate or interpret positional information is that the morphogen concentration *causes* them to behave in certain ways, for example, by activating or repressing some target genes. I see no problem in translating all the statements that contain the terms "information," "interpretation," and so on into strictly causal language.[28]

Wolpert (1989) distinguishes between a strong and a weak sense of positional information. According to the strong sense, there is "a cell parameter, positional value, which is related to position as in a coordinate system and which determines cell differentiation" (1989, 3). The weaker version, by contrast, "merely emphasizes position as a key determinant in cell development and differentiation" (ibid.). The difference between the two senses of positional information is not quite clear. Presumably, the weaker notion of positional information does not presuppose the existence of positional value. But what exactly does "value" mean here? And what should we make of the "coordinate system"?

One possibility is to read the strong concept of positional information in a sense that involves *representational content* or *intentionality*. This possibility should be briefly explored here.

It is often said that the most telltale characteristic of intentional states is that they can be said to be in *error* or to misrepresent some other state of affairs (e.g., Griffiths 2001, 397). Misrepresentation and error are cognitive categories that are not applicable to information in the weak sense (discussed in Section 8.1). Some signal either does or does not causally correlate with the state of its source, or it may do so with a certain probability $p < 1$ (in cases where there is noise), but it does not make sense to say that a signal is erroneous or that it misrepresents some source state. By contrast, a message in the full intentional sense (like the text you are reading now) may be erroneous.

By this criterion, if it is any good distinguishing weak from intentional information, we should conclude that positional information in developmental biology is intentional. For it seems to make perfect sense to say that a cell misreads a positional signal, or that the positional signal itself is false. In fact, many of the known *Drosophila* mutants where development is affected may be described in this manner. For example, the homeotic mutant *Nasobemia* that carries legs on its head instead of antennae (due to a mutation in the gene *Antennapedia*; see Gehring 1998, 28–30) can be described as a case where the homeotic selector genes misread the positional information provided by the egg-polarity and segmentation genes, which will induce some cells in the head to "think" that they are in the thorax, where their job would have been to make legs. Indeed, developmental biologists frequently speak of the "appropriate" or "right" behavior of cells within the embryo (see, e.g., Gurdon and Bourillot 2001, 797), suggesting that positional information is meaningfully said to be right or wrong.

At this point, a critical question must be raised: What are the notions of positional information and value really needed for? What explanatory burden do they carry? Wolpert can be read as claiming that a cell's positional value somehow represents its position in a coordinate system that is defined by one or several morphogens. But to bring in the concept of representation here is a highly problematic move, for it forces us to come up with an account of intentionality that does not attribute cognitive states to cells (which I take to be absurd).

One possibility is to apply *teleosemantics*, a well-known attempt to give a naturalistic account of meaning and related phenomena (e.g., Millikan 1989). According to this approach, meaning is tied to natural selection. Interestingly, the eminent theoretical biologist John Maynard Smith has taken such an approach in order to give an analysis of the concept of information in biology (Maynard Smith 2000).

Before we can turn to Maynard Smith's account, we must note a few things about the concept of intentionality. In philosophy of mind, this term is used to describe a characteristic of certain mental states, namely that such states can have an object. The object of a thought is that which the thought is about, for example, the future precipitation in my surrounds when I am thinking "it's going to rain." The meaning or sense of a statement is also said to be intentional; it bears some referential relation to an object (see Chapter 7). We sometimes say that a message received from a fellow human contains important information, thus referring to some part of the sense of the message (the sense is that which the message says). Thus, the concept of information can be used intentionally, in which case it refers to the sense or meaning of a message. This concept is also referred to as *semantic* information.

Maynard Smith argues that the biological information concept is basically intentional. Of course, he is not suggesting that cells can think or that they have mental states. What he is suggesting is that there is a strong analogy between an organism that expresses its genome and a computer that executes an algorithm that has been programmed by an engineer. A computer program, Maynard Smith argues, does contain information in the strong, intentional sense. An experienced programmer could use it to infer the intentions of the engineer. Therefore, the program carries meaning. Had the program been composed by a random selection algorithm instead of an intelligent engineer (which is technologically possible), it could still be said to have meaning, for it is not in principle different from a designed program. In a fully analogous way, biological organisms have been programmed. There has been no engineer. Instead, the genetic program is the result of many rounds of natural selection. Over many generations, alterations to the program that allow its carriers to survive and breed successfully have accumulated. Thus, the genetic program – which is transmitted in the form of DNA sequences and executed with the help of RNA and protein molecules – contains information. The information is about protein molecules that help the organism survive and reproduce. Thus, according to Maynard Smith's reasoning, the concept of genetic information is intentional.

There are two major difficulties with Maynard Smith's account. The first difficulty is structurally similar to a problem that we have noted in the etiological theory of biological functions (see Section 2.4). Consider a gene that has just suffered a random mutation that happens to be beneficial for its carrier. This new gene sequence has not arisen by natural selection but by some nondirected genetic mutation. However, it seems to make just as much sense to attribute genetic information to such a gene. If other genes contain information in an intentional sense, then so does this one; for it functions just the same as genes that have been selected for. But the teleosemantic account cannot be applied in this case because there has been no natural selection.[29]

But there is a more general, second difficulty. This becomes evident if we take Maynard Smith's engineering analogy seriously. Maynard Smith seems to think that a computer program is intentional because we can infer the intentions of the engineer who wrote it. This may be correct, but note that the intentional states are *in the engineer*, not in the computer. A computer program is a string of symbols that acquires a meaning only in the context of a community of engineers who understand what the program does and what it can be used for.[30] In fact, nothing even qualifies as a computer except in relation to some intentional agents who interpret the input and output states of the device (Grush 2001). But the engineer is what we do not have in the biological world. An organism has not been built with the intent of carrying out some operations. Natural selection may be a substitute for explaining adaptation or for attributing functions (perhaps, but see Section 2.4), but it cannot substitute for the intentional states of an intelligent designer when it comes to attributing meaning to any part of the system.

Thus, the teleosemantic approach fails to make good the thesis than a biomolecule can contain information in the strong, intentional sense. All meaning and intentionality are lost when we substitute natural selection for intelligent design in explaining the origin of a system.[31]

Maynard Smith could be defended on the grounds that he only proposed an *analogy* between biological organisms and a purposefully designed system like a computer that executes an algorithm. This analogy may be good, and it may have been extremely useful heuristically,[32] but it cannot justify the attribution of meaning and intentionality to DNA (nor any other part of an organism). For the intentional aspect is exactly the *disanalogous* part of the comparison of the genetic material with a computer program. Any analogy is accompanied by disanalogies, or else it would not be an analogy but a bona fide application of a concept. Maynard Smith's text is a powerful argument for a strong analogy between genetic material and computer programs, but an analogy nevertheless.[33]

After this critique, we can now return to the notion of positional information. Given the above considerations, it is not clear what work notions like positional information or morphogen gradient interpretation are doing in explaining pattern formation. Why can't we just say that the morphogen concentration *determines* the subsequent behavior of a cell? Even if there is an internal parameter that responds to morphogen concentration and that controls cell differentiation, why is it not sufficient to state the causal dependence of this parameter on the morphogen concentration, without saying that it specifies or represents the cell's position in a coordinate system?

The problem can perhaps be stated even more pointedly with the help of an analogy to nonliving systems. Consider an example of pattern formation that does not involve living cells, for example, ripples on a creek. These ripples arise out of hydrodynamic pressures caused by the water flowing unevenly over the rocks in the riverbed. Now, would we say that a particular volume of water contains positional information for the formation of ripples at the surface? Does each volume of water have a positional value that specifies or represents its position with respect to the rocks? Surely, this would not make any sense.

By contrast, Wolpert's use of these notions does not strike us as being completely senseless. This becomes especially evident when we consider how some embryos are capable of buffering considerable disturbances without losing the capacity for forming their species-specific patterns. Marine invertebrates are especially renowned for their regenerative capabilities. Some of them, like sea urchins, can form fully developed animals from substantially fragmented embryos.[34] Does this not show that embryonic cells can orient with respect to some kind of coordinate system or morphogenetic field, and that they therefore contain some sort of representation of the organic forms to be developed?

I think that this suggestion does not resolve the issue, as these regulative phenomena can also be explained in strictly causal terms: It is enough if the causal networks that lead from morphogen concentrations to cell differentiation are relatively resilient or robust to external disturbances. A creek may exhibit the same or similar ripples even if some of the rocks on the ground are pushed around. Even if the constancy of the behavior under a variety of external conditions is particularly astounding in some developing organisms, this is not a difference in *kind*. The introduction of intentional notions requires a difference in kind to a purely physical process of pattern formation (where we are content with a strictly causal explanation).

My suspicion is that the attractiveness of intentional notions such as positional value in developmental biology arises from a very strong predilection of the human understanding: namely, to view organic development as a *goal-directed process*. Embryonic development always seems to be drawn towards the finished product, which is usually represented by the adult animal (e.g., the imago in the case of the fruit fly). The adult form is the goal of development. Developmental biologists try to show how various processes that take place in the embryonic stages contribute to this goal. A morphogen gradient is unlike the ripple-causing rocks in a creek because the latter do not have the *function* of causing ripples. By contrast, it is just very difficult for us *not* to view some of the events that occur in a developing embryo as having the function of generating some specific form. Teleology thus comes back with a vengeance in the midst of reductionist molecular biology.

The main evidence that positional information is a cryptically teleological notion comes from the fact that it makes perfect sense to say that a cell *misreads* or *misinterprets* positional information. By contrast, we would not say that some unusual ripples on the water "misread" the positional information contained in some volumes of water. The reason is that, in the latter case, we do not assume that there is a way in which the water *ought* to behave. But in a developing organism, it is hard to fence off the notion that there are correct and incorrect ways for an embryo to develop. Correct ways are those that create the final form of the organism as we know it. In such individuals, all the components of the developmental system have performed their function properly.

As an example, let us return to the role of the *bicoid* gradient in the development of the fruit fly. In Section 8.2, I have described its role by saying that it specifies positional information that is then interpreted by some target genes like the segmentation genes and the homeotic selector genes. Another way of describing this is by simply saying that the *bicoid* gradient causes a cascade of gene expression events that generate the segmented structure of the insect body. The first if these accounts makes only sense if we attribute to the *bicoid* gradient the biological function of eliciting a graded response in the cell nuclei of the early *Drosophila* embryo. As we remember from Section 2.4, the attribution of biological functions requires the specification of some ultimate system goal. In the present example, the ultimate system goal for attributing a function to the *bicoid* gradient is the form of the *Drosophila* imago.

Thus, explanations involving notions such as positional information or morphogen gradient interpretation can be analyzed as a species of functional explanation.[35] Like most functional explanations in experimental (proximate) biology, such explanations analyze a complex system by showing how the components of this system contribute to some overall capacity of the whole system (see Section 2.4). In developmental biology, this capacity is posited from the outset: It is the capacity for forming the patterns or form of the adult organism. Developmental biologists analyze the processes occurring inside an embryo and its cells with this finished form in mind, and the notion of positional information states a relation between some early cell differentiation events and the later fate of these cells in the adult organism.

Thus, my diagnosis is that those who think that the concept of information in biology is intentional have gone too far. The concept may be *teleological* or *functional*, but intentional it is not. Biological molecules like morphogens carry no meaning, any way we cut the cake. What they may carry is *function*. The relevant sense of function in experimental biology, as I have argued in Section 2.4, is the causal role function, which picks out certain capacities of the parts of complex systems independently of selection history.

To conclude, explanations involving positional information and morphogen gradient interpretation can be rendered as causal explanations just like ordinary functional explanations. The passing on of morphogen gradients from mother to offspring in *Drosophila* does not differ fundamentally from the passing on of other cellular structures that are required for embryonic development (e.g., cell membranes).[36] Morphogen gradients do not constitute a species of genetic information in the strong, intentional sense. As a consequence, morphogen gradients *qua* information carriers do not support DST. What my analysis shows is that morphogen gradients at best carry information in the *weak* sense. But it is absolutely uncontroversial that the weak concept of information applies to various developmental systems components other than DNA.

The other important conclusion to be drawn from this section is that current explanations of ontogenetic processes are fully compatible with the kind of reductionism defended in Chapter 2. There, I have analyzed the mechanistic explanations of modern experimental biology as applications of principles from physics and chemistry to special types of systems. In addition, I have shown that functional explanations deploy such principles in a specific way, namely by exhibiting the contributions that some mechanism makes towards some system capacity of the larger system in which it is embedded. The above considerations suggest that developmental biologists take the same basic approach to explain pattern formation. The mechanisms behind these ontogenetic processes are partly composed of gradient-forming protein molecules that activate or repress genes to generate specific spatiotemporal patterns of gene expression. The concentration profiles of these gradients and the specific protein–DNA interactions in which these molecules engage are explained by physical–chemical laws, including thermodynamic laws and laws that determine macromolecular interactions in aqueous solutions.

8.4 IS DNA A MASTER MOLECULE?

So far, I have established that genes play a special role in inheritance (see Section 8.1). Although this in itself strongly suggests that genes must also play some special role in *development*, this role remains to be clarified. Could we say, for example, that they somehow "direct" or "orchestrate" development?

In replying to this question, we must first ask what exactly a "directing" or "orchestrating" role for genes would amount to. This is far from clear. What *is* a director or orchestrator? If we take the director of an orchestra as a model, his or her job (in concert) is to synchronize the different voices and to adjust their volume and mood to achieve a harmonious balance of the whole sounding body. The director must have an *idea* in his or her mind as to how the orchestra should sound. This idea takes a central role in the orchestration: what eventually comes out must in some way be a realization of this idea. If the director has no such idea but, instead, randomly intervenes into the playing of some oeuvre, we would certainly not say that he or she had "directed" or "orchestrated" the performance. Thus, the notion of directing or orchestrating is originally highly *intentionally laden*; it signifies more than a causal role. The director in its original sense belongs to the world of ideas and intentions, not to the world of causal relations.

Is there an analogous role in the world of causal relations, something like a causal director? It is not obvious to me what this should be. Things that come to mind include thermostats and automatic pilots. These devices autoregulate some process or a whole complex of processes with the help of some mechanisms that feed the state of the system back into the device. If this is a purely causal director (i.e., one without intentional states), then it is not a good model for the role of DNA. Even though feedback loops with regulatory functions exist in living organisms, for example, in the regulation of metabolism, the role of DNA is different. DNA functions as a template for making protein molecules that may be involved in autoregulatory mechanisms, but this does not exhaust the role of DNA.

As was already mentioned in Section 6.1, there exist a special class of developmental genes called *master control genes* that might be said to orchestrate development.[37] One of them, *Pax6*, activates a cascade of genes that leads to the formation of eyes (Halder et al. 1995; Gehring and Ikeo 1999). This class also includes the homeotic selector genes of the *Antennapedia* and *Bithorax* loci in *Drosophila* that we have already encountered several times in this book (see Sections 6.1 and 8.2). Such genes are referred to as "master control genes" because their protein products (mostly transcription factors; see Section 8.2) affect the expression of a large number of other genes. Of course, their activity is also influenced by many other genes; they sit at some nodal point in a complex web of gene regulatory relations. Perhaps these genes could be said to play a "directing" or "orchestrating" role. The function of their products is to interact specifically with a large number of other molecular structures (mostly gene regulatory regions on DNA). By contrast, most other proteins interact specifically only with a small number of other molecular structures (protein or DNA). However, this difference should not be overdone. First, the difference is purely quantitative. Second, even the protein products of the so-called master control genes can differentiate only between a few molecular structures. Thus, much of the complexity in the action of these proteins lies in the diversity of their regulatory targets. But this means that we should really view the master control genes as parts of complex causal networks that, as a whole, perform a series of regulatory functions (which is, I think, how most biologists see the role of these genes).

These considerations suggest that the idea of DNA as a "master molecule" is highly metaphorical. As an attempt to specify the special role of DNA in development, it was a nonstarter to begin with.

Similar problems beset the idea of a DNA-based developmental program. If "program" is interpreted as a set of instructions, then the same kind of criticism applies: instructions are *intentional*. They have no proper place in the world of causal relations. Comparing DNA to a computer program does not help either, because this concept is also intentional and, therefore, not legitimately applied to a purely causal mechanism (see my critique of Maynard Smith in Section 8.3).

The idea of a genetic program has also been criticized by some molecular biologists, most notably by Gunther Stent (1980, 1985). Stent objects that for something to be a program, there must be an isomorphic relation between the program and the process that it instructs. This is the case for protein synthesis. By contrast, there is no isomorphism between DNA and phenotype. Therefore, according to Stent, DNA contains no program for development. However, Stent's objection may be based on an unnecessarily restrictive notion of a program. I am not sure that a program needs to be isomorphic with the process that it instructs. Therefore, the most severe difficulty with the idea of a program probably remains the intentional load of such notions.

What needs to be examined next is whether there is a sense in which DNA could be said to play a special role in ontogeny on purely *causal* grounds. Consider the following list of causal roles of DNA:

(1) It functions as a template in the synthesis of linear biomolecules.
(2) It determines the linear sequence of RNA molecules.
(3) It determines the linear sequence of protein molecules.
(4) Its linear sequence can be determined either by the linear sequence of another DNA molecule or by the linear sequence of an RNA molecule, but never by the linear sequence of a protein molecule (Crick's Central Dogma).
(5) It is expressed by all-purpose mechanisms (protein synthesis and RNA synthesis) that treat all input states the same.
(6) The relationship between its sequence and the primary structure of its final products (protein) is chemically contingent or arbitrary.[38]
(7) It does not turn over during the life of an individual cell.
(8) It is subject to repair of the primary structure (as opposed to tertiary structure repair as carried out by molecular chaperones).
(9) The copy number per cell is usually one (haploid cell) or two (diploid cell) for each type of nuclear DNA molecule (=chromosome).
(10) It is passed on from each parent as a single molecule-token per type of molecule in gametes.
(11) It is passed on from one cell to another as a single molecule-token per type of molecule in somatic cell divisions.
(12) It is derived from both parents in approximately the same amount.
(13) It is copied only once per cell division (exceptions exist, e.g., in an embryonic syncytium).
(14) It is no chemical catalyst.
(15) A change in a single molecule-token can have heritable phenotypic effects.
(16) The destruction of a single molecule-token can be lethal to a cell.
(17) It is subject to heritable random variation in primary structure.

(18) It interacts specifically with a large number of protein molecules (transcription factors, histones, polymerases, etc.).

Some of these roles are not specific to DNA. For example, RNA shares roles (1)–(6). Roles (9)–(15) are shared by some epigenetic inheritance mechanisms. Role (18) may be shared by other classes of biomolecules, even if the number of *specific* interactants in the cell is probably unmatched for most DNA molecules. But this is merely a quantitative difference. The only causal roles that are possibly unique to DNA are (8), (16), and (17), but we cannot be sure of this at present (some epigenetic mechanisms might have these features).

Causal roles (1)–(3) deserve some special attention. For these causal roles are exerted exactly by those DNA regions that are defined as *genes* according to the molecular gene concept (see Section 7.1). Genes have a special kind of *causal priority* in the synthesis of certain macromolecules.[39] This priority is such that sequence *differences* between certain types of molecules are accounted for by a difference in the corresponding genes, but not vice versa. In the case of an ordinary protein-coding gene, variation in the gene's nucleotide sequence is causally responsible for sequence variation in the protein product, whereas sequence variation in the protein is not causally responsible for sequence variation in the gene. This kind of causal priority can be expressed by the following counterfactual claim: Consider an ordinary protein-coding gene G_1 that determines the linear sequence for protein product P_1 in cellular environment E. Let G_2 be an allele of G_1 and P_2 the corresponding variant of the protein. Now, the following counterfactual is true: If G_1 were replaced by G_2, then P_2 would be synthesized instead of P_1 in cellular environment E. The sequence difference between P_1 and P_2 in E is accounted for by the sequence difference between G_1 and G_2. The converse, however, is not true: If P_1 were to be replaced by P_2, then G_1 would not be replaced by G_2 in E (by virtue of Crick's central dogma).

Counterfactuals are often a good way of establishing causal priority, and this case is no different. The causal priority of genes over other linear biomolecules is clearly something that DS theorists have overlooked in their zeal for making all cellular constituents equal.

At this point, the following objection could be made. Genes, if understood as DNA sequences, have no causal priority over other biomolecules, since at least RNA can play the same role. RNA has the same causal priority over protein, and in some cases, even over DNA. The latter is the case, for example, in retroviruses with an RNA genome (for example, HIV). This objection can be countered as follows: First, yes, RNA sequences have causal priority over amino acid sequences. But in all the cases where the relevant RNA is a messenger (mRNA), there exist DNA sequences that have causal priority over this RNA. In these cases, RNA is just an intermediary. Second, those cases where RNA has causal priority over DNA (e.g., in the life-cycles of retroviruses) are exactly those cases where RNA will be said to contain *genes*.[40] Thus, it remains the case that genes are distinguished by their causal priority in the synthesis of linear biomolecules.

What my list also shows is that DNA has a unique *combination* of causal roles in the cell, even if for each causal role that it has there is some other cellular structure that can play it as well (although not necessarily in the same way). Clearly, no other cellular component has all the eighteen causal roles that I have listed.[41]

Thus, my analysis so far establishes that DNA has a unique combination of causal roles and that genes have a special kind of causal priority over other biomolecules. We can now ask whether this is not enough to refute DST's causal parity thesis.

At this point, DST could be defended as follows.[42] DST's thesis of causal parity, surely, does not mean that the role of DNA differs in no way whatsoever from that of other cellular components. What DST challenges is that there is some *categorical* difference in the role of DNA; something like informational content or a programming, directing, or orchestrating role.

In replying to this defense, it is necessary to distinguish between a strong and a weak version of the parity thesis. The *strong* version says that DNA and genes play no causal role that sets them apart from other developmental systems components. In other words, there is no causal role difference whatsoever, according to the strong thesis. The *weak* version says only that there is no *categorical* difference, that is, even though there are some differences in causal roles between DNA and genes, the latter do not belong to a separate category of developmental causes.[43]

What is clear from my analysis is that the strong causal parity thesis is false. DNA has a unique combination of causal properties that no other developmental system component possesses. In addition, in a sense that I have defined, genes have causal priority in the synthesis of linear biomolecules. These features distinguish DNA and genes from all other developmental systems components.

The weak causal parity thesis is more difficult to assess, since it is not obvious what a categorical difference in causal role would be. Being the only carrier of genetic information, presumably, would qualify as a categorical difference; however, we have seen that this notion (genetic information) is indeed ridden with difficulties, at least if understood in one of the two conventional senses of "information" (the so-called causal and intentional information concepts). Concerning my list of eighteen causal roles, DS theorists could say that this is all very well, but that none of this qualifies as a *categorical* difference in causal role, nor does the fact that DNA has a unique combination of causal roles.

This reply raises the question of what could, in principle, constitute a categorical difference in the causal role of DNA in development. In order to answer this question, it must be asked whether causes are the sorts of things that come in different categories. This is a difficult metaphysical question. Causes may be categorized into physical and mental, deterministic and probabilistic, local and nonlocal, and perhaps other categories of causes. But none of these seems to be the sort of difference that we expect in a developmental system. Such a system is purely physical; its causal interactions are governed by the laws of physics and chemistry (compare Chapter 2). The causal interactions are, so far as we know, deterministic.[44] They are almost certainly local. So why would we expect there to be a different category of causes in biological systems?[45]

As John Mackie has shown, all causes are insufficient but nonredundant parts of unnecessary but sufficient complexes of conditions (Mackie 1980, 62). Mackie called them "INUS-conditions" (from "*i*nsufficient but *n*ecessary part of an *u*nnecessary but *s*ufficient condition"). The glowing match in the trash receptacle was the cause of the office fire, even though inflammable materials, oxygen, and a dry environment were also necessary conditions. Thus, the match was only a sufficient cause in conjunction with the inflammable materials, the oxygen, and the dry environment. However, this complex of conditions was not necessary, as an unsafe electrical device might also have started the fire.

Mackie's analysis can be applied to causation in any field of inquiry, including even the law. Thus, we are dealing here with some very general metaphysical properties of causation.

Applying this analysis to biological systems, we can note that all the components of a developmental system make their causal contribution as INUS-conditions (compare Gannett 1999). There is no factor that can exert its effects all by itself. DNA is no different in this respect from all the other things that make a cell come alive. But if all causes have this property, what other category of causes could there be that could make the weak causal parity thesis false?

This question is puzzling, suggesting that what DST attacks was a vague idea to begin with; namely, the idea that development is a directed process like an orchestral performance. But metaphysical considerations such as the ones that I have just presented show that there are no analogs to a director in the world of causal interactions. Causal networks may differ in connectivity, but this difference is merely quantitative. Even if DNA as part of the causal network of a developing organism has the highest number of causal connections with specific interactants (which it probably does), this is a quantitative difference, not one that sets it categorically apart from other interactants. Furthermore, the causal priority of genes over other biomolecules that I have argued above does not constitute a categorical difference either. For any cause may have priority over another set of causes.[46]

Such considerations suggest that the weak causal parity thesis is true. However, DS theorists have accepted it for the wrong reasons. For the weak parity thesis is a consequence of any adequate metaphysical theory of causation. The epistemic norms of modern science require that the developmental process be explained in a reductionistic way. To give a reductionistic explanation means to analyze the causal roles that various components of a system play with the help of principles from physics and chemistry (see Chapter 2). A categorially distinct kind of causes is not to be expected within such a reductionistic framework. Therefore, although the strong parity thesis is plainly false, the weak version of DST's causal parity thesis is sort of built in in modern science. It is thus not a result of empirical research in developmental biology.

To conclude, let us go back to the notion of a DNA-based genetic program mentioned at the beginning. Both kinds of objections to this notion – the intentionality objection and DST – have their merits. The genetic program *is* a metaphor, and so is the idea that DNA has informational content. The same is true if informational content is attributed to other developmental systems components such as morphogenetic gradients. From a truly reductionistic standpoint, explaining ontogeny in developmental biology is a matter of describing the causal interactions between the various components of the developmental system and of showing how these interactions generate a species' characteristic patterns. This is what modern developmental biologists do; they just express it differently sometimes.

Once this is accepted, the question arises of how the role of DNA and genes in development is conceptualized correctly. I have argued that the idea of a "directing" or "orchestrating" role is a nonstarter; these notions make no sense in the world of causal relations. Thus, DS theorists are right to reject these notions. However, they reject them for the wrong reasons: namely, for *empirical* reasons. But the correct reasons for rejecting this are *metaphysical* in nature; they are prior to empirical biological knowledge. The weak causal parity thesis, as I have called it, is a constraint imposed by the underlying metaphysics of modern science.

The strong causal parity thesis, by contrast, is empirical, but false. What empirical research has shown is that DNA and genes play quite a special role in developmental systems: they exhibit a unique combination of causal roles as well as causal priority in determining the sequence of other linear biomolecules. This explains at the same time why DNA is the hereditary replicator with the greatest range of potential phenotypic effects and why biologists spend so much time studying it.

SUMMARY

The idea that DNA contains a genetic program or genetic information that instructs ontogenetic processes faces two main difficulties, namely (1) the intentionality implied by these notions and (2) the possibly undue privileging of DNA with respect to other factors that are necessary for development. Concerning difficulty (1), the intentional aspect is not easily accommodated by the basic reductionist explanatory approach that characterizes other areas in modern experimental biology. With respect to difficulty (2), developmental systems theory (DST) claims that any account of genetic information will fail to bring out a sense in which DNA plays a privileged role in development. Furthermore, there is no other way of attributing a special role to genes and/or DNA according to DST. Environmental factors, as well as non-nucleic-acid structures inherited by a cell, are claimed to be just as important.

I have criticized developmental systems thinkers for an imprecise use of well-defined biological concepts, in particular the concepts of inheritance, epigenetic inheritance, and replication. For example, epigenetic inheritance and the cultural transmission of traits must be conceptually distinguished from both genetic inheritance and the mere passing on and copying of intracellular structures such as membranes, peroxisomes, or centrioles. These distinctions have been hard earned in the history of biology. Nothing is gained by tearing them down.

Furthermore, I have argued against DST that DNA is a hereditary replicator with a far greater potential to cause phenotypic differences than other replicators. With the exception of a few epigenetic inheritance mechanisms, all developmental systems components other than DNA known to current biology are not *hereditary* replicators (for example, peroxisomes and other membranous structures, or morphogen gradients). A review of what is known today about epigenetic mechanisms (genomic imprinting, histone modification) provides no evidence for anything like the enormous phenotypic plasticity that can be caused by DNA sequence variation. There may be environmental factors that can have effects as drastic as those of DNA-based genetic variation; however, they are not heritable. Thus, DS theorists have unduly exaggerated the potential effects of nongenetic factors in heredity. However, this does not imply that there exist no traits that depend more strongly on nongenetic than on genetic factors, for example, certain behavioral or psychological dispositions in humans. Acknowledging a special role of DNA in inheritance requires no commitment to psychological genetic determinism.

In order to better understand how contemporary biology explains developmental processes, I have examined a case study featuring the well-understood molecular mechanisms that generate the segmented body pattern of *Drosophila*. Standard accounts of these mechanisms raise additional conceptual problems, especially by introducing yet another form of information, namely positional information. Positional information is said to reside in the concentration gradients of a special class of molecules called morphogens. I have shown that the notion of positional information only makes sense if development is viewed as a goal-directed process. Nonetheless, explanations involving positional information can be rendered as a species of functional explanation, where function is applied in its causal role sense as a subsystem capacity that contributes to an overall capacity of the entire system (in this case the capacity for pattern formation). Thus, morphogen gradients in an embryo do not carry information in an intentional sense; all they carry is causal role functions.

Furthermore, I have argued that all intentional or semantic information concepts in biology are problematic. The teleosemantic account of information has fatal defects. Semantic information in biology is at best an analogy, even if an important and heuristically fruitful one. Thus, DS theorists are right that informational properties do not privilege genes or DNA (because the latter do not really have informational properties, nor does any other biological structure).

I have then shown that DST's causal parity thesis (the claim that the causal role of DNA is on a par with those of other developmental resources) is conceptually independent from its claims about genetic information. Thus, even if the claim that the concept of genetic information cannot be clarified in a way that accords a special role to DNA in development is correct (which it is), the claim that there is no difference in causal role could be false. Thus, the thesis of causal parity must be assessed by its own merits.

In order to do so, I have distinguished between two senses of causal parity: First, is there *any* causal role specific to DNA/genes? To deny that there is one is to endorse a strong version of DST's causal parity thesis. Second, is the causal role of DNA/genes *categorically* different? To answer this question in the negative amounts to a weak version of the causal parity thesis. By reviewing some of the known causal roles that DNA play in developmental processes, I have refuted the strong causal parity thesis. There is a combination of causal roles that is clearly unique to DNA. Furthermore, genes exhibit a special kind of causal priority in the synthesis of biomolecules. As far as the weak parity thesis is concerned, I have argued that it is a consequence of a metaphysically adequate theory of causation that is not specific to biology. From a thoroughly reductionist point of view, we would not expect a special category of causes in a purely causal mechanism. Therefore, DS theorists rightly reject the notion that DNA directs or programs development, but they do so for the wrong reasons.

9

Scientific Realism: In Search of the Truth

Scientific realists believe that highly confirmed scientific theories and explanations should be accepted as at least approximately true. The concept of truth, in the view of most contemporary realists, should be interpreted as some kind of correspondence relation between the scientific statements that are said to be true and the states of affairs obtaining in a mind-independent, objective reality. Thus, realists take well-tested scientific theories (models, explanations) to be representations of a reality that exists independent of the categories and concepts of the human mind. They must therefore defend a combination of metaphysical and epistemological theses. The metaphysical thesis says that there is a mind-independent reality, and that it is somehow structured independent of scientists' concepts. The epistemological thesis is that the human mind has access to this reality.[1]

Today, scientific realism has to face three different kinds of antirealist opponents. First, *instrumentalists* hold that the pragmatic purpose of scientific theories is not to represent a mind-independent world, but to allow us to orient in and interact with the world in which we find ourselves, for example, by predicting future experiences. In this spirit, Bas van Fraassen (1980) has defended the view that to accept a scientific theory means to accept it as *empirically adequate*, not as true (this claim is normative, not descriptive of actual scientific practice). For somewhat different reasons, Alex Rosenberg (who is a realist about physical theories) has developed an instrumentalist account of biology (Rosenberg 1994).

Second, the highly influential historian of science Thomas Kuhn (1970) has developed an account of scientific change according to which science does not progress toward truth. According to Kuhn, convergence on truth is impossible, as the world that scientists describe in their theories does not exist independent of the scientists' beliefs. Instead, it is always co-constituted by subject-sided and object-sided moments (Hoyningen-Huene 1993,

Chapter 2). On this view, which may be described as "Neo-Kantian," there *is* a mind-independent ("noumenal") world and it does have some impact on our knowledge. However, the mind makes a contribution to the world we experience, and it is not possible to subtract out this contribution in order to know something about the mind-independent world. Thus, the world that science describes is a *phenomenal* world.[2] By contrast, scientific realism takes the world that mature scientific theories represent to be the noumenal, mind-independent world; therefore, Kuhn must be classified as an antirealist.

Third, sociologists of science have argued that scientific knowledge is socially constructed. What this means is that the theoretical choices that scientists make are constrained by their interests, social networks, local laboratory settings, etc., but not by nature (e.g., Latour and Woolgar 1979; Latour 1987; Knorr Cetina 1981, 1999; Shapin and Schaffer 1985; Pickering 1999). On this view, there is no reason to believe that scientific theories represent mind-independent facts. Indeed, for the social constructivist, facts are *made*, not discovered. Thus, instrumentalists, Kuhnians, and social constructivists – a rather mixed lot in many respects[3] – are united in rejecting the main claims of scientific realism as presented above.

It will not be possible, in the final chapter of this book, to do full justice to these three positions opposed to realism and the reasoning that led to them. Furthermore, for reasons that become clear in Section 9.1, it cannot be my goal here to provide a realist rebuttal to these three positions. The goal of this chapter is more modest: I want to discuss some strengths and weaknesses of different arguments *for* scientific realism with special attention to experimental biology. Thus, I hope to strengthen the case for scientific realism indirectly, rather than giving a full-fledged defense, which would require a rebuttal to all the different antirealist positions.[4]

In Section 9.1, I determine the relationship between scientific realism in experimental biology and some more general metaphysical issues. Furthermore, I discuss the question of what we could sensibly be realists about in the context of experimental biology. Next, I critically review three different strategies that have been pursued in defense of scientific realism. Section 9.2 treats the so-called miracle argument (sometimes referred to as the "no miracles argument"). Section 9.3 discusses the experimentalist argument, and Section 9.4 the argument from independent determinations. In Section 9.5, I discuss a case study from the history of microbiology that features a case where a biological structure that appeared in electron microscopes was shown not to be real, that is, to be an experimental artifact. This case harbors some important lessons for the aspiring realist.

9.1   REALISM ABOUT WHAT?

What can and should we be realists about in the context of experimental biology? Any answer this question must begin by taking into consideration the main objects of experimental biology, namely organisms.

Living organisms are processes in space and time. As such, they can only be real to the extent that space and time are. Some classical philosophers have rejected the reality of space and time. Most famously, Kant viewed them as forms of intuition (*Anschauungsformen*) that the human understanding uses to organize its sensory data.[5] According to Kant, we are unable to have experience of any material objects outside of space and time, but space and time do not exist independent of us. What there is in the outside world independent of the human understanding is unknowable in principle.

Therefore for a Kantian, a living organism, insofar as it is a process in space and time and thus accessible to our senses, must be an appearance that is in an important sense *created* by the human understanding. On such a view, it is not possible to be a realist about biology, if realism is understood in the sense that I have alluded to in the beginning. However, the reasons have nothing to do with biology specifically. A Kantian cannot take *any* material object as real in an absolute sense (though Kant himself would accept them as *empirically real*), and this includes organisms, since they are material objects.[6]

There is nothing that I can do here to counter this kind of antirealism. Of course, Kant's philosophy of material nature has its own difficulties and, therefore, ought to be ultimately rejected. But even if Kant is rejected, it remains true that realism about biology is dependent upon realism about objects in space and time in general. To defend such a form of realism is clearly beyond the scope of this book. This topic belongs to the philosophy of physics. Thus, the realism that I defend is *conditional*. It holds only under the condition that realism about material objects in space and time in general is justifiable.

There is one objection that could be made to my strategy of conditionalizing realism in biology on the reality of space and time. Is it not, an objector might ask, a *prerequisite* for there being a phenomenal world (in Kant's sense) that there be real organisms? If the world as we experience it is in some sense a construction of the human mind, who or what is doing the constructing? Surely, it is *organisms* that construct worlds, even if it is true that space and time are constructs of the human understanding. This could provide an argument for the reality of organisms. In other words, we might have here a lever for arguing for realism in biology that is not accessible if we focus on physics alone. It might actually turn out that the realism issue in biology is actually more fundamental than the realism issue in physics after all.

I do not want to exclude the possibility that this is actually the case, nor that such a strategy might succeed in countering philosophical doubts concerning the reality of material objects. However, the issue is far beyond the scope of this book. It is a more a general problem of metaphysics, not so much of the philosophy of experimental biology.

But now we must confront the following impasse. If, as I have argued, realism about biology presupposes realism about space and time, and if the possible biological foundations of realism are something that I do not want to discuss here, what issues are left? In other words, is there any substantial realism issue about experimental biology *as such*, a problem that is independent of the reality of material objects in general, or the reality of space and time?[7] If there is no such issue, we are done with discussing scientific realism in experimental biology.

However, I think that there is indeed such an issue. It has to do with something that philosophers of science call *theoretical entities*. These are things that play an important role in explaining natural phenomena, but that are unobservable to the unaided senses. In physics, theoretical entities include things such as atoms, elementary particles, fields, and forces. In biology, an important class of theoretical entities is genes. Originally, genes were introduced to explain certain observed inheritance patterns. Their material substrate was not known. Later, genes were identified as DNA segments (see Chapter 7). Even though DNA can be visualized in electron microscopy (even in an active state, i.e., while being expressed), genes remain nevertheless theoretical entities. For what we see under an electron microscope is always a *theoretical interpretation* of some raw data.[8] Raw data in electron microscopy are provided by the image in the ocular or on a photographic film. Biologists need a lot of theoretical knowledge to ensure that an image represents what they think it does. Whether we can really *see* through a microscope is controversial (see Hacking 1985), and it is especially debatable in the case of structures that are smaller than the wavelength of visible light.[9]

For these reasons, genes remain theoretical entities even though we have electron micrographic images of them.[10] This does not change if we consider them as chemical entities, that is, as stretches of DNA molecules. First, chemical entities (molecules) are also unobservable. Second, even if they have the specific nucleotide sequence of a stretch of DNA in front of them, biologists cannot be absolutely sure that it contains a gene. There are still no absolutely reliable methods of predicting genes from DNA sequences alone. Thus, genes are not directly observable; they are theoretical entities.[11]

269

Other theoretical entities that are important in experimental biology include membranes, organelles (e.g., mitochondria), macromolecules like proteins, complex aggregates of protein and nucleic acid like ribosomes or chromatin, complexes of protein and polysaccharides, chemiosmotic gradients (see Chapter 4), and morphogen gradients (see Chapter 8). These entities and structures are too small to be viewed with our unaided senses. Furthermore, some of them are extremely fragile and thus difficult to isolate. Complex preparation procedures are necessary to view some of them by light or electron microscopy (see Section 9.5).

The history of biology shows several instances in which some theoretical entity that was thought to exist later turned out to be an artifact of some experimental procedure. I discuss an example in Section 9.5. For now, it is enough to note that biologists deal with entities and structures that cannot be observed directly and whose existence can therefore be subject to doubt. Such doubts would prevail even if we had no doubts concerning the reality of material objects in general. In other words, even if I accepted that material objects in time and space are real, we can have reasonable doubts as to the existence of some biological entities.

These considerations show that there *is* a philosophically interesting and substantial realism issue in experimental biology: a problem that is independent of the more fundamental question of whether space and time and other physical entities are real. It is this problem that the present chapter is about. If we suspend philosophical doubts concerning the reality of material objects in general, what reasons can we have for thinking that some entity or structure observed in biological experiments actually exists? In other words, what could give us confidence that the tiny structures and mechanisms described by biologists are as the scientists say they are? In the following three sections, I discuss the different kinds of arguments that could be used to establish this in the subsequent sections.

A further question concerns what kinds of things a realist in experimental biology might want to be a realist about. So far, I have mentioned realism about theoretical entities such as genes and very small biological structures such as organelles or chromosomes, and a realist in experimental biology might certainly want to be a realist about these. But there are other things that come to mind, in particular theories, laws, natural kinds, functions, and information. I discuss each of these in turn.

Realism about theories is the standard brand of realism. It claims that well-confirmed theories should be accepted as true or as approximately true.[12] The origins of this debate lie in the philosophy of physics, where it is comparatively clear what a theory is (for example, Newton's three laws of mechanics constitute a physical theory). By contrast, as we have seen in Section 2.2, it is not quite as clear what a theory is in experimental biology. At any rate, they are very different from physical theories (with the possible exception of classical genetics). But no harm will be done if, for the purposes of this discussion, we refer to any *descriptions of biological processes and mechanisms* as theories. The purpose of this stipulation is to ensure that our discussion here latches onto the more general realism debates in the philosophy of science. Realism about such theories, then, will consist in the claim that they are *true* descriptions of the biological processes and mechanisms that they describe. Alternatively, they may be said to be approximately true.[13]

Some philosophers are realists about theoretical entities, but antirealists with respect to theories. I discuss this position in detail in Section 9.3.

Realism about laws combines two claims. First, statements describing laws of nature *can* be true (whether or not we can be sure about their truth). Second, what *makes* law statements true is some state of affairs in an objective reality. The first claim has been challenged, for example, by Nancy Cartwright (1983). She has argued that law statements in physics must be hedged by so-called *ceteris paribus* clauses that state the absence of disturbing factors (cf. Earman and Roberts 1999). Since not all the possible disturbing factors can be listed, no law statement can be strictly true. The laws of physics "lie" according to Cartwright. Concerning the second claim, there are several options as to how it can be developed. For example, Dretske (1977), Armstrong (1983), and Tooley (1987) have argued that laws are relations between universals. Since they hold universals to be real, they can also be realists about relations between universals, and hence realists about laws.

Another possibility is to view laws, not as relations between universals, but as powers or dispositions that are grounded in the essential properties of things (Ellis 2001). A comparison of what the two positions have to say about the necessity of laws best brings about the main difference between this view and the universals account of laws. For the universals account, laws are contingent parts of the world. This means that there might be possible worlds with exactly the same kinds of things with the same kinds of properties, but that are governed by different laws of nature. By contrast, on the essentialist account of laws advanced by Brian Ellis and some others, a possible world with different laws of nature would also have to be a possible world with different sorts of things and properties. For example, a world in which there are electrically charged objects with the properties they have in our world is also a world in which Coulomb's law holds. Conversely, if a law different from Coulomb's governed electrostatic interactions, then the property of being charged would have to be different.

271

I take it that Machamer et al. (2000), who think that laws are not necessary in biological explanations (see Section 2.3), identify the concept of a law of nature with the universals account, since what they refer to as "activities" are highly similar to what Ellis calls powers or dispositions.

It is difficult to adjudicate between these two realist accounts of laws of nature. Fortunately, for the purposes of our discussion, it does not matter which one is preferred. For I have argued in Chapter 2 that all genuine laws of nature that play a role in experimental biology are really laws of physics and chemistry (which is not the case in population biology; see Weber 1999a). Thus, the problem of realism about laws properly belongs in the philosophy of physics and chemistry, and the philosophy of population biology.

Realism about natural kinds, the question of whether there are genuinely biological natural kinds, is an old one. For Aristotle, biological species were the paradigms of natural kinds. Since then, biologists and philosophers of biology have argued for centuries about whether or not biological species are real (Ereshefsky 1992). Ernst Mayr (1982) has argued at length that species are real, but that they have no essences. This implies that species are not natural *kinds*, a view that is widely accepted today. But if species are not natural kinds, what other biological natural kinds could there be? In Chapter 7, I have referred to certain kinds in classical genetics as natural kinds. Some of the objects of genetics (genes, chromosomes, crossing-over events, etc.) often share a specific causal role that may be viewed as constitutive for natural kinds.

At this point, a difficulty must be noted. I have argued in Chapter 2 that there are no autonomous biological laws, at least not in those areas of biology that are the subject of this book. There is some tension between this claim and my thesis that there are higher-level natural kinds such as genes, chromosomes, and so on. The reason is that laws and natural kinds are thought to be closely related. Namely, it is often argued that natural kinds are exactly those kinds of things that are bound together by laws of nature (e.g., Fodor 1974). Thus, by claiming that there are higher-level biological natural kinds, am I not committed to there also being biological laws?

The way out of this conundrum will be to allow for the possibility that not all natural kinds are alike (see Waters 1998, 2000). Specifically, in experimental biology, we must recognize *functional* kinds such as genes, chromosomes, and classes of enzymes, as well as *kinds of processes* such as transcription, endocytosis, and DNA replication. I see no reason that functional kinds and kinds of processes should be denied the status of natural kinds, as long as it is understood that these natural kinds differ from the natural kinds found in physics and chemistry. Brian Ellis (2002, 27f.) differentiates between *fixed* and *variable* natural kinds and argues that the former are only found in physics and chemistry. In Section 2.3, I have basically accepted this view and exhibited its strong relation to John Beatty's evolutionary contingency thesis. Electrons and the element calcium are examples of physicochemical natural kinds, and it would be foolish to think that there are natural kinds of the same sort in biology. In particular, I do not see why there should be a law of nature for each and every natural kind. To give an example, chromosomes can be viewed as belonging to a biological natural kind. They form a natural kind because there are some robust generalizations about them. For example, they contain either DNA or RNA, they are replicated by the cell, they mutate when irradiated, and so on. However, these generalizations either are not laws of nature in the strict sense or are physical or chemical laws (Beatty's evolutionary contingency thesis, discussed in Section 2.3). In the latter case, the laws range over physical or chemical natural kinds, that is, fixed natural kinds. In the former case, the kinds are biological, and hence variable. The generalizations in which they are involved lack nomic necessity. But I see no reason to deny them the status of natural kinds.

Realism about biological functions is the question of whether biological functions should be regarded as real. Clearly, when biologists attribute a function or a set of functions to some biological entity (for example, the mitochondrion), they are implying that the function is really there, independent of their recognizing it as a function. Whether such a form of realism is justifiable will depend on what is meant by a biological function. I have adopted a modified version of Cummins's causal role function as the relevant sense of function for experimental biology (see Section 2.4). This account poses some difficulties for a realist. This is because, as we have seen, causal role functions are individuated by the contribution that some item makes to an overall system capacity. The problem is that the choice of the relevant system capacity is to some extent up to us. I have proposed to choose self-reproduction as the ultimate system capacity with respect to which all other functions are picked out, because self-reproduction (i.e., the maintenance of an organism's form) is the most conspicuous feature of living organisms that we want to understand. But it must be admitted that this may be a mere predilection of the human understanding, which arises because we are organisms ourselves. Individual organisms are somewhat fuzzy processes. But the ontology of living beings is something that would need to be worked out further, which I cannot do here. For the time being, realism about functions must be tempered.

Realism about information asserts that information is a basic ontological category. In other words, it is possible to claim that information would be in the world, even if no one were to interpret it. Presumably, the genetic information that is said to be encoded in the nucleotide sequences of DNA was there before intelligent human beings evolved to discover it. However, I have shown in Chapter 8 that the information concept in biology is problematic. Even though there is an unproblematic sense of information, namely the pragmatic information concept, this sense of information is not open to a realist interpretation. In this weak sense, anything can be information for anything, provided that there is some reliable covariation between the states of the information source and the signal. However, this kind of information is not in the world. It is always a three-place relation between a cognitive agent, the information source, and the signal. Thus, a realist interpretation of this kind of information would be mistaken.

Concerning the stronger, intentional sense of information, I have argued that it is not legitimately attributed to biological entities (be it DNA sequences, morphogen gradients, or whatever). Thus, realism about information in this sense is not warranted in biology either.

Having thus identified the legitimate objects of scientific realism in experimental biology, I now turn to the question of how such realism could be justified. I shall concentrate on realism about biological theories (in the sense outlined above) and theoretical entities. Realism about laws, natural kinds, and functions would probably have to be defended in a similar way; at least I am not aware of any alternatives.

9.2   THE "MIRACLE ARGUMENT"

The most widely discussed argument for scientific realism is known as the "miracle argument" (or the "no miracles argument"). The basic idea behind this argument is that if realism were false, then it would be a miracle if science were successful. Since there are no miracles and science *is* successful, realism must be true. More formal presentations of this argument usually reconstruct it as a so-called inference to the best explanation or IBE (also known as "abduction"). According to this inference rule, it is justifiable to infer from a statement that describes some phenomenon to another statement that provides the best explanation for this phenomenon, even where such an inference is not permissible according to the rules of deductive logic. According to the "miracle" argument, the best explanation for the success of scientific theories is their approximate truth; therefore, we can infer the approximate truth of a scientific theory from its empirical success.[14]

There are various problems with this strategy of arguing for realism. First, Bas van Fraassen (1989, 142–143) has argued that IBE is unsound as an inference rule. For the best explanation available for some given phenomenon at some given time could still be a very, very bad one. It could be the "best of a bad lot." Thus, we cannot generally have reasons to believe that the best explanation is the true one. Second, it has been argued that, if used to defend scientific realism, IBE arguments are circular (e.g., Laudan 1984; Fine 1984; Lipton 1991, 158ff.; Oberheim and Hoyningen-Huene 1997). According to this criticism, to presuppose the soundness of IBE (as realists must in order to argue for realism) begs the question for the antirealist, as the soundness of IBE as a basis for the "miracle" argument is exactly what is up for debate in the first place. Third, the "miracle" argument has been questioned on the ground that there have been scientific theories that were once empirically successful but nevertheless turned out to be false (Laudan 1984; Carrier 1991). There were even theories with nonreferring terms[15] that enjoyed spectacular empirical successes, for example, ether theories of light propagation or thermodynamic theories that viewed heat as a substance (caloric). But if false theories whose terms did not even refer can still make successful novel predictions, then the central premise of the miracle argument is undermined. Empirical success comes cheap. A "cosmic coincidence" is no longer required to explain it.

The abundance of successful historical theories that were later shown to be false has also been used to challenge scientific realism directly, by means of a so-called pessimistic induction (Laudan 1984). In this antirealist move, the past failures of scientific theories are used as an induction base to infer that our best current theories are also likely to fail. There is some irony in this move, as antirealists are not typically the kind of people who would rest their bets on inductivism as a sound methodology. Nevertheless, this is an important challenge that must be met by realists (see Psillos 1996, 1999 for an attempt).

The difficulties associated with past successful theories that were later refuted seem to be the ones that have worried scientific realists the most. Various authors have tried to solve this problem by arguing that if a theory later shown to be false is successful, then the defective parts of the theory cannot have been involved in the prediction (e.g., Kitcher 1993, 141–149; Psillos 1994; Leplin 1997, 146–151). However, this strategy suffers from the difficulty that both the identification of defective parts of theories *and* the identification of those parts that are responsible for their empirical success rely on assuming present-day scientific theories to be true (Stanford, 2003).

Another approach to making the "miracle" argument good is reducing the burden it has to bear. For example, Martin Carrier (1993) argues that, in order to be empirically successful, theories must at least classify the phenomena correctly. Therefore, the success of a theory shows that it correctly classifies the phenomena. This is a much weaker form of realism than the one that is under consideration here.

Additional difficulties loom when we ask what exactly the success of scientific theories consists in. Clearly, it would be a mistake to mean by "success" the theoretical success of a science, that is, its success in producing true theories, for this would beg the question for the antirealist. The most serious contender for scientific success is *predictive* success, that is, a theory's ability to make accurate predictions. Thus, the success in question is a theory's ability to predict possible *observations* that can then verified.

But prediction *simpliciter* is not enough. Many authors have argued that a scientific prediction must not predict just any experimentally verified fact if it is to carry any relevance for scientific realism (see Leplin 1997, Chapter 2 for a review). Obviously, if a theory was deliberately tailored to accommodate some observable fact, then the prediction of that fact by this theory is no miracle. Consequently, it requires no further explanation and the "miracle" argument for realism has no force. Only if a theory successfully predicts *novel* facts can the realist legitimately request an explanation for this success. A theory, as it were, must see further than its inventor – only then does its success ask for an explanation.

The basic intuition behind this argument is this. Take a theory such as general relativity theory (GTR). This theory allowed Einstein to predict the deflection of starlight in the sun's gravitational field, which can be observed during solar eclipses. This prediction was in no way built into GTR from the outset. Now, if GTR was false, how could it have gotten some facts right that its author did not build into the theory? Surely, it would be a cosmic coincidence if a false theory correctly predicted the numerical value for the deflection of light near a large mass.[16] Anything but the approximate truth of general relativity theory would make this prediction a miracle. Hence, the theory must be approximately true.

Even though this argument has a strong intuitive appeal, there are some further difficulties with this suggestion, as it turns out to be difficult to spell out what exactly "novel" should be taken to mean.[17] Most philosophers of science agree that it cannot mean that a fact was *unknown* at the time the theoretical prediction was made. For example, the anomalous (from the standpoint of Newtonian theory) precession of Mercury was known before Einstein invented the general relativity theory, yet its success in explaining this known fact was viewed as strong evidence for the theory. Surely, the temporal order of the prediction and the observation cannot be of any epistemological importance. In order to fix this problem, some people have suggested use-novelty as the salient sense of novelty. This means that a fact must not have been *used* in the construction of a theory, should a prediction of this fact by the theory furnish any kind of warrant for its truth. There are some difficulties with this suggestion, too, but it might be possible to fix them (see Leplin 1997, Chapters 2, 3).

Can we apply such a criterion of novelty in predictions in experimental biology? At first sight, this avenue seems promising, as there are positive examples. Take, for example, Peter Mitchell's prediction that uncouplers of oxidative phosphorylation affect the proton conductivity of membranes (see Section 4.2). No biochemist would have expected this to be true on the basis of the rival chemical theory alone. And indeed, this prediction of Mitchell's was viewed as strong evidence for the theory. Another example is Watson and Crick's famous 1953 prediction of a semiconservative replication mechanism for DNA, which was confirmed by Meselson and Stahl about four years later (Holmes 2001). These predictions seem no less spectacular than the famous predictions of Einstein and other physicists. Thus, a realist might want to use these successful novel predictions to argue for scientific realism about the theories that predicted these facts (i.e., the chemiosmotic theory and the Watson–Crick model of DNA, respectively).

There are, however, several difficulties with such a strategy. First, novel predictions seem to be an exception rather than the rule in experimental biology. A systematic search for them in the history of the field has yet to be made, but I am not aware of too many good examples. If this were true, it would limit scientific realism to just the few examples where novel predictions have been made by biologists. This is a consequence that some scientific realists might be willing to swallow (Eric Oberheim, personal communication), but most scientists will probably want to be realists about a larger part of modern biology.

A second difficulty arises from the fact that predictions in experimental biology are rarely, if ever, quantitative. But the prediction of a novel *qualitative* fact takes less of a miracle to explain in nonrealist terms. This is particularly obvious in the case of semi-conservative replication. All biochemists at the time agreed that there are basically three possibilities as to how double-stranded DNA could replicate: conservative, semiconservative, and dispersive (Holmes 2001, 419). Thus, it might be argued that Watson and Crick had a one-in-three chance to get it right. But this requires no cosmic coincidence of the kind that scientific realists invoke in order to proffer their preferred explanation of predictive success.[18]

In the case of Mitchell's prediction concerning the effect of uncouplers on the proton conductance, this seems less likely to have been a coincidence, even though the prediction itself was qualitative. But nevertheless, qualitative predictions of false theories seem to require much less of a cosmic coincidence than quantitative predictions with a high degree of precision.[19]

These considerations show that the miracle argument is at best of limited value for defending scientific realism in experimental biology. Therefore, I move on to another strategy.

9.3 THE EXPERIMENTALIST ARGUMENT

Ian Hacking has attempted to develop a different approach for defending realism (Hacking 1983, Chapter 16). His argument is intended to establish only realism about *entities*, not about theories. Thus, according to Hacking, we are justified in accepting the existence of things like electrons, but not the truth (not even approximate truth) of theories about electrons (compare Cartwright 1983). Hacking's position vis-à-vis theoretical entities is encapsulated in his claim that "if you can spray them, then they are real" (1983, 23). The idea is that theoretical entities are not merely the subject of scientific theorizing. After a while, they may also become "tools" (1983, 263). According to Hacking, this means that the causal properties of some entities like electrons are used in scientific practice to *do* things like experimentally inducing certain measurable effects.[20] Presumably, this would not be possible if these entities and their causal properties used for such experiments did not exit. Hacking writes: "We are completely convinced of the reality of electrons when we regularly set out to build – and often enough succeed in building – new kinds of device that use various well-understood causal properties of electrons to interfere in other more hypothetical parts of nature" (1983, 265).

Thus, it is not enough for Hacking if a theoretical entity is used as a part of a theory to make successful predictions. Only once the causal powers of these entities are also used to intervene with physical processes are we justified in accepting them as real.

The main problem with Hacking's experimentalist argument is to show that it is not just the "miracle argument" in new guises. What exactly is the difference between predictive success according to the miracle argument and instrumental success according to the experimentalist argument? When a scientist uses knowledge about a theoretical entity to build a new apparatus, does he or she not simply rely on the predictive power of the theory that postulates this entity? After all, scientists cannot "spray" electrons; they can only manipulate some visible apparatus about which they *think* that it sprays electrons.

Hacking sees a difference here. For him, scientists do not need a *complete* theory about some entity such as the electron in order to use it in experimental practice. All they have to rely on is knowledge about some "well-understood low-level causal properties" (1983, 274). These are the properties that are deployed in generating experimental effects. Thus, it is Hacking's contention that if such knowledge of low-level causal properties gives rise to successful experimental deployment of the entities in question, this justifies a kind of confidence in the existence of these entities that goes beyond simple predictive success. However, the kind of realism supported by this argument is only entity realism, not theory realism.

Unfortunately, Hacking does not give us any criteria for distinguishing between low-level causal properties that are sufficient for using the entity for manipulation and high-level properties that require a full-blown theory. In the main example Hacking uses to illustrate his point, physicists used electrons to detect a certain kind of particle decay: namely, so-called neutral currents (events that involve the weak force). The theoretical physicists Steven Weinberg and Abdus Salam had predicted that such currents should violate the principle of parity, which states that equal numbers of left- and right-handed particles must be produced. One way of detecting this is by measuring the polarization of scattered electrons. In the 1970s, scientists at the Stanford Linear Accelerator were able to confirm parity nonconservation, which was greeted as an outstanding success for the Weinberg–Salam model. According to Hacking, this successful prediction does not provide sufficient grounds for realism about the Weinberg–Salam model. However, the fact that the physicists successfully used electrons for this experiment is supposed to be grounds for the real existence of *electrons*. What then, we may ask, were the low-level causal properties of electrons used for this experiment? It turns out that the crucial property used was the electron spin. But in what sense is spin a "low-level" property? After all, spin is a quantum-mechanical property (not to be identified with classical angular momentum).

Hacking claims that there is a "rough and ready quantum understanding" (1983, 268) of why certain materials emit linearly polarized electrons of the kind that were used in the Stanford Linear Accelerator. Furthermore, he speaks about "home truths" (265) about electrons that are supposed to be sufficient in order to use them for designing experimental setups. But in what sense are the knowledge that electrons can have exactly two different spin states, and that there exist linearly as well as circularly polarized electron beams, "home truths" or knowledge of "low-level causal properties"? It seems to me that this is rather full-blown theoretical knowledge from quantum mechanics – the kind of thing that Hacking does not allow us to be realists about.

Hacking must show that the successful use of some low-level causal properties of an entity to predict the behavior of an apparatus provides reasons
to believe that these entities exist *and* actually have the low-level properties
used. The "and" here is important, for to believe in the existence of electrons
necessarily means to believe in the existence of a class of objects with certain causal properties (see the introduction to Chapter 7), not merely in the
existence of some entity with unspecified properties. A form of realism that
only said "there exist some entities the properties of which we do not know"
would not be a form of realism at all (in fact, it is compatible with Kant's or
Kuhn's antirealism). In Hacking's main example, physicists used quantum-
mechanical properties of electrons to design an experimental apparatus. But
it is hard to see how Hacking's position differs from full-blown realism about
quantum mechanics that is supported by the latter's predictive success, which
is exactly what Hacking wanted to avoid. It seems that Hacking owes us a
more precise specification of what "low-level causal properties" are and how
they should allow us to detach theoretical entities from the theories in which
they feature.

These difficulties of Hacking's strategy are known (see especially Reiner
and Pierson 1995 and Psillos 1999, 256–258). I now show that they also arise
when Hacking's argument is applied to the entities of experimental biology.

At first, it seems easy to apply Hacking's argument to experimental biology. After all, experimental biologists have long become engineers. They use
biological entities such as genes (see Sections 6.1 and 6.4) or proteins (e.g.,
antibodies; see Section 4.7) not just as research tools, but also for medical
diagnosis and treatment. We might indeed be more inclined to believe that
genes exist once we have seen them being used to save lives, for example, by
inserting a human gene into a bacterium in order to produce a therapeutically
important protein such as insulin. According to Hacking, such instrumental
success in the deployment of theoretical entities justifies belief in the mind-
independent existence of genes, but not belief in the *theory* that describes
genes. Unfortunately, I do not see how Hacking's distinction between entity
realism and theory realism can be meaningfully drawn in biology (nor in
physics; see above). As we have already seen, a realism worthy of its name
needs to claim that we can have knowledge that certain entities exist *and* that
they have certain properties, for instance, a set of causal powers. In other
words, even an entity realist must ascribe some causal powers to the entities
he or she believes in. Now, the general problem is that this admission threatens to collapse the distinction between entity realism and theory realism. To
ascribe certain causal powers to an entity is tantamount to propounding a
theory about this entity. In other words, to claim that an entity exists *and* has

certain properties is the same as claiming that a certain theory is true, namely the theory that describes the causal powers of the entity in question. But this reduces Hacking's experimentalist argument to the "miracle" argument for theory realism.

It now seems to me that in experimental biology it is especially difficult to dissociate theoretical entities from the theories in which they feature. Take the example of the gene. Would it be meaningful to be an entity realist about genes but a theory antirealist with respect to genetic theory? Genes are defined as DNA segments that determine the linear structure of a protein or RNA molecule in a certain cellular environment (see Section 7.1). To accept that a class of entities satisfying this description exists means to accept the truth of some theoretical postulates: namely those postulates that state the function of genes within the cell. Thus, entity realism collapses into theory realism. Now, it could be objected that genes can also be characterized in a way that requires less theoretical knowledge, for example, by applying the classical gene concept (Section 7.1). Thus, perhaps it would be coherent to be an entity realist about classical genes while rejecting theory realism with respect to the molecular theory of the gene. However, in order to support this kind of entity realism with Hacking's argument, it would have to be shown that *classical* genes have been successfully deployed as research tools. Perhaps the widespread practice of breeding genetically defined *Drosophila* strains carrying specific marker genes for genetic experiments would even qualify as "using" a theoretical entity in Hacking's sense. However, this practice hardly works without the theoretical assumptions that Morgan and his students referred to as "the theory of the gene" (see Section 3.2). But then we are again thrown back into trying to support the truth of a theory on the basis of its empirical success, which is just the "miracle" argument all over again. Thus, we are thrown back into all the difficulties with this argument that were discussed in the previous section and that Hacking wanted to avoid.

9.4 THE ARGUMENT FROM INDEPENDENT DETERMINATIONS

Wesley Salmon (1984, 213–227) has pointed out that historically one of the most successful realist arguments was the argument for the reality of atoms from independent determinations of Avogadro's number (which gives, for example, the number of hydrogen molecules in 2 grams of hydrogen gas). In the 1910s, the physicist Jean Perrin showed that thirteen different experimental approaches to determining Avogadro's number had produced approximately the same figure. Especially striking is the fact that these determinations used *different physical processes* for measurement. Therefore, the agreement of the values for Avogadro's number must be attributed either to a chance coincidence or to the real existence of atoms (but see below; this is just reporting the argument). Because such a coincidence is extremely unlikely, atoms must be real. Perrin's arguments were historically highly influential and contributed to the end of positivism in the physical sciences (Nye 1972).

Salmon has analyzed this argument in terms of a *common cause* of the independent determinations of Avogadro's number. In Salmon's reconstruction, each independent determination is viewed as an event with a certain probability. The probability that all the independent determinations coincide is extremely low, unless it is assumed that all the events share a common cause. This common cause, according to Salmon, is the reality of the atoms. Salmon thinks that he can infer the reality of atoms just by applying Reichenbach's principle of common cause, which is an application of the probability calculus. Thus, even though there is a strong resemblance to the "miracle" argument, Salmon's version does not rely on inference to the best explanation, just on a seemingly innocuous application of the theory of probabilities.

There are a number of difficulties with Salmon's construal of the argument for the reality of atoms. The first difficulty concerns his notion of the probability that a given determination of Avogadro's number returns a certain value (or lies within a certain range of values). In order for Salmon's argument to work, this probability must be interpreted as an *objective* probability, for example, as a frequency of outcomes in a set of possible outcomes (the reference class). But what is the set of possible outcomes of a certain determination of Avogadro's number? It seems to me that, once again (compare Section 4.6), an appeal to the concept of probability raises more questions than it answers. The second difficulty has to do with the question of whether atoms are properly viewed as the common cause of the outcomes of the different determinations of Avogadro's number. Since these determinations were done on different samples in different laboratories, no sample of individual atoms qualifies as the common cause-*token* of all these different event-tokens. At best, atoms are a common *type* of cause for a certain type of events. But does the common cause principle also apply to types of causes of types of events? This requires at least further justification. Finally, van Fraassen (1980, 123) has pointed out that the principle of common cause is known not to be strictly true. It is violated by quantum-mechanical effects such as the Einstein–Podolski–Rosen (EPR) correlations.

The difficulties of Salmon's reconstruction of Perrin's argument do not mean that the argument is fundamentally flawed. It could be the reconstruction that is at fault, not the argument in its original version. Indeed, there is something intuitively plausible about the idea that if sets of different experimental techniques that make use of quite different physical processes still yield corresponding results, then the entities or processes responsible must be real.

An alternative to Salmon's appeal to the principle of common cause is to reconstruct the argument as an inference to the best explanation (see Section 9.2). This might work as follows: The agreement of the independent determinations of Avogadro's number cries out for an explanation. For if the physical processes deployed in the different experimental determinations are truly independent, why should the results agree? There are three types of possible explanations that can be given: (1) The agreement can be attributed to chance. (2) It can be explained by the actual existence of atoms and molecules. (3) It can be explained by postulating that all the determinations had something in common, some unknown similarity. This last explanation thus denies that the determinations are, in fact, independent. Let us further assume that (1) is not an option, since it amounts to invoking a miracle. Then we are left with the realist explanation (2) and an explanation that is compatible with antirealism (3). If it can be shown that (2) is the better explanation than (3), and if we accept IBE as a sound inference rule, then we have established realism about atoms by a variant of the miracle argument.

The critical problem, obviously, is to show that the realist explanation of the concurrence of the independent determinations is truly superior to any antirealist explanations. Antirealists did have their share of difficulties with explaining novel predictive success of scientific theories when countering the traditional version of the "miracle" argument; however, there are also some promising new attempts (Stanford 2000; Lyons 2003).

To give an antirealist explanation of the agreement of independent determinations is far from being a trivial task.[21] The antirealist can try to attribute the agreement to a hidden similarity in all the different experiments. But so long as there is no *additional* evidence for such a similarity, it remains a purely hypothetical posit that the antirealist must introduce *ad hoc* in order to save antirealism from the empirical evidence. To make things worse, these similarities, if they are supposed to reside in the objects and processes *in themselves*, cannot be *known* according to the antirealist (for this is what antirealism says). Thus, the antirealist is in a weak position. Alternatively, the antirealist could appeal to something like a unity of the phenomenal world, which, again, puts him in a weak position vis-à-vis the realist, who has a clearer and simpler explanation for the agreement. This suggests that an IBE-reconstruction of the argument from independent determinations is perhaps the strongest argument for scientific realism that can be mustered, because it represents the most difficult challenge to antirealism in any of its three variants mentioned in the introduction.

The case is further strengthened by noting that this variant does not appeal to novel predictions and thus avoids some of the difficulties of the traditional version of the miracle argument mentioned in Section 9.2, namely exactly those difficulties that are created by relying on the predictive success of theories. It should also be noted that, in this variant of the miracle argument, it is not the success of scientific *theories* that is to be explained by realism, but the agreement of some *empirical* results. This version thus also avoids the problems with past successful theories mentioned in Section 9.2.

It must be admitted, however, that there is a problem analogous to the problem of past successful theories. For the independent determinations variant of the "miracle" argument is weakened by cases where a set of independent determinations at first showed that some entity was real, but it was later found that this concurrence was an experimental *artifact* (see the next section). This would allow antirealists to issue a pessimistic induction analogous to the cases discussed in Section 9.2. In fact, I discuss an example in the following section. Here, what remains to be shown is that this kind of argument has applications in experimental biology.

As we have seen in several places, *Drosophila* geneticists had developed a technique called linkage mapping that allowed them to localize genes to specific positions on a chromosome (we have encountered this technique in Sections 3.2, 6.1, and 7.1). In this mapping technique, geneticists measured the frequency with which crossing-over occurs between two genetic loci. Under the theoretical assumption that the frequency of crossing-over is roughly proportional to the physical separation of the loci on the chromosome, it follows that the frequency of crossing-over can be used to measure that distance. But how could geneticists be sure that this theoretical assumption was justified? In other words, how could they know that a linkage map truly reflects physical distances?[22] The history of how this was established is very interesting for our present purposes (Weber 1998b).

If crossing-over were strictly proportional to physical distance, then a linkage map should always be *additive*. In other words, the relative frequency of crossing-over between two loci A and B and between B and C should add up to the frequency measured between A and C. The first genetic maps produced in Thomas Hunt Morgan's laboratory in the 1910s did exhibit additivity, but only for loci that were located very close together. For loci that were located at a greater distance, the crossover frequency for remote factors was smaller than predicted from adding up the frequencies measured from loci located in between. For a while, there were two competing interpretations of this deviation from additivity. According to one interpretation, the deviations were evidence that the chromosome is not a linear structure after all. Instead, it is a three-dimensional structure. This interpretation was due to William Castle and was generally known as the "rat trap" model, because his genetic maps looked like little wire cages (Castle 1919). Hermann Muller argued against the rat trap model (Muller 1920). His preferred explanation for nonadditivity was the occasional occurrence of double crossing-over between two loci. Since a double crossover mutant would phenotypically look like a fly in which no crossing-over had occurred, double crossing-over would lead to an apparent deficiency in recombinants over large map distances, which is what was observed.

So far, both Castle's and Muller's interpretations explained the deviations from additivity. However, Muller had an additional argument: He was able to show that the deviation from additivity *itself* followed a certain regularity. On his interpretation, the deviation from additivity should increase with increasing map distance, because the frequency of double crossing-over should increase with larger distance between two loci. This was actually observed. By contrast, in the rat trap model, we would expect no such regularity in the deviation from additivity. Muller's argument appears to have carried the day, as Castle adopted the linear interpretation of the genetic map the same year (Castle 1920). Thus, there were reasons to believe that linkage mapping faithfully represents the arrangement of genes on chromosomes before other mapping techniques came on the scene.

In 1933, T. S. Painter published a remarkable new technique for mapping *Drosophila* genes. This method was based on the giant salivary gland chromosomes found in the fly larvae (see Section 6.1). After suitable staining, these giant chromosomes display a characteristic banding pattern that is visible under an ordinary light microscope. An examination of various mutants revealed that some mutants carry large structural alterations in their chromosomes, such as deletions, inversions, duplications, or translocations (where one piece of a chromosome was attached to a different one). By correlating many different chromosomal rearrangements with the known phenotypes for various mutations, it was possible to construct entire maps that were based on this cytological technique only. Strikingly, these maps were *colinear* with the older linkage maps (Weber 1998b). This is not to say that the two kinds of maps corresponded exactly with respect to the distances between the genes. Rather, what turned out to be invariant in these two kinds of maps was the *linear ordering* of the genes (Figure 9.1).

I suggest that the fact that different techniques produced the same results provides a powerful argument that the linear arrangement of genes on

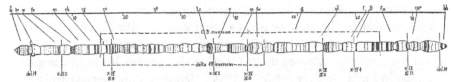

Figure 9.1. Colinearity of cytological and genetic maps. The thin line on top represents a genetic map of the X-chromosome of *Drosophila melanogaster*. This map is based on the frequencies of crossing-over between the genes shown. The image below is a drawing of a giant chromosome, where the position of the genes was determined with the help of the characteristic bands. The linear ordering of the genes is the same in both maps, which provided evidence that these maps represent real structures. Reprinted with permission from T. S. Painter (1933): "A New Method for the Study of Chromosome Rearrangements and the Plotting of Chromosome Maps". *Science* 78: 585–586. © 1933 American Association for the Advancement of Science.

chromosomes is a real property of these chromosomes. However, for this agreement to carry any significance, we must require the antecedently established reliability of *both* techniques. As we have seen, geneticists had good reasons to think that linear genetic maps represent real structures before cytological mapping of giant chromosomes was invented. Furthermore, cytological mapping needed its own checks to ensure the reliability of maps. One obvious way was to double-check map distances against several so-called markers (genetic loci used for mapping). The concurrence of independent evidence is only convincing with respect to the realism question if the results from at least two different methods, which are *both* reliable in their own right, agree, and only if the ways in which these methods produce their results are well understood. The reason is that we want to avoid being impressed by a concurrence that has a *trivial* explanation, e.g., something about the experimental technique that we do not fully understand. In the case of Perrin's independent determinations of Avogadro's number or in the present example of genetic mapping, I can see no trivial explanation for the concurrence of the results. To undermine this argument, antirealists need to come up with one. Only in such cases are we justified in inferring the reality of the structures and entities involved by an IBE-type argument.

To conclude, there is a variant of the "miracle" argument that works well in experimental biology. According to this variant, we are justified in accepting some structures or entities (and some of their properties) as real if two or several sets of experimental techniques that rely on different processes, which have antecedently been shown to be reliable in their own right, return corresponding results.[23] These criteria are necessary to rule out trivial explanations of the correspondence, for example, experimental artifacts.

In the next section, I discuss a historical example of such an experimental artifact.

9.5 EXPERIMENTAL ARTIFACTS: LESSONS FROM THE MESOSOME

In laboratory slang, an "artifact" is a phenomenon or appearance that is thought not to represent a real biological structure or process, but one created by the experimental method or instrument used (e.g., Ownby 1988, vii). In the context of microscopy,[24] we can distinguish between *optical* artifacts and *preparation* artifacts. The former are caused by the microscope or by an interaction of the specimen with the microscope. The latter arise when the biological specimen is fixed, cut, stained, or decorated for light or electron microscopy. Optical artifacts have become rare with technologically advanced microscopes, but preparation artifacts are probably still one of the most frequent forms of error in biological laboratories. Of course, an artifact only constitutes an error as long as it is not *recognized* as an artifact, that is, if it is thought to represent a real structure.

Further reflection shows that the notion of artifact is epistemologically trickier than one might have thought. Consider the stereotypic characterization that artifacts are "created by the experimental procedure." This natural way of speaking requires analysis. Are not *all* experimental results in a sense created by the experimental procedure? Or, in the context of microscopy, are not *all* microscopic images caused by an interaction of the specimen, the preparation procedure, and the microscope? Apparently, what distinguishes artifacts from nonartifacts is that the latter arise from *the right kind* of interaction between object and instrument: namely an interaction that produces a faithful image of what is really there in the specimen. But this suggests that the judgment of whether or not something is real or an artifact involves sophisticated *epistemic norms*. Perhaps we can put it like this: A microscopic image requires more *interpretation* than an ordinary image. The biologist who interprets, say, an electron micrograph forms a *hypothesis* (or a set of hypotheses) about what the different lines, dots, and splotches on the image represent. Thus, what eventually appears in a microscopic image is cognitively processed by the scientist. This processing involves a lot of theoretical and practical knowledge as well as epistemic norms. The recognition of something as an artifact, then, is a judgment to the effect that some elements of an image do not represent a real biological structure. The considerations and arguments behind such a judgment can be quite complex. This raises the question of what the epistemological grounds are for such judgments. As usual, the best way to approach this question is by examining actual biological practice.

In the history of biology, we mainly find examples of structures that were initially thought to be real and were then recognized as artifacts (see below). But the reverse is also on record. For example, until the 1950s, many cell biologists thought that a certain stack-shaped intracellular structure first described by C. Golgi in 1898 was an artifact. Belief in the reality of the "Golgi apparatus," as it came to be known, hinged on the question of whether fixation of cells with silver or osmium tetroxide (OsO_4) can lead to heavy metal depositions in the cell. However, with the introduction of electron microscopy in the 1950s, the Golgi apparatus was increasingly considered to be a real structure. Since then, the Golgi apparatus has had a spectacular career (Farquhar and Palade 1981). Today, it is subject to intense research in biochemistry and cell biology, as it turned out to be responsible for the posttranslational processing and intracellular sorting of proteins (i.e., dispatching proteins after synthesis to their final destination in the cell).

The more typical case is that some structure that was once thought to be real is suddenly (or gradually) judged to be an artifact. One of the most famous and long-lived artifacts in the history of biological electron microscopy was the bacterial "mesosome," which was thought to be a membrane-enclosed intracellular organelle in many different species of bacteria. Mesosomes are thought to be artifacts today. In other words, according to contemporary microbiologists, there is no such structure in bacterial cells. This is quite remarkable given the fact that, from ca. 1960 until the mid-1980s, mesosomes were actively studied in many microbiological and biochemical laboratories around the world. Several hundred articles were published in the scientific literature reporting on the structure and possible functions of mesosomes.[25] How is it possible that some entity that had been the subject of so much research all of a sudden ceased to exist?

The history of the mesosome is closely tied to that of the transmission electron microscope in biology. In this technique, an electron beam is passed through a biological sample and leaves an image on a fluorescent or photographic plate. As electrons do not pass through matter easily, the specimen must be extremely thin and it must be placed in a vacuum. Both of these conditions are extremely hard on biological materials. In order to cut very thin slices out of biological material, it is necessary to somehow fixate it. Furthermore, water rapidly evaporates in a vacuum, which makes it necessary to dehydrate the material beforehand. Obviously, the problem is to fix and dehydrate it in such a way as not to damage the fragile biological structures. Finally, the radiation energy of the electron beam has a strong destructive impact on biological material. Care must be taken not to simply burn the material away immediately. I want to give the reader a feeling for the formidable difficulties of observing intact biological ultrastructures in a transmission electron microscope.

A standard preparation technique in biological electron microscopy that was developed in the 1950s is known as the Ryter–Kellenberger technique. It works as follows. Bacterial cells are chemically fixed by adding 0.1% OsO_4 (osmium tetroxide) to the culture medium while they are still growing. After the medium is removed, the cells are fixed in 1% OsO_4. After the sample is dehydrated, it is embedded in a polyester resin. Subsequently, the resin is mechanically cut into very thin slices that can be inserted into the microscope. Fitz-James (1960) observed what were initially known as "peripheral bodies" in such preparations and gave them the name "mesosomes." They appeared to be invaginations of the bacterial plasma membrane that frequently appeared near the septum of a dividing cell. For these reasons, mesosomes were thought to be involved in cell wall synthesis.

In 1968, N. Nanninga reported on mesosomes observed with a new preparation technique known as "freeze-etching" (Nanninga 1968). In this technique, bacterial cells are rapidly frozen. In order to protect the biological structures from damage by growing ice crystals, a so-called cryoprotectant is added, for example, a 20% glycerol solution (cryoprotectants make water solidify amorphously without forming crystals). Instead of the frozen sample being cut into thin slices, it is cracked apart at a random plane. It had been shown that frozen biological samples break preferably along certain structures such as membranes. (This provided biologists with the first opportunity to inspect the surfaces of biological membranes under the electron microscope, instead of just being able to look at cross-sections). In the freeze-etching technique, the ice is then partly removed by freeze-drying. Finally, the specimen is shadowed from the side with heavy metal in order to make the surface structure detectable for the electron microscope. Alternatively, the freeze-drying step can be omitted. The latter technique is known as "freeze-fracturing."

Using these new techniques, Nanninga examined the effect of varying experimental conditions on the appearance of mesosomes (Nanninga 1971). If the cells were not prefixed with 0.1% OsO_4, as in the Ryter–Kellenberger method, for fewer mesosomes were observed in freeze-fractured preparations. Nanninga offered the possible explanation that the absence of chemical fixation destabilizes the mesosomes with respect to the freeze-fracture technique. In retrospect, one might suggest that Nanninga could or even should have suspected a fixation artifact already at this stage. However, he did not question the reality of mesosomes until 1973, at least not in print (Rasmussen 1993,

251). When he eventually did, other laboratories tried to produce evidence for Nanninga's initial hypothesis that unfixed samples fracture right across the mesosomes in such a way as to make them invisible to this technique (e.g., Higgins and Daneo-Moore 1974). Thus, at this stage, microbiologists still blamed their *methods*, rather than questioning the existence of the mesosomes if they failed to show up. This indicates how established mesosomes were in the early 1970s. Indeed, the 1975 issue of the *Annual Review of Microbiology* still carried a review article specifically on mesosomes (Greenawalt and Whiteside 1975).

But now suspicions were mounting. Silva et al. (1976) directly explored the possibility that mesosomes are an artifact created by chemical fixation. Also using the freeze-fracturing technique, they showed that large, complex mesosomes are indeed only found when mesosomes are subjected to prefixation by adding 0.1% OsO_4 to the culture broth. Suspecting that the fixant causes membrane damage during this phase, Silva et al. examined the kinetics of mesosome formation during the fixation phase. They found a linear increase in the number of mesosomes up to a saturation point. Furthermore, Silva and associates showed that OsO_4 has membrane-damaging properties. For example, it was found to lyse protoplasts (cells from which the cell wall had been removed) and cause K^+ leakage from cells. Finally, they showed that known membrane-damaging substances such as certain anesthetics caused the appearance of mesosome-like structures.

The final deathblow for the mesosome came with new preparation techniques in the 1980s. One such technique was the use of sectioned, frozen-hydrated material that has been cryoprotected with the help of glucose. This technique avoided three major possible causes of damage to fragile biological structures: chemical fixation, staining, and dehydration. Dubochet et al. (1983) carefully examined the residual possible sources of artifacts in sections of frozen-hydrated bacteria. Residual sources include (i) the freezing process, (ii) the sectioning (cutting) of thin slices, and (iii) the irradiation by the electron beam. Dubochet and co-workers tried to control for all three possible sources of error. They ruled out (i) by examining the behavior of frozen material while it was drying (by sublimation). Areas where ice crystals had formed were clearly differentiable from amorphous ice. Thus, freezing artifacts are avoidable by strictly using amorphously frozen materials, which can be checked. Artifacts of type (ii) can also be controlled because damage inflicted by the microtome (a mechanical cutting device) is recognizable by its orientation with respect to the cutting direction. Finally, artifacts of type (iii) can be controlled by varying the strength and duration of the irradiation and examining its effects on the biological materials in question. Using this carefully controlled technique, Dubochet confirmed that no mesosomes were present in bacterial membranes. In the 1980s, a consensus formed that mesosomes do not exist.

Different commentators on this historical episode have disagreed on the correct reconstruction of why mesosomes were suddenly judged to be artifacts, after so many years of enjoying a high status as not only real, but also important bacterial organelles. The historian Nicolas Rasmussen (1993, 2001) has given a largely sociological explanation. He suggested that mesosomes were dropped as real entities because biochemists lost interest in them (1993, 255–256). For many years, biochemists had tried to characterize the mesosomal membrane, in particular with respect to biochemical differences from the plasma membrane (e.g., the presence of specific enzymes). Even though several laboratories claimed to have found such differences, the results obtained by different laboratories varied wildly. For this reason, biochemists eventually moved on to other projects. The mesosome, as it were, lost its biochemical "constituency" without which it was not acceptable as a real entity.

Rasmussen's account is unconvincing. What is lacking in this account is good historical evidence that mesosomes were really judged to be artifacts *because* biochemists lost interest in them. To show this would require a much more detailed historical investigation. As it stands, the causal relation between the biochemists' interest in mesosomes and the microscopists' judgment that they are artifacts could well be reversed: Biochemists could have lost interest in mesosomes because their existence was open to doubt as the result of the microscopists' work. Recall that the first hints that mesosomes were fixation artifacts emerged as early as 1971, and this could have been sufficient to turn biochemists away.

Rasmussen's conclusions have been opposed, in different ways, by two philosophers: Sylvia Culp (1994) and Robert Hudson (1999). Culp has suggested that microbiologists used *robustness* as a criterion to decide whether or not mesosomes were artifacts. The concept of robustness was first introduced by Wimsatt (1981). Wimsatt used the term in a very general sense: namely as a kind of invariance of results under different methods, interventions, or assumptions. The results in question may be experimental data or the results of theoretical calculations. Culp (1994) has a narrower notion of robustness. Leaving out the details of Culp's account, the basic idea behind robustness is this.[26] If some structure is observed by a variety of different experimental techniques, then this counts as evidence for the nonartifactuality of this structure (see also Hacking 1983, 200–202). Thus, the basic intuition seems to be the same as that behind Perrin's argument for the existence of atoms and the argument from independent determinations discussed in Section 9.4. I shall try to give this intuition some additional rationale.

A result that has been obtained with several independent techniques is more reliable than one that is the product of just a single technique. Presumably, the reason is that if scientists use just a single technique to measure or observe something, then there is always the possibility that the observation is the result of a flaw in the technique. But when a variety of techniques are used and the results agree, then scientists have a reason to believe that the techniques are *not* flawed. If they were flawed, then they would be unlikely to produce the same result. Instead, we would expect them to produce wildly divergent results.

Culp's main thesis is that mesosomes were judged to be artifacts once several independent methods failed to indicate their presence in bacterial cells. These independent methods, according to Culp, include freeze-fracturing, cryosubstitution (freezing followed by embedding and sectioning), and ultrarapid freezing without dehydration followed by sectioning. By contrast, techniques that did show well-defined mesosomes included only the Ryter–Kellenberger prefixation technique. Thus, the evidence against mesosomes was more robust than the evidence for mesosomes. Therefore, microbiologists were justified in rejecting them as real biological structures (Culp 1994, 53).

Robert Hudson has challenged Culp's conclusion by claiming that considerations of robustness played no role in the rise or fall of the mesosome. In fact, robustness could have been used to argue that mesosomes exist (Hudson 1999, 297). What is more, the biologists themselves did not appear to use any kind of robustness criteria when debating whether or not mesosomes were artifacts. Instead, what they were using is something Hudson refers to as "reliable process reasoning." This form of reasoning proceeds by providing evidence that some experimental technique is reliable, that is, does not produce artifacts. Hudson does not attempt to analyze this form of reasoning in detail.

I would like to propose yet another interpretation of this historical episode (of course, this will be the correct one). My suggestion is that demonstrating the artifactual status of mesosomes was simply a matter of directly showing that mesosomes are produced by chemical fixation. This was shown with the help of scientific evidence that is not different in kind from the standard evidence for some theoretical mechanism (see Chapter 4). In other words, mesosomes were rejected as real biological structures once there was experimental *evidence* that they were formed by the membrane-damaging effect of certain chemicals such as the OsO_4 used for chemical fixation. What made this evidence convincing and thus the judgment of the artifactuality of mesosomes sound was basically the same experimental strategy that we have encountered in Chapter 4. There, I have argued that a good experimental test of a theory or hypothesis is one in which scientists have systematically varied the experimental conditions to enable them to control for possible errors (see Section 4.7). Here, biologists were testing the hypothesis that mesosomes are produced by chemical fixation. This hypothesis is quite different from the theoretical hypothesis I examined in Chapter 4 (the chemiosmotic hypothesis). Nevertheless, the approach to testing it was essentially the same. For example, we have seen that Silva et al. (1976) carried out various experiments to establish a causal relationship between chemical (OsO_4) fixation and the number of mesosomes observed in freeze-fractured preparations. The finding that mesosomes only occurred after chemical fixation and the kinetic measurements provided evidence that such a causal connection existed. However, additional tests were necessary in order to rule out possible errors, that is, false causal attributions. For example, one possible mistake was that chemical fixation did not cause mesosomes to *form*; but that it *preserved* them during the freeze-fracturing of the cells and that they are destroyed during preparation unless they are chemically fixated. The findings that OsO_4 exhibited membrane-damaging properties and that known membrane-damaging substances also caused more mesosomes to appear spoke against this possibility. Yet it did not rule it out.

A more stringent test was provided by Dubochet et al. (1983), who also observed a correlation between chemical fixation and the appearance of mesosomes. But in addition, this group was able to rule out the possible error of taking OsO_4 to be the cause of mesosome appearance, whereas in fact mesosomes disappeared in the absence of the fixant. As we have seen, Dubochet et al. conducted an inquiry into the possible sources of preparative damage and were able to show that the samples were not damaged by freezing, sectioning, or irradiation. Thus, they were able to control for the factors that might cause mesosomes to disappear from the cell. This gave them reason to believe that such factors were absent and that, therefore, there were no such things as mesosomes in the bacterial cells. This is exactly the kind of error-aware reasoning that also governs the testing of theories according to Chapter 4 (see Sections 4.6 and 4.7).

To sum up, I agree with Hudson that the judgment concerning the artifactuality of mesosomes had nothing to do with robustness. Scientists rest their belief in the existence of some entity in the reliability of the experimental procedures used. This reliability is tested with the same kind of methods that are used to test theoretical hypotheses (see Chapter 4), that is, by systematically altering experimental conditions to control for possible errors.[27] If an experimental procedure fails this test, then an entity is no longer accepted as real.

The mesosome case also confirms my thesis of the previous section, namely that agreement of results alone is insufficient grounds for the reality of some structure. There, I argued that such agreement carries little epistemic weight by itself as long as it could be due to an experimental artifact. In the mesosome case, there was a trivial explanation for why mesosomes turned up under a variety of different preparation methods: All these methods had in common that they inflicted physical injury to bacterial membranes. Of course, defenders of the robustness criterion could argue that, in this case, the methods were not really independent and that robustness is only a reliable criterion if applied to truly independent methods. But how can we be sure that two methods are truly independent? Only an examination of the conditions under which experimental procedures are reliable in the sense of being able to control for possible errors (see Section 4.6) can show that nothing spurious is going on when experimental results agree.

SUMMARY

I have first established that, even though realism about the objects of biology presupposes realism about the material world in general, there is a philosophically interesting realism issue in experimental biology. This problem arises because the objects of experimental biology are epistemically not very accessible (due to their small size and their fragility). Doubts of the reality of biological structures and entities are thus possible even if the existence of material objects as such is taken for granted. The proper objects of scientific realism in experimental biology are theories (in a wide sense of the term), theoretical entities and not-directly-observable structures, laws, natural kinds, and possibly functions. Realism about biological information is not warranted, as I have shown in the previous chapter. In this chapter, I have focused on realism about theoretical entities and not directly observable structures.

I have next critically examined three argumentative strategies that have been pursued in defense of scientific realism, the "miracle" argument, the experimentalist argument, and the argument from independent determinations. The first strategy faces the difficulty that novel predictions (which are thought to be necessary to support realism) are rare in experimental biology. Furthermore, if there are novel predictions, they are usually qualitative. The second strategy suffers from the impossibility of distinguishing between theory realism and entity realism in biology. The third strategy appears to be the strongest one if it is reconstructed as an inference to the best explanation (as opposed to an instance of the principle of common cause) and augmented by the requirement that the different methods involved in the independent determinations must have been antecedently shown to be reliable.

The historical case of mesosomes reveals some pitfalls of the argument from independent determinations, for mesosomes were observed under a wide variety of experimental conditions. The mesosome case shows that biologists base their confidence about the existence of unobservable structures on the reliability of the methods they use. This reliability must be established by suitable investigations into the possible errors of the experimental techniques. The reasoning deployed in such investigations is basically the same as when theories and hypotheses are being tested.

Notes

1. An anonymous reader for Cambridge University Press suggested that this work is not a study in the philosophy of biology, but in the philosophy of science as applied to biology. This is partly how it should be, provided that my view that there is a certain unity in all of modern science is correct. However, issues that are specific to experimental biology are also discussed in this book (see especially Chapters 6 and 8). Therefore, this book is both: a study in the philosophy of biology and a study in the philosophy of science as applied to biology.
2. The designation "experimental biology" is not entirely accurate, as evolutionary biology and ecology are also partly experimental sciences today. There are still learned societies carrying "experimental biology" in their name; however, this is a historical remnant from times when a strong distinction was made between natural history and experimental science. I have chosen the designation for the title of this book partly for want of a better one ("philosophy of proximate biology" just doesn't work), but also to emphasize the strong focus on experimental practice.

CHAPTER 2

1. For example, Oppenheim and Putnam (1958); Nagel (1961); Schaffner (1967, 1969, 1974b, 1976, 1993); Ruse (1973, 1976); Hull (1972, 1974, 1976); Fodor (1974); Ayala (1974, 1985); Wimsatt (1974, 1976a, 1976b); Causey (1977); Darden and Maull (1977); Roll-Hansen (1979); Hooker (1981); Mayr (1982, 59–63); Kitcher (1984); Bechtel (1984); Rosenberg (1985); Balzer and Dawe (1986); Hoyningen-Huene (1989, 1992, 1997); Beatty (1990); Waters (1990, 1994, 2000); Dupré (1993); Bechtel and Richardson (1993); Sarkar (1998); Weber (1998a, 2001c); Carrier (2000).
2. For example, Schaffner (1969, 1993); Hull (1972, 1974, 1976); Ruse (1973, 1976); Goosens (1978); Kimbrough (1979); Kitcher (1984); Rosenberg (1985); Balzer and Dawe (1986); Waters (1990, 1994, 2000); Weber (1998b); Gasper (1992); Vance (1996); Sarkar (1998). The term "classical genetics" denotes the body of knowledge and investigative techniques that studies genes and chromosomes with

the help of crossing experiments in the manner that Gregor Mendel invented in the nineteenth century. The classical period is usually identified with the work of Thomas Hunt Morgan and his group in the early decades of the twentieth century (see also Chapters 3, 6, and 7).

3. There have been debates as to whether there might not be a disjunction of properties that could be identified with dominance. However, the general feeling in the philosophical community has been that, even if this disjunction were known, it would be too unmanageable to provide useable bridge principles. Furthermore, some have argued that bridge principles need to establish a one-one relation between natural kinds (Fodor 1974; cf. Hoyningen-Huene 1997). Presumably, the molecular realizers of a genetic property like dominance do not form a natural kind at the molecular level.

4. See Wimsatt (1976b); Schaffner (1993, 287); Machamer et al. (2000); Craver and Darden (2001); Weber (2001c); Glennan (1996, 2002); Woodward (2002).

5. Since the membrane is only about 5 nm thick, the electric field gradient in the resting potential has the enormous value of about 100,000 V/cm.

6. Note that many biological processes are *cyclical*, such that set-up conditions and termination conditions are identical. Other examples are muscle contraction, metabolic pathways such as the urea cycle (see Section 3.3), the operation of the respiratory chain (see Section 4.1), and DNA replication.

7. The transitions between the different states of these voltage-gated channels seem not to occur in a deterministic fashion. It was shown that a population of such channels has a characteristic relaxation time for each type of transition. These relaxation times are like half-lives in radioactive decay. To my knowledge, it is currently unknown whether individual transition events could be subject to quantum measurement uncertainty, in other words, whether these channels could have quantum superposition states. This could have interesting implications for the old philosophical problem of determinism and free will, but this is not the place to explore these.

8. See, for example, Kitcher (1989); Woodward (1997, 2000, 2001); Schaffner (1993, Chapter 6); van Fraassen (1980); Kitcher and Salmon (1987).

9. This might explain why experimental biologists (unlike evolutionary biologists and ecologists) rarely use the term "theory." When they use it, they mean something that has an unproven, provisional character.

10. I thank Paul Hoyningen-Huene for pointing this difference out to me.

11. The thesis of explanatory heteronomy is related to what Sarkar (1998, 43) calls "fundamentalism." By this, he means that "the explanation of a feature of a system invokes factors from a different realm (. . .) and the feature to be explained is a result of the rules operative in that realm" (ibid.). I go even further than this. First, I claim that some biological explanations like the one discussed in Section 2.1 are *nothing but* applications of theories and principles from the more fundamental realm. Second, I defend the metaphysical claim that the "rules" from the physical realm are laws of nature and that there are no genuine laws that are neither physical nor chemical (see Section 2.3).

12. This claim seems to be based on a particular account of laws. An account of laws of nature such as Brian Ellis's (2001, 2002) could perhaps be made consistent with Machamer et al.'s views (see also Section 9.1).

13. Machamer et al. do not provide much analysis of their concept of activities; all they are saying is that "activities are types of causes" (2000, 6), which does not really help. In addition, they endorse a kind of metaphysical dualism according to which both entities and activities are to be seen as ontologically basic. Thus, Machamer et al. reject both substantivalism, according to which entities and their states are ontologically prior and all change is derivative from them, and process ontology, which holds change to be ontologically prior and entities as well as states to be derivative. Unfortunately, Machamer et al.'s metaphysical excursion does not accomplish much toward clarification of these issues. In particular, it remains unclear how activities and entities are supposed to be related.

14. Several authors have proposed to replace the old distinction between contingent and necessary regularities by a continuum. For example, regularities could be classified according to their *stability* or *invariance* under a variety of conditions or interventions (e.g., Mitchell 2000; Woodward 2001). Indeed, there seems to be no reason to think that there should only be two kinds of regularities. In our present example, the degree of nomic necessity could be said to be high for the regularities in the behavior of the mechanism's constituent entities, since the regular behavior of, for example, ion channels can be observed under a variety of different experimental conditions (such as patch-clamp measurements, where a membrane patch containing channel proteins is directly mounted on a pipette-shaped electrode in order to measure electrochemical currents). The regularities shown by entire neurons could be said to show a lower degree of nomic necessity, as they are quite fragile to external conditions and experimental interventions.

15. The difference between capacities and causal dispositions, according to Cartwright (1999, 59), is that the latter are tied to a single type of manifestation. For example, brittleness is a disposition, because it manifests itself only as breaking. By contrast, capacities may produce different manifestations in different circumstances. For example, aspirin has a capacity that can relieve headaches or, in other circumstances, prevent thrombosis or cause bleeding.

16. There are authors who do not see laws and capacities as distinct. On such a view, law statements are analyzed as ascriptions of capacities (Glennan 1997). The issue is not merely terminological. Whether *all* natural laws can be explicated as capacities of individual entities (as Cartwright and others argue) is a substantive question.

17. Daniel Sirtes (personal communication) objects that my criticism affects only Cartwright's notion of capacities, but not Machamer et al.'s Craver's notion of activity. Activities need not be identified with the exercise of capacities possessed by *individual* entities. Even though Machamer et al. (2000, 5) say that they are not aware of any activities that are not activities *of* an entity, their dualist ontology (see note 13) allows this in principle. My counterobjection is that laws are still necessary, namely to explain why certain constellation of entities behave in a certain way, e.g., in the way predicted by the Nernst equation.

18. Beatty, of course, is no reductionist (Beatty 1990). However, there is no conflict here with his views, since I do not claim that evolutionary and population biology are reductionistic in the sense outlined here. Beatty would strongly emphasize that these parts of biology are not in the business of discovering laws, but of reconstructing natural history, without thereby being less "scientific." This does not contradict the position I am defending here for nonevolutionary biology.

19. See Allen, Bekoff, and Lauder (1998) for a recent collection of essays on the topic, and McLaughlin (2001) for an excellent book-length study with a rich bibliography.

20. However, there remains an analogue to the problem of first appearance, namely the so-called problem of "first-cycle functions" (McLaughlin 2001, 168).

21. Craver (2001) does not seem to note this, even though he gives a nice account of how Cummins-functions are applied in biological practice.

22. I wish to thank Martin Carrier for pointing this out to me.

23. Note that my proposal differs from McLaughlin's in that I am not adopting an etiological account of function, according to which functions must explain the presence of the function bearers. On my account, the capacity for self-reproduction does not do any causal-explanatory work with respect to the function bearers; it merely serves as an ultimate system goal with respect to which role functions at different levels of the biological organization acquire their status as functions. This also avoids the main difficulty of Cummins's account of role functions, but at lower metaphysical costs than McLaughlin's etiological account.

24. An anonymous referee has asked why we should stop at the level of individual organisms for identifying role functions. Why not populations, species, or even ecosystems? Indeed, the idea that communities of organisms or ecosystems have functions has been floating around in environmental science for a long time. However, this amounts to postulating superorganisms, an idea that has been controversial throughout the history of biology (see Weber and Esfeld 2003). The salient question is whether entities that are more inclusive than individual organisms are sufficiently cohesive in space and time and whether they really display the kind of homeostatic properties that distinguishes organismic individuals.

25. Roberta Millstein has pointed out to me that astronomers talk about the "birth" and "death" of stars, suggesting that stars admit of functional ascriptions in a sense similar to that in biology. I do not want to exclude this; however, this would require a closer study of the explanatory practices of astronomy.

26. The term "function" is used in a somewhat broader sense here than in the previous section (Section 2.4). There, we were analyzing quasi-teleological statements of the form "the function of X in System S is to ϕ." Here, we are using the term "function" to denote a special class of properties, namely properties that are individuated by their causal role. Since I have concluded that functions in the sense of Section 2.4 (for short, function$_{2.4}$) are also causal roles, the two different senses of "function" are clearly related. Namely, an ascription of function$_{2.4}$ implies an ascription of function$_{2.5}$, but not vice versa; see Mahner and Bunge (2001). The main difference is that, here, some capacity does not have to contribute to an ultimate system goal such as self-reproduction in order to qualify as a function$_{2.5}$. For example, "x is an electric conductor" is a functional predicate in the sense of the present discussion (i.e., a function$_{2.5}$), but not in the sense of Section 2.4. The concept of dominance in genetics is a borderline case, since it is not clear whether it has a biological function$_{2.4}$.

27. The same point could be made with symmetries instead of natural laws (van Fraassen 1989).

28. Another possible exception is the clonal selection theory of antibody generation.

CHAPTER 3

1. Note that this claim is different from the claim advanced by some sociologists of science, namely that scientific *objects* such as quarks are constructed. A scientific realist has no problem accepting that theories are constructed, for this in no way precludes that these theories express states of affairs that exist independently of the human mind.

2. In the previous chapter, I have argued that experimental biologists do not construct self-contained theories, but just explanations that are applications of physical and chemical theories. Thus, the relevant sense of the term "theory" used there is that of a description of laws of nature. In this chapter (and in Chapter 9), I use the term "theory" in a broader sense, namely as any description of a biological mechanism.

3. E.g., Simon (1977); Nickles (1980b, 1980c); Langley et al. (1987); Thagard (1988); Darden (1997).

4. The repressor was later shown to be a protein molecule that specifically recognizes a short DNA sequence element and binds to it. However, the repressor model in its original form was not committed to the repressor being made of protein; in fact, the early evidence considered by Jacob and Monod indicated that it was not a protein (Jacob and Monod 1961, 333). Instead, they speculated that the cytoplasmic repressor consists of RNA (1961, 352).

5. In fact, the same experiments that were used by Jacob and Monod to establish the repressor model also played an important role in the discovery of mRNA, but I shall ignore this part of the story for simplicity.

6. This hypothesis resonated well with the so-called instructional theory of antibody formation, according to which antigens somehow impart structural information to antibodies. By analogy, an inducer could impart structure to newly synthesized enzymes and thus be necessary for the production of functional enzymes. In his Nobel Lecture delivered in 1965, Jacques Monod relates that it was psychologically very difficult for him to give up this idea (Monod 1966, 479). The final deathblow for this idea came in 1960, when Monod and co-workers showed that induction works in β-galactosidase mutants that lack any affinity for galactosides.

7. Probably the best historical account is still Judson (1979, Chapter 7). See also Morange (1998, Chapter 14).

8. Pardee, Jacob, and Monod actually used this terminology; see, e.g., Pardee et al. (1959, 171, 174).

9. Schaffner did not spell out the deductive argument explicitly, but I find it helpful to do so.

10. In fact, Pardee had previously been working on the feedback inhibition of an enzyme by its end product, namely aspartate transcarbamylase (Judson 1979, 405).

11. Arthur Pardee does not recall talking to Szilard in December 1957, but he told Judson (1979, 410) that "repression was very much in all of our thinking at that time."

12. Following (Darden 1991, 248–251), Schaffner's suggestion concerning the genesis of the repressor model could be described as "invoking a theory type." This special kind of analogical reasoning involves taking an abstract structure from a pre-existing theory from the same or a different domain and instantiating this structure in a new

way. For example, Darwin's theory of natural selection may provide such an abstract structure for a certain type of theory, which was used to construct selection theories in domains such as immunology (clonal selection theory). Similarly, there might be an abstract theory structure for derepression mechanisms (Schaffner's "formal derepression hypothesis," mentioned in the text), which was instantiated by Jacob and Monod to give the repressor model of the *lac* region.

13. Schaffner's proposal nicely resolves the uncertainty concerning Szilard's role that Judson (1979, 409) pondered in his book. Judson was aware of Schaffner's paper, but he found it "hard to follow" (654).

14. There might be a similarity here to Clark Glymour's (1980) "bootstrap" theory of confirmation (see the introduction to Chapter 4).

15. Schaffner (1993, 18) has since abandoned this thesis and criticized his own 1974 paper for failing to differentiate between the *generation* and *primary evaluation* of scientific theories. I agree with this diagnosis.

16. The Morgan group did not interpret this linear ordering as an abstract calculus that could be used to predict the outcome of crossing-over experiments. Rather, they thought that the linear order of the genetic map represented the actual physical arrangement of genes on the chromosomes (see Weber 1998b).

17. Recall that the same question arose in the previous section, in relation to Schaffner's claim that the repressor model was deduced from the PaJaMo experiment.

18. Curiously, this echoes the older distinction between context of discovery and context of justification mentioned in the introduction to this chapter. According to the classical form of this distinction, the philosophy of science is not to be concerned with psychological aspects of scientific research (i.e., the scientists' private mental life), only with the logical aspects.

19. Frames consist of a name and a number of slots for entering properties and the values that these properties take. The computer then receives instructions to compare the individual slots of different frames and to form conclusions about the relations between the objects represented by the frames (see Thagard 1988, 16).

20. An exception is the strategies for diagnosing faults and fixing theories; their source seems to be the repair analogy. After completing her book, Darden found some evidence for the relevance of these strategies in contemporary research while conducting a participant study in Joshua Lederberg's laboratory (Darden and Cook 1994).

21. The Krebs notebooks and some other documents relevant to the discovery of the urea cycle have recently been published (Grasshoff and Nickelsen 2001).

22. Kulkarni and Simon (1988) have designed a computer program to simulate the experimental strategy followed by Krebs and Henseleit, as well as the construction of the ornithine cycle. However, the historical adequacy of this account is doubtful (see Grasshoff 1995). The main problem is that Kulkarni's and Simon's AI program KEKADA (the name is a Hindi translation of the German word *Krebs*) contains a heuristic rule that makes it consider catalytic action as soon as it is given the results of the ornithine experiments. (KEKADA also considered catalytic action for ammonia, which is nonsensical given Krebs's state of knowledge in 1932.) Grasshoff argues that the catalytic action of ornithine was not immediately obvious to Krebs; this idea was rather the *result* of a major intellectual effort. Because Kulkarni and Simon simply wrote the catalyst hypothesis into KEKADA's code,

their simulation trivializes Krebs's intellectual achievement – which is considered to be one of the most brilliant in the history of biochemistry.

23. This suggestion conforms well to Krebs's own description that his solution was an instance of "pattern recognition." According to Kuhn, normal science consists in solving puzzles in a way similar to that used for the exemplary problem solutions (exemplars) that are constitutive for paradigms. A newly solved puzzle is related to the exemplars by a web of *similarity relations*. Thus, solving a new puzzle consists in recognizing a pattern of similarities (see Hoyningen-Huene 1993, Section 3.2).

24. T. S. Kuhn has characterized anomalies as instances where "nature has violated the paradigm-induced expectations that govern normal science" (Kuhn 1970, 52). According to Kuhn, such anomalies may eventually trigger a scientific revolution, where an existing paradigm is abandoned and a new one takes its place. Alternatively, anomalies may be resolved and added to the stock of exemplary problem solutions of the ruling paradigm. None of the three examples discussed here seems to fit Kuhn's distinctions entirely. The revisions in the theories made in response to the anomalies hardly qualify as scientific revolutions in Kuhn's sense (at best they are "mini-revolutions" in the sense of Andersen 1996). On the other hand, these cases seem to be more than just new exemplars of the kind that characterize successful normal science, since they necessitated considerable modification of existing theories.

25. See also Schaffner's (1993, 37–43) account of the genesis of M. F. Burnet's clonal selection theory of antibody formation.

26. Another possible example of the role of analogies is Krebs's discovery of the citric acid cycle (which later became known as the "Krebs cycle") in 1937. However, Holmes presented evidence that Krebs was not guided by the concept of a cyclic reaction scheme when he devised the citric acid cycle (Holmes 1993a, 428–429). This may seem remarkable, given that the urea cycle was Krebs's first major achievement. But Holmes's analysis shows that Krebs basically arrived at the citric acid cycle in the same way as he devised the urea cycle.

27. Note that this is also true of Darden's strategies for diagnosing problems and fixing faults in theories. At first sight, this heuristic appears to be quite generally applicable; however, just to say that scientists proceed like repairpersons does not offer much guidance in solving scientific problems. Of course, this strategy can be made more effective or even sufficient. For example, Lindley Darden has collaborated with an AI laboratory to implement this strategy on a computer. The most sophisticated version of their program TRANSGENE employs a backtracking procedure in diagnosing the source of failed predictions of genetic theory (Darden 1998). But this backtracking procedure works by simulating genetic crossing experiments and thus assumes a highly discipline-specific character.

CHAPTER 4

1. Bayesians are not claiming that scientists actually use Bayes's theorem when they are assessing evidence (except, perhaps, in special cases). Most of them view the "Bayesianismo" as some sort of an idealized model of actual scientific reasoning (see, e.g., Franklin 1990, 100).

2. For other historical accounts see Allchin (1991, 1994, 1996, 1997); Weber (1991); Prebble and Weber (2003); Robinson (1997); Rowen (1986); Weber (2002a; 2002b).
3. The phlogiston analogy is fruitfully explored in Allchin (1997).
4. Paul Feyerabend has argued (against Thomas Kuhn) that it is always rational for scientists to pursue alternative theories, no matter whether the established theory is in a state of crisis or not (see Feyerabend's comments on a draft of Kuhn's *Structure*, published by Hoyningen-Huene 1995). The ox-phos story supports Feyerabend on this point, since the chemical theory was not in a crisis when Mitchell proposed his alternative.
5. For someone like Mitchell to receive a Nobel Prize was quite an extraordinary event in the highly professional field of biochemistry. When he wrote his now famous article of 1961, he was an outsider to the field of oxidative phosphorylation. His hypothesis challenged all the major authorities in the field. Due to illness, Mitchell did no scientific work in the years following publication of his article. After leaving a post at the University of Edinburgh, he built his own, private lab in rural Cornwall, named Glynn Research Laboratories. From 1965 on, he published a series of experimental findings from his own lab, which he thought supported the chemiosmotic hypothesis. Most of these experiments were actually done by his only co-worker, Jennifer Moyle, while Mitchell was busy doing theoretical calculations. In contrast to Mitchell's main opponents in their large, prestigious biochemistry departments with substantial research grants, Glynn's financial situation was less than ideal (see Weber 1991).
6. Novel predictions, i.e., predictions of facts that were not used in the construction of the theory that predicted these facts, are thought to provide strong evidence for that theory (see Section 9.2). My case supports this thesis to some extent; however, the evidence was not strong enough to convince Mitchell's opponents.
7. Biochemists also used intact tissue slices to elucidate the order of metabolic reactions (Holmes 1992). However, a biochemical pathway was only considered to be fully understood once all the enzymes responsible had been identified and purified in an active state (see Section 5.3). This approach is known as "reconstitution."
8. Possibly, earlier results that seemed to demonstrate ox-phos in soluble systems contained submitochondrial particles.
9. See the highly ironic ending in Greville (1969, 72): "The suggestion might be made that, since the chemical mechanism involves multiplication of hypotheses, it should be removed by Occam's razor (...). This would be a pity, for in these days beards and dialogue are characteristic of youth and iconoclasm."
10. The situation here is actually more reminiscent of Kuhnian *incommensurability* (see the introduction to Chapter 7) than of underdetermination. In Weber (2002a), I argue that the chemical and chemiosmotic theories indeed exhibit the main features of incommensurability (taking up a suggestion originally due to Allchin 1994), but that this did not preclude an empirical comparison. This aspect is omitted from the present account because this chapter is already too long.
11. Of course, it could still be the case that there exist possible alternative theories that are empirically equivalent or equally well supported by the evidence. However, such alternatives were simply not available.
12. The concept of an experimental artifact will be discussed in Chapter 9 (Section 9.5).

13. Racker's reconstitution experiments were not the only new results that contributed to the resolution of the controversy. Boyer (1981) writes that the chemiosmotic theory was only acceptable to him once he had a plausible mechanistic explanation of how a membrane potential could drive the synthesis of ATP. He proposed such an explanation in 1973, on the basis of work in his laboratory on the exchange of phosphate-bound water molecules by the ATPase. This work later won him a Nobel Prize, too (Allchin 2002).

14. Boyer et al. (1977). This remarkable publication still reveals a considerable divergence of views on several matters among the main players of the ox-phos controversy. Furthermore, the review was partly written with the intention of improving the reputation and hence the funding prospects of bioenergetics within biochemistry (Prebble and Weber 2003, 195f.). However, all of the authors seemed to accept chemiosmotic coupling more or less. A few years earlier, this would probably not have been possible.

15. Gottfried Schatz, personal communication.

16. Severity, of course, is not an end in itself. The intended epistemic function of severity is that a highly severe test of a hypothesis provides a reason for holding the hypothesis to be true. In other words, a severe test on the basis of some evidence allows scientists to ampliatively (i.e., nondeductively) infer from the evidence to the hypothesis. The rationale behind this kind of inference backed by severe tests is an argument of the form: Had the hypothesis been false, it would have been very unlikely to pass the test. Hence, the hypothesis passing the test supports the view that it is true. Of course, such inferences from severe tests are in principle fallible; that is, they cannot guarantee the truth of the conclusion.

17. Antibodies, also known as immunoglobulins, are large, Y-shaped protein molecules produced by the immune systems of vertebrates for defense against pathogens. They are also important tools for biochemical research. Antibodies directed against almost any substance can be experimentally produced by repeatedly injecting rabbits with this substance (in this case ATPase) and taking the rabbit's blood serum a couple of weeks later.

18. The P:O ratio measures the amount of phosphate incorporated into ATP per oxygen atom consumed by respiration.

19. Later, the laboratory established a reconstitution procedure that gave rise to unidirectional incorporation of enzymes (Eytan, Matheson, and Racker 1976).

20. An example where this was possible is provided by the bromthymol blue experiments done in Chance's and Mitchell's laboratories (see Section 4.3).

CHAPTER 5

1. It was never denied that experience can play other roles in scientific research, for example, in the genesis of theories or in exploring new phenomena that are not yet covered by any theories. However, on the received view, this falls into the context of discovery and, therefore, outside the purview of epistemology and philosophy of science.

2. Historians moved away from the theory-centered approach before the philosophers, and for their own reasons. This trend in the historiography of science was mainly due to the emergence of anthropologically oriented laboratory studies (e.g., Latour

and Woolgar 1979; Knorr Cetina 1981), as well as the shift from "internalist" to "externalist" historiography in the wake of the so-called strong program (e.g., Shapin and Schaffer 1985).

3. I was actually involved in such a project during my first training as a molecular biologist (Verner and Weber 1989).

4. Rheinberger's work has been viewed as a radical departure from Anglo-American empiricist philosophy of science (e.g., Burian 1995, 124). This would seem unsurprising, given Rheinberger's background in French postmodernist philosophy. This chapter may be viewed as an attempt to show how this approach to the analysis of scientific practice could be integrated with empiricist philosophy of science.

5. Rheinberger also cites the work of Ludwik Fleck as a point of departure.

6. Note the stark contrast to Popper (1959, 107): "The theoretician puts certain definite questions to the experimenter, and the latter, by his experiments, tries to elicit a decisive answer to these questions, and to no others. All other questions he tries hard to exclude."

7. Uridine is one of the four nucleotide bases that make up ribonucleic acid (RNA). Whereas DNA contains the bases adenosine (A), thymidine (T), guanosine (G), and cytosine (C), RNA contains uridine (U) instead of thymidine. One of the roles of RNA in protein synthesis is that of a "messenger" that transfers information from DNA to protein. RNA is made by the cell by a process known as "transcription," in which an RNA copy is made from the DNA template. This "messenger RNA" (mRNA) directs the synthesis of proteins. Historians of biology (and biologists) disagree on when precisely mRNA was discovered, thus confirming an old insight due to Thomas Kuhn (1970, 54–62), namely that scientific discovery is more like a gradual process. At any rate, the in vitro system discussed by Rheinberger also played an important role in the discovery of mRNA as well as other species of RNA molecules involved in protein synthesis ("transfer" or tRNA and "ribosomal" or rRNA).

8. Because the availability of such radioisotopes as well as the techniques for counting radioactive decay came out of military research done during World War II and in the Cold War period, this constitutes an example of how the larger sociotechnological context can have a strong impact on scientific research.

9. Some philosophers have argued that the concept of information is used metaphorically in biology (see Section 8.1 and Section 8.3). Even if this is correct, it does not imply that the concept of information was historically unimportant. Metaphors and analogies have a high heuristic potential in scientific problem solving (see Chapter 3).

10. Mitchell himself was a confessing Popperian (e.g., Mitchell 1981, 29). For details on the research style of the Glynn laboratory, see Weber (1991); Prebble and Weber (2003).

11. See Prebble and Weber (2003) for an engaging biography of Mitchell.

12. Danielli has been credited by G. N. Cohen and J. Monod as the originator of the idea that organic molecules are transported through cell membranes by specific, membrane-bound enzymes (Cohen and Monod 1957). An example of such an enzyme is galactoside permease, the product of the y gene of the *lac* operon (see Section 3.2).

13. John Prebble has argued that Mitchell's chemiosmotic hypothesis was also in part a product of Mitchell's Heraclitean philosophy of nature (Prebble 2001). Mitchell recognized two fundamentally different kinds of entities in nature, namely *statids* and *fluctids*. The former are exemplified by a crystal, the latter by a flame. The characteristic of fluctids is that matter is flowing through a distinct region of space while maintaining a certain form. Furthermore, Mitchell defined *fluctoids* as objects that are composed of both static and flowing elements, for example, a Bunsen burner. Of course, Mitchell thought that living cells are the most complex fluctoids. While these ideas show Mitchell as an original and independent philosophical thinker, the specific connection of these ideas to the chemiosmotic hypothesis is not entirely clear. His achievement may also be viewed as the solution of a structured problem under a set of constraints (stoichometric, thermodynamical, cell-biological, etc.) in the manner discussed in Chapter 3 (see Section 3.3).

14. As Hans-Jörg Rheinberger pointed out to me in conversation, the term "bifurcation" is somewhat misleading, because in the theory of nonlinear dynamics it usually denotes a point where a system can take one of two possible trajectories in the state space. A more appropriate term in the present context might be "branching."

15. It is possible that Racker retrospectively exaggerates the extent to which the experiments with artificial vesicles were inspired by Mitchell's theory. It could be that they started to use phospholipids simply in order to give the enzyme a more "natural" environment.

16. This was pointed out to me by Douglas Allchin.

17. The most recent addition to mitochondrial functions is that they are involved in apoptosis (programmed cell death).

18. Rheinberger (personal communication) rejects this kind of social reductionism, as do the proponents of the "Paris School" in sociology of science, e.g., Latour (1999).

19. This alternative explanation was suggested to me by Hans-Jörg Rheinberger.

20. Rheinberger (ibid.) also writes that "[t]his vagueness is inevitable because, paradoxically, epistemic things embody what one does not yet know." Perhaps, what he means is that epistemic things become objects of research before there is a fully worked out and experimentally tested theory or mechanism that specifies all the properties and causal interactions of this object. I maintain that, the moment a research object enters into the scientists' thinking, in whatever speculative and vague a way, it becomes a theoretical entity.

21. It could be objected that these are questions that only arise within a *realist* framework, and that it is actually a strength of Rheinberger's approach that it requires no reference to objects that exist independent of scientific practice. However, I think it is rather a weakness of the nonrealist stance that it does not really allow us to say that scientists studied the *same* objects with *different* techniques. I understand Burian's criticism as directing us exactly to this problem. In Chapter 9, I try to defend a realist position by arguing that the grounds for realism in experimental biology lie exactly in the fact that some research objects can be identified by different experimental approaches.

22. Neither of these questions has really been dealt with in connection with philosophical and AI research on "discovery" (see Chapter 3), where experimental methods and research problems are usually treated as simply given (Maull 1980).

CHAPTER 6

1. Burian (1992, 1993a, 1993b, 1996); de Chadarevian (1998); Ankeny (2000); Clarke and Fujimura (1992); Clause (1993); Creager (2002); Geison and Creager (1999); Geison and Laubichler (2001); Holmes (1993b); Kohler (1991; 1993; 1994); Lederman and Burian (1993); Lederman and Tolin (1993); Mitman and Fausto-Sterling (1992); Rader (1999); Rheinberger (2000a); Summers (1993); Zallen (1993).
2. For general histories of classical *Drosophila* genetics see Sturtevant (1965); Allen (1978); Carlson (1981).
3. Restriction enzymes or restriction endonucleases are bacterial enzymes that cut DNA, some of them at specific sequence motifs. These enzymes, which were first described in 1965 by Arber, Nathans and Smith, are part of a bacterial cell's defense against viral DNA.
4. "Hybridization" refers to the experimental production of double-stranded nucleic acid molecules where the two strands are derived from different sources. The technique makes use of the strong disposition of complementary nucleic acid molecules (which is also crucial for their informational properties) to anneal to a double helix. Under certain conditions, it is possible to hybridize single-stranded RNA to complementary DNA; this forms the basis of colony hybridization and many other molecular techniques.
5. This is another important technique, which uses the unique distribution of cleavage sites for different restriction enzymes in order to physically map DNA molecules. The DNA fragment to be mapped is cut with a number of restriction enzymes and the resulting fragments are separated by molecular weight by gel electrophoresis. Since the molecular weight of the fragments is proportional to their length, the specific pattern of restriction fragments can be used to construct a map of the restriction sites on the DNA fragment.
6. Actually, the products of these genes were known from classical studies in biochemical genetics (xanthine dehydrogenase and acetylcholinesterase, respectively). However, this knowledge could not then be used for cloning because the concentrations of the corresponding mRNAs were too low to permit isolation of specific RNAs for colony hybridization. By contrast, some other "household genes" like *Adh* (alcohol dehydrogenase) were cloned by prior identification of an mRNA (Benyajati, Wang, Reddy, Weinberg, and Sofer 1980).
7. "Fushi tarazu" is Japanese and means "not enough segments," because *fushi tarazu* mutants show a reduced number of segments at an early embryonic stage.
8. The source of this idea is Bruno Latour and Steve Woolgar's pioneering ethnographic study of laboratory life (Latour and Woolgar 1979).
9. An important aspect of the moral economy of the *Drosophila* community was the habit of freely exchanging strains between competing laboratories.
10. In genetic engineering, the term "vector" designates a DNA molecule (typically a plasmid) that can be used to insert DNA into other organisms.
11. Another example of preparative experimentation is chemical synthesis.
12. Note that the idea that experimental knowledge sometimes endures changes at the theoretical level is one of the claims associated with the New Experimentalism (see Chapter 5).

13. See Burian (1993a); Clause (1993); Holmes (1993b); Kohler (1993); Lederman and Burian (1993); Lederman and Tolin (1993); Summers (1993); Zallen (1993); Geison and Creager (1999); Rader (1999).
14. This was pointed out to me by Angela Creager.
15. A counterexample might be provided by Sydney Brenner's choice of the nematode *Caenorhabditis elegans* as a model organism for studying development and the genetic control of behavior in the 1960s. (Today, Brenner works on yet another model organism, the fish *Fugu*.) But even in this case it can be argued that the research questions were thoroughly transformed by the constructive work of shaping *C. elegans* into a powerful model system (see de Chadarevian 1998).
16. Some viruses, as well as prions (infectious agents causing diseases like BSE), do not contain DNA. However, these "organisms" are not self-reproducing.
17. "Induction" in general refers to inference from logically weaker premises, for example, inferring a general conclusion from a few individual cases. "Enumerative" induction designates an inferential procedure by which a conclusion about the properties of all the members of a set is drawn by simply generalizing (without additional information) from the properties of the members of a proper subset. An example of enumerative induction would be to infer from the observation of some finite number of white swans that all swans are white. Famously, Karl Popper has rejected inductive inference. Today, philosophers of science agree that *enumerative* induction plays no role in science. However, there are other forms of inductive inference that are thought to be sound. We are about to examine an example.
18. Eukaryotic genes are interspersed with noncoding sequences known as "introns." The protein-coding parts of genes are called "exons." In a process known as "splicing," the introns are enzymatically removed from the primary transcript (the first RNA molecule that is produced when a gene is expressed). In some genes, there are different ways in which the exons can be combined. Thus, the mechanism of alternative splicing allows the cell to synthesize more than one functional protein from a single gene.

CHAPTER 7

1. I wish to thank Howard Sankey for pointing this out to me.
2. It can be objected that whether some way of referring to genes is correct or not can only be judged from today's perspective, assuming the truth of our contemporary knowledge. Kitcher fully accepts this charge (1993, 100–101, fn. 13). However, he defends this "whiggish" approach by claiming that historical interpretation always has to adopt a certain standpoint, and taking the contemporary standpoint is preferable because the latter is superior with respect to its knowledge about nature. I agree with this at least insofar as considerations of reference are concerned, which make no sense without taking a realist stance on contemporary scientific knowledge.
3. Olby (1979, 1985) also argues that Mendel's concern was not with heredity but with the origin of new species by hybridization. Since his experiments provided no new insights into this (presumed) phenomenon, this might provide some explanation of why Mendel's work was ignored at first.

4. Initially, Bateson's use of this term was not consistent; he sometimes meant the visible, phenotypic traits and sometimes the genotypic elements that determine these traits (Darden 1991, 179). The genotype/phenotype-distinction is usually attributed to Wilhelm Johannsen, who formulated it in 1909 on the basis of his concept of a "pure line." Elof Carlson has argued that failure to clearly distinguish between phenotypic characters and genotypic factors led several geneticists astray, for example, the eminent W. E. Castle (Carlson 1966, Chapter 4).

5. The classic complementation test does not work with dominant mutations. There are ways around this problem, which I will not go into here.

6. Raphael Falk pointed out to me that all cases of intragenic recombination before 1955 are only clear in retrospect, that is, after Benzer's discovery of genetic fine structure (to be discussed below).

7. This was probably a specific feature of Benzer's phage system, not a general characteristic of the recombination process.

8. The additional genetic complexities recently found by molecular biologists has led some commentators to suggest that the gene concept is in a state of "dissolution" (Fogle 2000), or even that there is no unified gene concept at the molecular level (Beurton 2000). These authors argue that it has become increasingly difficult to decide which DNA sequences are part of a gene, and that, therefore, the molecular gene concept should be abandoned. It seems to me that these claims are predicated on the assumption that genes should be like the simple entities they were thought to be back in the 1960s, namely uninterrupted pieces of DNA that are colinear with protein. However, nothing in the molecular gene concept says that genes should be simple structures.

9. In the extensive debates on reduction in genetics, several authors have argued that genes cannot be defined in purely physicochemical terms (e.g., Hull 1974, Chapter 1; Kitcher 1984; Rosenberg 1978). I think that this is essentially correct; however, as Waters (1994, 2000) shows, the molecular gene concept is partly functional (i.e., it picks out its referents by their causal role in the cell, not just by their molecular structure). Therefore, for the time being, the molecular gene concept does not support a reduction of genetics to physics and chemistry. What it does support is some form of reduction of genetics to molecular biology. Waters views the latter discipline as distinct from organic chemistry; according to him, it is a proper *biological* discipline with its own concepts, investigative procedures, and explanatory patterns. As far as Hull's, Kitcher's, and Rosenberg's claims are concerned, they might ultimately be proven wrong by the rapid advances in genomics. Bioinformatic methods for predicting genes from DNA sequences alone are being improved at a rapid pace.

10. This is supported by the following quotation from Morgan's preface to *The Mechanism of Mendelian Heredity*: "[W]e have made no assumption concerning heredity that cannot also be made abstractly without the chromosomes as bearers of the postulated hereditary factors" (Morgan et al. 1915, viii).

11. It could be suggested that the family resemblance of the different gene concepts that I have mentioned suffices for accounting for some kind of coherence between the different historical versions. However, this would be a rather weak kind of coherence. It is worth examining whether different gene concepts had more in common, e.g., a common reference.

12. I want to leave open the question of the possible reference of Mendel's gene concept, since we do not know for sure how Mendel envisioned his *Faktoren*, especially with regard the causal powers that he attributed to them.

13. For example, the American geneticist W. E. Castle called the "purity of gametes" into question (Castle 1915). He thought that he had evidence for some kind of "contamination" of the factors in heterozygotes. Later, this effect was shown to be due to multiple factors.

14. It will be noted that the expression on the right-hand side also contains the term "gene"; thus circularity seems to loom here. However, this would only be a problem if (2) were a *definition*, which it is not. Note that the term "gene" is only *mentioned* on the left-hand side, while it is *used* on the right-hand side.

15. Recently, this protein was shown to be a membrane transporter responsible for the transport of guanine and tryptophan, which are precursors of the red eye pigment of *Drosophila* (Mackenzie et al. 1999). This is a nice example of how unexpected the molecular function of classical genes can turn out to be; classical geneticists had always assumed that the *white* gene was involved in the synthesis of eye pigments.

16. The defining characteristic of such a series is that different alleles produce, in *trans* heterozygotes, a phenotype in which the phenotypic manifestations common to both alleles are expressed, whereas the traits with respect to which the two alleles show different manifestations are in the wild-type condition. Hence, these alleles are nonallelic according to some traits (those traits which are normal in the heterozygote) and allelic for some other traits (those traits showing the mutant phenotype). These "step-alleles" can be arranged in a series ranging from alleles being almost fully nonallelomorphic with respect to each other (i.e., the heterozygous flies are almost normal) to alleles that seem to be allelomorphs in that the heterozygous show the full mutant phenotype. Because of this discrete array of degrees of allelomorphism, this phenomenon was termed "step-allelomorphism."

17. This resolved an old puzzle, which had surrounded the *achaete–scute* locus since the 1930s. Whereas N. P. Dubinin (1932) and some other Russian geneticists had thought that the step-allelomorphism shown by mutants at this locus could be explained by the existence of several "subgenes," Muller and Prokofyeva (1935) first argued that the locus contains three independent genes (*achaete, scute*, and *lethal of scute*). However, Raffel and Muller (1940) had to admit that the locus was probably more complex than this.

18. Of course, mass needs to be distinguished into Newtonian mass, relativistic mass, and relativistic rest mass. When I say "mass doesn't come in different flavors," what I mean is that the property of mass *within* a given theoretical framework (Newtonian or relativistic) is the same property regardless of the nature of the body that has a mass – at least according to the current state of physics (if I understand it correctly).

19. More precisely, some *tokens* of the term "gene" referred successfully.

20. Hartry Field has argued that the reference of the Newtonian concept of mass was *indeterminate*; that is, there is no fact of the matter as to whether this concept referred to relativistic mass or to rest mass (Field 1973). However, according to Field, the term "mass" in Newtonian mechanics did refer to *something*; it just lacked a fundamental differentiating capacity that only Einstein's concept of mass introduced. Field called this phenomenon "partial denotation." I think the referential

relationship between the different gene concepts is not adequately captured by
Field's account (even though Field suggested that it is; see 1973, 477), because
there *is* a fact of the matter as to what the classical gene concept referred to, even
if the relationship to the contemporary concept is complex.

21. Howard Sankey has pointed out to me that the extensions of the Ptolemaïc and
Copernican concepts of planet also overlap partially. There is indeed some similarity
to the case of the gene, in that "planet" also denotes a class of distinct individuals
that can be classified in different ways. However, I don't think we could say that the
Ptolemaïc concept of planet referred to a natural kind, since the reason for excluding
the Earth from the planets was lost in the Copernican Revolution (namely, its being
the center of the universe). By contrast, I have argued that even older concepts
of the gene referred to natural kinds, even if these kinds do not correspond to the
natural kind picked out by the molecular gene concept.

22. Hans-Jörg Rheinberger has suggested that the case of the gene shows that we need
an "epistemology of the imprecise" (Rheinberger 2000b).

CHAPTER 8

1. For example, Delbrück (1971) and Holland (2001). Of course, Aristotle knew
nothing about DNA. He thought that the male semen contains an organism's form
and its "principle of motion," where "motion" is akin to what we call "development"
today (Aristotle 1943, 729a).

2. See Sarkar (1996), Godfrey-Smith (2000c), Weber (2001b), Kitcher (2001), and
Holland (2001). The historian of biology Lily Kay has based some of the main
conclusions in her monumental study of the history of the genetic code (Kay 2000)
on the assumption that information in genetics is just a metaphor. The theoretical
biologist John Maynard Smith defends the use of the information concept in biology (Maynard Smith 2000; cf. Sarkar 2000; Godfrey-Smith 2000b; Winnie 2000;
Sterelny 2000b); so does Jablonka (2002). I shall discuss attempts to defend the
concept of information in genetics in Section 8.3. Evelyn Fox Keller (2002) shows
that theorizing in developmental biology has been strongly influenced by various
metaphors.

3. See Oyama (1985, 2000); Griffiths and Gray (1994, 1997); Griffiths and Knight
(1998); Oyama, Griffiths, and Gray (2001); Smith (1992); Griesemer (2000);
Griffiths (forthcoming). For critical discussions see Sterelny, Smith, and Dickison
(1996); Kitcher (2001); Godfrey-Smith (2000a); Schaffner (1998); Morgan (2001);
Waters (forthcoming).

4. DS theorists differ in particular in how useful they find information talk in biology.
Some, like Oyama (1985), seem to think that it is useful; they just argue that
information does not distinguish DNA from other developmental resources. Others,
like Griffiths and Gray (1994), doubt that information is an appropriate category in
biology.

5. Sterelny does not go all the way with the DS theorists. He thinks that genes are
distinct because they are replicators and represent aspects of the phenotype (Sterelny
et al. 1996; Sterelny 2000a, forthcoming). However, according to these authors,
genes are not the only replicators. DST, by contrast, rejects the replicator/interactor
distinction altogether (see below).

6. Such mechanisms could be responsible for the difficulty of cloning animals from differentiated cells (Reik and Dean 2003).
7. Note that, here, the methyl group is attached to a protein (histone) that is bound to DNA. By contrast, in genomic imprinting the methyl group is attached to the DNA molecule directly.
8. I wish to thank Dani Sirtes for pointing this out to me.
9. There are some exceptions to this, for example, a bacterial gene regulatory mechanism called attenuation. In this mechanism, protein synthesis stalls when the amino acid tryptophan is low in the cell. The reason for the stall is that the gene codes for a stretch of protein that is high in tryptophan. This stall prevents the premature abort of the transcription process (transcription and translation are coupled in bacterial cells). This mechanism seems to affect mainly the rate of translation, not the identity of the gene product.
10. Bird songs might be viewed as *memes* in the sense of Dawkins (1989).
11. The best introduction to gene selectionism is still Dawkins (1989). Sober and Wilson (1994) provide a guide to the literature. An elegant resolution of the problem can be found in Waters (1991).
12. See Purdue and Lazarow (2001) for a review of the current state of research on the biogenesis of peroxisomes.
13. See Godfrey-Smith (2000d) for a recent discussion of the concept of replicator. Replicators are usually contrasted with interactors (Hull 1981) or vehicles (Dawkins 1982), which are the entities that interact with the environment, whereas replicators are the entities that are being copied from one generation to the next.
14. For example, it is debatable whether DNA methylation patterns could be said to be replicators. I might be willing to concede that they are; however, the evolutionary role of DNA methylation is far from clear at the time of this writing. For example, it is not clear at present for how many generations such modifications can persist. Another class of potential candidates for true replicators is the prions, the causative agents of various diseases such as bovine spongiforme encephalopathy (BSE) and Creutzfeldt-Jacob disease. According to current thinking, these proteins are capable of transmitting a misfolded state to other molecules of the same kind. The misfolded state – which also causes the disease symptoms – is thus infectious. I wish to thank Renato Paro for calling these examples to my attention.
15. See, for example, Dawkins' reply to Gould in Dawkins (1989, 271–272).
16. Sterelny (forthcoming) has noted the mistake and uses the term "predictive information" instead.
17. Epigenetic inheritance mechanisms such as the ones discussed under Step One are not known, at present, to contain any proofreading mechanisms (Renato Paro, personal communication).
18. Griffiths and Gray (1994) consider developmental *processes* to be more fundamental than developmental *systems*. A developmental process token is capable of initiating more token-processes of the kind to which it belongs. A developmental system is whatever exists from the beginning to the end point of the cyclical process, where the exact location of these points is yours to choose. DS theorists sometimes describe gametes (egg and sperm) as "bottlenecks" in the developmental process.
19. The following account is based on Lawrence (1992); Alberts et al. (1994); Gehring (1998); Mann and Morata (2000); and Wolpert et al. (2002). I had to simplify

the story. I think that the omissions do not affect the philosophical conclusions; however, skeptical readers should consult the sources cited above.

20. A separate system patterns the embryo along the dorsoventral (from back to front) axis. Similar mechanisms are at work there.

21. The notion of morphogen dates back to the late nineteenth century and is usually attributed to Thomas Hunt Morgan (e.g., Lawrence 1992, 204–206). The idea has recurrently attracted the interest of theoreticians, as it seemed to provide a physically and chemically realistic solution to the problem of pattern formation (e.g., Turing 1952; Wolpert 1969; Crick 1970). In spite of some early evidence for the existence of morphogens in various organisms like *Hydra* or leafhoppers, the idea remained controversial until the discovery of the *bicoid* gradient in *Drosophila*.

22. The name of this class of genes reflects the phenotypes of some mutations in these genes, which are characterized by gaps in the segment pattern of the early embryo (most of these mutations are lethal in the embryonic stage). The colorful names of the individual genes derive from attempts to describe the sometimes bizarre phenotypes of mutations in these genes. By convention, capitalized names indicate that typical mutations in these genes are dominant. Most of the genes were first identified exclusively by the methods of classical genetics and then cloned for molecular studies (see Chapter 6 for this approach). Some of the gene names are German because most of these mutants were found in the course of a large-scale search for developmental genes conducted by the German geneticist Christiane Nüsslein-Volhard (Nüsslein-Volhard and Wieschaus 1980), who shared a Nobel Prize in 1995 for this work. For historical accounts, see Keller (1996) and Morange (2000).

23. As their names suggest, the genes *even-skipped* and *fushi tarazu* are characterized by mutants that lack every other of the parasegments in the embryonic stage. "Fushi tarazu" is Japanese for "not enough segments."

24. Interestingly, in some insects the role of *bicoid* is played by the gene *hunchback*, which functions as a gap gene in *Drosophila*. This seems to be the ancestral condition (Schröder 2003). Thus, evolution has occasionally redeployed some of these genes for a different, but related function.

25. Some of these transcription factors are known to bind to the regulatory regions of their own genes, thus creating autoregulatory feedback loops (Schier and Gehring 1992).

26. The question might arise of how only eight homeotic selector genes (three in the Antennapedia and five in the Bithorax complex) can specify the identity of fourteen parasegments. The key to this apparent paradox lies in the combinatorial action of these genes, which was already predicted by Lewis (1978) on the basis of classical genetic experiments.

27. The highly influential British mathematician Alan Turing introduced an ingenious mathematical approach for modeling morphogenesis that also involved gradient-forming diffusible morphogens (Turing 1952). In his model, a system of oscillating chemical reactions generates wavelike patterns in a developing embryo. Even though Turing is sometimes credited as one source of the idea of a diffusible morphogen, his specific models are not thought to apply to real developmental systems today. None of the known morphogens forms a reaction-diffusion system as in Turing's model. However, some conceptual similarities between Turing's views and contemporary models may not be sufficiently appreciated.

28. Ken Waters has argued the same point about DNA-based genetic information (Waters 2000).

29. This problem was already noted by Sarkar (2000, 211), charging Maynard Smith with "hyperselectionism." In his reply to commentaries, Maynard Smith announces that he is "not worried" by this charge. However, his reply to Sarkar misses the point: "Of course there are accidental and even maladaptive features of organisms, but the function of a symbolic signaling system is not one of them" (Maynard Smith 2000, 216). The point is not that genetic systems could have arisen by chance or be maladaptive. The problem is that, for Maynard Smith's teleosemantic approach to work, only DNA sequences that are adaptive can be said to carry information.

30. Note that the same is true about linguistic expressions (spoken and written language).

31. This problem besets the whole teleosemantic program, but this is not the place to argue this point.

32. Recall that analogies have been proposed as important parts of problem-solving heuristics in scientific discovery; see Chapter 3. In her impressive historical account, the late Lily Kay argues that the prevalence of informational notions in twentieth century genetics and molecular biology is a result of the rise of information technology and the military use of this technology for cryptographic purposes (Kay 2000). This is a possible scenario, and it is fully compatible with my account of the concept of genetic information. Scientists may use any ideas that are floating around in their broader culture for problem solving. Maynard Smith's (2000) text may actually be considered to be evidence for Kay's thesis.

33. Curiously, Maynard Smith in some places talks about the "coding analogy" and the "information analogy" himself. I suspect that he did not think that DNA really carries meaning, but this is what he said.

34. Phenomena like these have led the great embryologist Hans Driesch to give up all attempts to explain development mechanically and to turn to vitalism around the turn of the nineteenth century (Weber 1999b).

35. Note that my analysis differs from the so-called teleosemantic accounts of intentional states (Millikan 1989) and of genetic information (Maynard Smith 2000), because I do not invoke natural selection.

36. Ulrich Stegmann (personal communication) has complained that my account does not explain why developmental biologists attribute information to morphogen gradients but not to other components. It is possible that this has to do with the strong dependence of the organism's form on the shape of the morphogen gradient; however, this problem requires further analysis. It is not my goal here to defend the biologists' use of informational notions.

37. The term "master control genes" is due to E. B. Lewis (see Lewis 1992 for a historical review).

38. The concept of arbitrariness is more complex than one might think. Ulrich Stegmann (forthcoming) provides a subtle analysis of this notion.

39. The notion of causal priority has, to my knowledge, not been applied in this debate. Some developmental systems theorists have considered whether there is some kind of causal *asymmetry* between genes and other cellular constituents (see Sterelny and Griffith's Step Five, discussed in Section 8.1) and, of course, have concluded that

there is none. However, they have not made the relevant sense of causal asymmetry very clear.

40. I am indebted to Ken Waters for pointing this out to me.
41. I am not claiming that the list is complete. If some known causal role is missing, please call it to my attention.
42. This possible defense of DST was suggested to me by Paul Hoyningen-Huene.
43. My use of the attributes "weak" and "strong" follows the convention that the stronger thesis should logically entail the weaker one. This is the case here: If there is no causal difference between DNA and other components, this logically entails that there is no categorical causal difference. The converse does not hold; thus the thesis that claims no categorical difference is logically weaker.
44. This is still subject to debate; see, for example, Horan (1994), Brandon and Carson (1996), Weber (2001a).
45. Some vitalists such as Hans Driesch thought that there is a special kind of causation involved in biological organisms (Weber 1999b), but this is not a viable option today.
46. Ken Waters (personal communication) thinks that there is a categorical difference in causal role between genes and other developmental resources. He spells out this difference mainly in terms of the investigative reach that genes offer to researchers (compare Schaffner 1998), but he plans to substantiate this claim further.

CHAPTER 9

1. Such a form of realism is defended, for example, by Richard Boyd (1981, 1983, 1985, 1990, 1992); Howard Sankey (2001, 2002); Stathis Psillos (1999); Jarrett Leplin (1997); Alan Musgrave (1988); Wesley Salmon (1984); and Michael Devitt (1984). Ronald Giere (1985, 1988) has defended a model-theoretic form of realism that does not involve truth. In addition, there are several other weaker variants of realism that cannot be discussed here, for example, structural realism (Maxwell 1970), "schoolbook" realism (Rescher 1987), or natural kind realism (Carrier 1993). A specific weaker version of realism, namely entity realism, will be discussed in Section 9.3.
2. For Kuhnian antirealists, the relation of reference (see Chapter 7) holds between theoretical terms and the phenomenal world. By contrast, for realists that relation is seen as obtaining between theoretical terms and the world in itself (Andersen 2001).
3. It is important to keep antirealism separate from *relativism*. According to relativists, there can be no rational way of choosing a set of methods or a theoretical framework. This is not identical to antirealism. For example, Kuhn has always eschewed relativism, and so do instrumentalists like van Fraassen. By contrast, social constructivism clearly is a form of relativism.
4. For some rebuttals see Musgrave (1988), Hacking (1999), Kitcher (1993), Psillos (1999).
5. See the *Critique of Pure Reason* (Kant 1998, B37/A22–B58/A41).
6. Of course, there are thinkers who have denied that organisms are purely material objects, namely vitalists. But the last great vitalist, Hans Driesch, was himself a

Kantian. He introduced a vital factor termed "entelechy" to explain pattern-forming and self-regulating developmental processes. However, entelechy, for Driesch, is an a priori concept formed under a specific category of reason, namely teleology (an idea that Kant himself would reject). Thus, curiously, Driesch, although clearly a vitalist, was not a *realist* about vital forces (Weber 1999b).

7. Here, I am indebted to Paul Hoyningen-Huene.

8. Some philosophers have argued that *all* our perceptions are theory-laden, even our direct sense perceptions (Kuhn 1970, see especially Chapter X; Feyerabend 1988, see especially Chapter 6). If this is taken seriously, there can be no difference in principle between direct observations and theoretical interpretations of data generated by a scientific instrument. This leads to some difficult philosophical issues that cannot be discussed here. For reasons that I have mentioned above, the scientific realism to be defended here is conditional on realism about the objects that we can perceive directly.

9. In order to be visible under a microscope, a structure must have a size that is greater than the wavelength of the detector beam. Electron microscopes have a much higher resolution than light microscopes because they use an electron beam, which has an extremely short wavelength compared to visible light.

10. Readers who don't believe that genes can be photographed should see electron micrographs of bacterial DNA molecules with nascent RNA molecules attached to them, which show a Christmas tree-like appearance. The gene reveals its location on the DNA by the attached RNA molecules (e.g., Stryer 1981, 606).

11. Doubts that genes exist independent of their changing conceptualizations in genetic theory can be found throughout the history of genetics (e.g., Goldschmidt 1946, Stadler 1954, Fogle 2000).

12. The notion of approximate truth, of course, poses some logical difficulties. It is controversial whether this should be admitted as a logical category that is on a par with truth/falsity. A possible way of avoiding this conundrum is saying that an approximately true theory is strictly false, but that there exists a similar theory that is strictly true.

13. An additional difficulty is posed by the fact that biologists often have to idealize and/or simplify things in order to see the wood for the trees. In these cases, realism may be maintained by claiming that a model captures some *salient* aspect of the process or mechanism under study. For example, realism about chemiosmotic theory (see Chapter 4) may entail nothing more than the claim that a chemiosmotic gradient rather than a chemical intermediate couples oxidation and phosphorylation, with all the details remaining open. If this sounds too weak and not realist enough, compare it to what an instrumentalist would have to say ("the chemiosmotic mechanism is empirically adequate").

14. Among others, this kind of argument has been used to defend realism by Smart (1963); Putnam (1975b, 73); Boyd (1981, 1983, 1985); Musgrave (1988); Leplin (1997); and Psillos (1999).

15. See Chapter 7 for the concept of reference.

16. The quantitative aspect of the prediction is important here, for the mere fact of deflection could also have been predicted under the Newtonian theory, provided that light was assumed to be made of corpuscles that have a mass.

17. See, e.g., Grünbaum (1964); Worrall (1989); Mayo (1996, Chapter 8); Zahar (1973); Leplin (1997).

18. This is not to deny the historical importance of the Meselson–Stahl experiment. When this experiment was done, the main issue at stake was not the Watson–Crick structure of DNA, which was already widely accepted. The main issue was the mechanism of replication. With respect to this mechanism, molecular biologists had grave difficulties imagining the semiconservative mechanism envisioned by Watson and Crick, as it implied that the two strands in a replicating molecule would have to rotate around each other at a mind-boggling angular velocity (which they actually do on the current understanding). Thus, semiconservative replication seemed implausible. The double-helical structure was also consistent with a conservative mechanism. Several important theoreticians including Gunter Stent and Max Delbrück were skeptical of semiconservative replication and proposed conservative models (Holmes 2001, Chapter 1).

19. Jarrett Leplin (1997, 77) thinks that only the prediction of *qualitative generalizations* by a theory, when no other theory predicts the same qualitative fact, counts as a *novel* prediction. However, he also takes quantitative accuracy as a measure of predictive success (he chooses to keep novelty and accuracy conceptually independent); thus his account is also difficult to apply to our cases.

20. A similar strategy to argue for scientific realism has been developed independently by Ronald Giere (1988, 124–127). Interestingly, both Hacking and Giere formed this idea while spending some time in high-energy physics laboratories.

21. The following defense of the argument from independent determinations owes much to some written comments by Eric Oberheim on a draft of this chapter.

22. As Bill Wimsatt formulated it: Map distance is a *theoretical* concept (Wimsatt 1987, 36).

23. See also Burian's criticism of Rheinberger, discussed in Section 5.3.

24. I will concentrate on microscopic artifacts in this section. However, most of what I say can be adapted to nonvisual experimental methods (see Section 4.7).

25. I have searched the National Library of Medicine database *PubMed* for articles containing the term "mesosome" in their titles, keywords, or abstracts. The search returned over 300 articles. *PubMed* carries articles in scientific journals from 1966 until the present. This figure probably underestimates the actual number of articles, as only the more recent ones are filed with electronic abstracts and the term "mesosome" is not always mentioned in the title or keywords.

26. Culp's main concern is not with realism, but with a problem that has been introduced by Harry Collins (1985, Chapter 4), namely the so-called experimenter's regress (see also Culp 1995). In essence, an experimenter's regress threatens if some experimental procedure is evaluated solely on the basis of the results it produces; yet there is no independent way of checking the correctness of these results. Collins's main example is a gravity wave detector. As the discussion below will show, in experimental biology there are ways in which the reliability of experimental procedures can be ascertained without relying on it's producing some expected result. Another concern of Culp's is the "theory-ladenness" of data, which I think is conceptually different from both the "experimenter's regress" and the problem of experimental artifacts. Note that the fact that scientists use a theory in interpreting experimental data need not affect the status of the interpreted data as evidence for

this same theory, so long as some kind of conflict with nature remains possible (Glymour 1980; Hoyningen-Huene 1988). For these reasons, I suggest that Culp's concern ought to have been realism.

27. Compare Hacking: "Practice – and I mean in general doing, not looking – creates the ability to distinguish between visible artifacts of the preparation or the instrument, and the real structure that is seen with the microscope" (1983, 191).

Bibliography

Ackermann, R. 1989. The New Experimentalism. *British Journal for the Philosophy of Science* 40: 185–90.

Adams, M. D., et al. 2000. The Genome Sequence of *Drosophila melanogaster*. *Science* 287: 2185–95.

Alberts, B., D. Bray, J. Lewis, M. Raff, K. Roberts, and J. D. Watson. 1983. *Molecular Biology of the Cell*. New York: Garland.

Alberts, B., D. Bray, J. Lewis, M. Raff, K. Roberts, and J. D. Watson. 1994. *Molecular Biology of the Cell*. 3d ed. New York: Garland.

Allchin, D. 1991. Disagreement in Science: The Ox-Phos Controversy, 1961–1977. Ph.D. thesis, University of Chicago.

Allchin, D. 1994. The Super Bowl and the Ox-Phos Controversy: "Winner-Take-All" Competition in Philosophy of Science. In *PSA 1994*, Vol. 1, edited by D. Hull, M. Forbes, and R. M. Burian. East Lansing: Philosophy of Science Association, pp. 22–33.

Allchin, D. 1996. Cellular and Theoretical Chimeras: Piecing Together How Cells Process Energy. *Studies in the History and Philosophy of Science* 27: 31–41.

Allchin, D. 1997. A Twentieth-Century Phlogiston: Constructing Error and Differentiating Domains. *Perspectives on Science* 5: 81–127.

Allchin, D. 2002. To Err and Win a Nobel Prize: Paul Boyer, ATP Synthase and the Emergence of Bioenergetics. *Journal of the History of Biology* 35: 149–72.

Allen, C., M. Bekoff, and G. V. Lauder (eds.) 1998. *Nature's Purposes: Analyses of Function and Design in Biology*. Cambridge, Mass.: MIT Press.

Allen, G. E. 1978. *Thomas Hunt Morgan: The Man and His Science*. Princeton: Princeton University Press.

Alonso, M. C., and C. V. Cabrera. 1988. The achaete-scute Gene Complex of *Drosophila melanogaster* Comprises Four Homologous Genes. *The EMBO Journal* 7: 2585–91.

Andersen, H. 1996. Categorization, Anomalies and the Discovery of Nuclear Fission. *Studies in History and Philosophy of Science* 27: 463–92.

Andersen, H. 2001. Reference and Resemblance. *Philosopy of Science* (Proceedings) 68: S50–S61.

Ankeny, R. A. 2000. Fashioning Descriptive Models in Biology: Of Worms and Wiring Diagrams. *Philosophy of Science* (Proceedings) 67: S260–S272.

Bibliography

Ariew, A. 2003. Ernst Mayr's 'Ultimate/Proximate' Distinction Reconsidered and Reconstructed. *Biology and Philosophy* 18: 553–65.

Aristotle. 1943. *Generation of Animals*. With a translation by A. L. Peck. London: William Heinemann.

Armstrong, D. M. 1983. *What Is a Law of Nature?* Cambridge: Cambridge University Press.

Ashburner, M., et al. 2000. Gene Ontology: Tool for the Unification of Biology. The Gene Ontology Consortium. *Nature Genetics* 25: 25–9.

Avery, O. T., C. M. McLeod, and M. McCarthy. 1944. Studies on the Chemical Nature of the Substance Inducing Transformation of Pneumococcal Types. Induction of Transformation by a Deoxyribonucleic Acid Fraction Isolated from Pneumococcus Type III. *Journal of Experimental Medicine* 79: 137–58.

Ayala, F. J. 1974. Introduction. In *Studies in the Philosophy of Biology. Reduction and Related Problems*, edited by F. J. Ayala and T. Dobzhansky. Berkeley: University of California Press, pp. VII–XVI.

Ayala, F. J. 1985. Reduction in Biology: A Recent Challenge. In *Evolution at a Crossroads: The New Biology and the New Philosophy of Science*, edited by D. Depew and B. Weber. Cambridge, Mass.: MIT Press, pp. 65–80.

Baltimore, D. 2001. Our Genome Unveiled. *Nature* 409: 814–16.

Balzer, W., and C. M. Dawe. 1986. Structure and Comparison of Genetic Theories: (2) The Reduction of Character-Factor Genetics to Molecular Genetics. *British Journal for the Philosophy of Science* 37: 177–91.

Bargmann, C. I. 1993. Genetic and Cellular Analysis of Behavior in *C. elegans*. *Annual Review of Neuroscience* 16: 47–51.

Barr, M. M., and P. W. Sternberg. 1999. A Polycystic Kidney-Disease Gene Homologue Required for Male Mating Behavior in *C. elegans*. *Nature* 401: 386–9.

Bateson, W. 1909. *Mendel's Principles of Heredity*. Cambridge: Cambridge University Press.

Beadle, G. W., and E. L. Tatum. 1941. Genetic Control of Biochemical Reactions in *Neurospora*. *Proceedings of the National Academy of Sciences of the United States of America* 27: 499–506.

Beatty, J. 1990. Evolutionary Anti-Reductionism: Historical Reflections. *Biology and Philosophy* 5: 199–210.

Beatty, J. 1995. The Evolutionary Contingency Thesis. In *Concepts, Theories, and Rationality in the Biological Sciences. The Second Pittsburgh–Konstanz Colloquium in the Philosophy of Science*, edited by G. Wolters, J. G. Lennox, and P. McLaughlin. Pittsburgh: University of Pittsburgh Press, pp. 45–81.

Bechtel, W. 1984. Reconceptualization and Interfield Connections: The Discovery of the Link between Vitamins and Co-enzymes. *Philosophy of Science* 51: 265–92.

Bechtel, W., and R. C. Richardson. 1993. *Discovering Complexity: Decomposition and Localization as Strategies in Scientific Research*. Princeton, N.J.: Princeton University Press.

Beisel, C., A. Imhof, J. Greene, E. Kremmer, and F. Sauer. 2002. Histone Methylation by the *Drosophila* Epigenetic Transcriptional Regulator ASH1. *Nature* 419: 857–62.

Bender, W., P. Spierer, and D. S. Hogness. 1983. Chromosomal Walking and Jumping to Isolate DNA from the Ace and rosy Loci and the Bithorax Complex in *Drosophila melanogaster*. *Journal of Molecular Biology* 168: 17–33.

322

Bibliography

Benyajati, C., N. Wang, A. Reddy, E. Weinberg, and W. Sofer 1980. Alcohol Dehydrogenase in *Drosophila*: Isolation and Characterization of Messenger RNA and cDNA Clone. *Nucleic Acids Research* 8: 5649–67.

Benzer, S. 1955. Fine Structure of a Genetic Region in Bacteriophage. *Proceedings of the National Academy of Sciences of the United States of America* 41: 344–54.

Beurton, P. 2000. A Unified View of the Gene, or How to Overcome Reductionism. In *The Concept of the Gene in Development and Evolution. Historical and Epistemological Perspectives*, edited by P. Beurton, R. Falk, and H.-J. Rheinberger. Cambridge: Cambridge University Press, pp. 286–314.

Beurton, P., R. Falk, and H.-J. Rheinberger (eds.) 2000. *The Concept of the Gene in Development and Evolution. Historical and Epistemological Perspectives*. Cambridge: Cambridge University Press.

Bigelow, J., and R. Pargetter. 1987. Functions. *Journal of Philosophy* 84: 181–96.

Billeter, M., Y. Q. Qian, G. Otting, M. Muller, W. Gehring, and K. Wüthrich. 1993. Determination of the Nuclear Magnetic Resonance Solution Structure of an Antennapedia Homeodomain-DNA complex. *Journal of Molecular Biology* 234: 1084–93.

Bingham, P. M., R. Levis, and G. M. Rubin. 1981. Cloning of DNA Sequences from the white Locus of *D. melanogaster* by a Novel and General Method. *Cell* 25: 693–704.

Boyd, R. N. 1981. Scientific Realism and Naturalistic Epistemology. In *PSA 1980*, Vol. 2, edited by P. D. Asquith and R. N. Giere. East Lansing: Philosophy of Science Association, pp. 613–62.

Boyd, R. N. 1983. On the Current Status of the Issue of Scientific Realism. *Erkenntnis* 19: 45–90.

Boyd, R. N. 1985. *Lex Orandi est Lex Credendi*. In *Images of Science. Essays on Realism and Empiricism, with a Reply from Bas C. Van Fraassen*, edited by P. M. Churchland and C. A. Hooker. Chicago: University of Chicago Press, pp. 3–34.

Boyd, R. N. 1990. Realism, Approximate Truth, and Philosophical Method. In *Scientific Theories*. Minnesota Studies in the Philosophy of Science, Vol. XIV, edited by C. W. Savage. Minneapolis: University of Minnesota Press, pp. 355–91.

Boyd, R. N. 1992. Constructivism, Realism, and Philosophical Method. In *Inference, Explanation and Other Frustrations. Essays in the Philosophy of Science*, edited by J. Earman. Berkeley: University of California Press, pp. 131–98.

Boyer, P. D. 1981. An Autobiographical Sketch Related to My Efforts to Understand Oxidative Phosphorylation. In *Of Oxygen, Fuels, and Living Matter*, edited by G. Semenza. Chichester: Wiley, pp. 229–44.

Boyer, P. D., B. Chance, L. Ernster, P. Mitchell, E. Racker, and E. C. Slater. 1977. Oxidative Phosphorylation and Photophosphorylation. *Annual Review of Biochemistry* 46: 955–1026.

Brandon, R. N., and S. Carson. 1996. The Indeterministic Character of Evolutionary Theory: No "No Hidden Variables Proof" but No Room for Determinism Either. *Philosophy of Science* 63: 315–37.

Bridges, C. B. 1914. Direct Proof through Non-disjunction That the Sex-Linked Genes of *Drosophila* are Borne by the X-Chromosome. *Science* 40: 107–9.

Bridges, C. B. 1916. Non-Disjunction as Proof of the Chromosome Theory of Heredity. *Genetics* 1: 1–52, 107–63.

Burian, R. M. 1985. On Conceptual Change in Biology: The Case of the Gene. In *Evolution at a Crossroads: The New Biology and the New Philosophy of*

Science, edited by D. Depew and B. Weber. Cambridge, Mass.: MIT Press, pp. 21–42.

Burian, R. M. 1992. How the Choice of Experimental Organism Matters: Biological Practices and Discipline Boundaries. *Synthese* 92: 151–66.

Burian, R. M. 1993a. How the Choice of Experimental Organism Matters: Epistemological Reflections on an Aspect of Biological Practice. *Journal of the History of Biology* 26: 351–67.

Burian, R. M. 1993b. Technique, Task Definition, and the Transition from Genetics to Molecular Genetics: Aspects of the Work on Protein Synthesis in the Laboratories of J. Monod and P. Zamecnik. *Journal of the History of Biology* 26: 387–407.

Burian, R. M. 1995. Comments on Rheinberger's "From Experimental Systems to Cultures of Experimentation." In *Concepts, Theories, and Rationality in the Biological Sciences. The Second Pittsburgh–Konstanz Colloquium in the Philosophy of Science*, edited by G. Wolters, J. G. Lennox, and P. McLaughlin. Pittsburgh: University of Pittsburgh Press, pp. 123–36.

Burian, R. M. 1996. "The Tools of the Discipline: Biochemists and Molecular Biologists": A Comment. *Journal of the History of Biology* 29: 451–62.

Burian, R. M., R. C. Richardson, and W. J. Van der Steen. 1996. Against Generality: Meaning in Genetics and Philosophy. *Studies in History and Philosophy of Science* 27: 1–30.

Cairns, J., G. S. Stent, and J. D. Watson (eds.) 1992. *Phage and the Origin of Molecular Biology*. Expanded ed. Cold Spring Harbor: CSHL Press.

Campuzano, S., Carramolino, L., Cabrern, C. V., Ruiz-Gomez, M., Villares, R., Boronat, A., and J. Modolell. 1985. Molecular Genetics of the achaete-scute Gene Complex of *D. melanogaster*. *Cell* 40: 327–38.

Campuzano, S., and J. Modolell. 1992. Patterning of the *Drosophila* Nervous System: The achaete-scute Gene Complex. *Trends in Genetics* 8: 202–8.

Carlson, E. A. 1966. *The Gene. A Critical History*. Philadelphia: Saunders.

Carlson, E. A. 1981. *Genes, Radiation, and Society: The Life and Work of H. J. Muller*. New York: Cornell University Press.

Carnap, R. 1938. Logical Foundations of the Unity of Science. In *International Encyclopedia of Unified Science*, Vol. 1. Chicago: University of Chicago Press, pp. 42–62.

Carrasco, A. E., W. McGinnis, W. J. Gehring, and E. M. de Robertis. 1984. Cloning of an *X. laevis* Gene Expressed during Early Embryogenesis Coding for a Peptide Region Homologous to *Drosophila* Homeotic Genes. *Cell* 37: 409–14.

Carrier, M. 1991. What Is Wrong with the Miracle Argument? *Studies in the History and Philosophy of Science* 22: 23–36.

Carrier, M. 1993. What Is Right With the Miracle Argument: Establishing a Taxonomy of Natural Kinds. *Studies in History and Philosophy of Science* 24: 391–409.

Carrier, M. 1995. Evolutionary Change and Lawlikeness, Beatty on Biological Generalizations. In *Concepts, Theories, and Rationality in the Biological Sciences. The Second Pittsburgh–Konstanz Colloquium in the Philosophy of Science*, edited by G. Wolters, J. G. Lennox, and P. McLaughlin. Pittsburgh: University of Pittsburgh Press, pp. 83–97.

Carrier, M. 1998. New Experimentalism and the Changing Significance of Experiments: On the Shortcomings of an Equipment-Centered Guide to History. In *Experimental*

Essays – Versuche zum Experiment, edited by M. Heidelberger and F. Steinle. Baden-Baden: Nomos, pp. 175–91.

Carrier, M. 2000. Multiplicity and Heterogeneity: On the Relations between Functions and Their Realizations. *Studies in History and Philosophy of Biological and Biomedical Sciences* 31: 179–91.

Cartwright, N. 1983. *How the Laws of Physics Lie*. Oxford: Clarendon.

Cartwright, N. 1989. *Nature's Capacities and Their Measurement*. Oxford: Oxford University Press.

Cartwright, N. 1999. *The Dappled World: A Study of the Boundaries of Science*. Cambridge: Cambridge University Press.

Castle, W. E. 1915. Some Experiments in Mass Selection. *American Naturalist* 49: 713–26.

Castle, W. E. 1919. Are Genes Linear or Non-linear in Arrangement? *Proceedings of the National Academy of Sciences of the United States of America* 5: 500–6.

Castle, W. E. 1920. Model of the Linkage System of Eleven Second Chromosome Genes of *Drosophila. Proceedings of the National Academy of Sciences of the United States of America* 6: 25–32.

Causey, R. 1972. Attribute-Identities in Microreductions. *The Journal of Philosophy* 69: 407–22.

Causey, R. 1977. *Unity of Science*. Dordrecht: Reidel.

Chalmers, A. 1999. Making Sense of Laws of Physics. In *Causation and Laws of Nature*, edited by H. Sankey. Dordrecht: Kluwer, pp. 3–16.

Chance, B. (ed.) 1963. *Energy-Linked Functions of Mitochondria*. New York: Academic Press.

Chance, B., and L. Mela. 1966a. A Hydrogen Concentration Gradient in a Mitochondrial Membrane. *Nature* 212: 369–72.

Chance, B. and L. Mela. 1966b. Proton Movements in Mitochondrial Membranes. *Nature* 212: 372–6.

Chance, B., and L. Mela. 1966c. Hydrogen Ion Concentration Changes in Mitochondrial Membranes. *Journal of Biological Chemistry* 241: 4588–99.

Chance, B., C. P. Lee, and L. Mela. 1967. Control and Conservation of Energy in the Cytochrome Chain. *Federation Proceedings* 26: 1341–54.

Chervitz, S. A. Aravind, L., Sherlock, G., Ball, C. A., Koonin, E. V., Dwight, S. S., Harris, M. A., Dolinski, K., Mohr, S., Smith, T., Weng, S., Chery, J. M., and D. Botstein. 1998. Comparison of the Complete Protein Sets of Worm and Yeast: Orthology and Divergence. *Science* 282: 2022–8.

Clarke, A. E., and J. H. Fujimura (eds.) 1992. What Tools? Which Jobs? Why Right? In *The Right Tools for the Job. At Work in Twentieth-Century Life Science*. Princeton, N.J.: Princeton, University Press, pp. 3–44.

Claude, A. 1946a. Fractionation of Mammalian Liver Cells by Differential Centrifugation. 1. Problems, Methods, and Preparation of Extract. *Journal of Experimental Medicine* 84: 51–9.

Claude, A. 1946b. Fractionation of Mammalian Liver Cells by Differential Centrifugation. 2. Experimental Procedures and Results. *Journal of Experimental Medicine* 84: 61–89.

Bibliography

Clause, B. T. 1993. The Wistar Rat as a Right Choice: Establishing Mammalian Standards and the Ideal of a Standardized Mammal. *Journal of the History of Biology* 26: 329–49.

Cohen, G. N., and J. Monod. 1957. Bacterial Permeases. *Bacteriological Reviews* 21: 169–94.

Collins, H. M. 1985. *Changing Order. Replication and Induction in Scientific Practice.* London: Sage.

Cook-Deegan, R. 1994. *Gene Wars.* New York: Norton.

Cooper, C., and A. L. Lehninger. 1956. Oxidative Phosphorylation by an Enzyme Complex from Extracts of Mitochondria. I. The Span β-Hydroxybutyrate to Oxygen. *Journal of Biological Chemistry* 219: 489–506.

Craver, C. F. 2001. Role Functions, Mechanisms, and Hierarchy. *Philosophy of Science* 68: 53–74.

Craver, C. F., and L. Darden. 2001. Discovering Mechanisms in Neurobiology: The Case of Spatial Memory. In *Theory and Method in the Neurosciences*. Pittsburgh–Konstanz Series in the Philosophy and History of Science, edited by P. K. Machamer, R. Grush, and P. McLaughlin. Pittsburgh: Pittsburgh University Press, pp. 112–37.

Creager, A. N. H. 2002. *The Life of a Virus: Tabacco Mosaic Virus as an Experimental Model, 1930–1965*. Chicago: University of Chicago Press.

Crichton, M. 1990. *Jurassic Park*. New York: Ballantine.

Crick, F. H. C. 1970. Diffusion in Embryogenesis. *Nature* 225: 420–2.

Crick, F. H. C., and L. Orgel. 1964. The Theory of Inter-allelic Complementation. *Journal of Molecular Biology* 8: 161–5.

Culp, S. 1994. Defending Robustness: The Bacterial Mesosome as a Test Case. In *PSA 1994, Vol. 1*, edited by D. Hull, M. Forbes and R. Burian. East Lansing: Philosophy of Science Association, pp. 47–57.

Culp, S. 1995. Objectivity in Experimental Inquiry: Breaking Data–Technique Circles. *Philosophy of Science* 62: 430–50.

Cummins, R. 1975. Functional Analysis. *Journal of Philosophy* 72: 741–65.

Dambly-Chaudière, C., and A. Ghysen. 1987. Independent Sub-patterns of Sense Organs Require Independent Genes of the achaete-scute Complex in *Drosophila* Larvae. *Genes and Development* 1: 297–306.

Darden, L. 1991. *Theory Change in Science: Strategies from Mendelian Genetics*. Oxford: Oxford University Press.

Darden, L. 1997. Recent Work in Computational Scientific Discovery. In *Proceedings of the Nineteenth Conference of the Cognitive Science Society*, edited by M. Shafto and P. Langley. Mahwah, N.J.: Lawrence Erlbaum, pp. 161–6.

Darden, L. 1998. Anomaly-Driven Theory Redesign: Computational Philosophy of Science Experiments. In *The Digital Phoenix: How Computers Are Changing Philosophy*, edited by T. W. Bynum and J. Moor. Oxford: Blackwell, pp. 62–78.

Darden, L., and M. Cook. 1994. Reasoning Strategies in Molecular Biology: Abstractions, Scans and Anomalies. In *PSA 1994, Vol. 2*, edited by D. L. Hull, M. Forbes, and R. M. Burian. East Lansing, Mich.: Philosophy of Science Association, pp. 179–91.

Darden, L., and N. Maull. 1977. Interfield Theories. *Philosophy of Science* 44: 43–63.

Darden, L., and R. Rada. 1988a. Hypothesis Formation Using Interrelations. In *Analogica*, edited by A. Prieditis. Los Altos, CA: Morgan Kaufmann, pp. 109–27.

Darden, L., and R. Rada. 1988b. Hypothesis Formation Using Part–Whole Interrelations. In *Analogical Reasoning*, edited by D. Helman. Dordrecht: Kluwer, pp. 341–75.

Bibliography

Darwin, C. 1868. *The Variation of Plants and Animals under Domestication*. London: John Murray.

Dawkins, R. 1982. Replicators and Vehicles. In *Current Problems in Sociobiology*, edited by the King's College Sociobiology Group. Cambridge: Cambridge University Press, pp. 45–64.

Dawkins, R. 1989. *The Selfish Gene*. New Edition. Oxford: Oxford University Press.

de Bono, M., and C. I. Bargmann. 1998. Natural Variation in a Neuropeptide Y Receptor Homolog Modifies Social Behavior and Food Response in *C. elegans*. *Cell* 94: 679–89.

de Chadarevian, S. 1998. Of Worms and Programmes: *Caenorhabditis elegans* and the Study of Development. *Studies in History and Philosophy of Biological and Biomedical Sciences* 29C: 81–106.

de Duve, C. 1975. Exploring Cells with a Centrifuge. *Science* 189: 186–94.

de Vries, H. 1889. *Intracelluläre Pangenesis*. Jena: Gustav Fischer.

Delbrück, M. 1971. Aristotle-totle-totle. In *Of Microbes and Life*, edited by J. Monod and E. Borek. New York: Columbia University Press, pp. 50–5.

Devitt, M. 1984. *Realism and Truth*. Oxford: Blackwell.

Dietrich, M. R. 2000. From Gene to Genetic Hierarchy: Richard Goldschmidt and the Problem of the Gene. In *The Concept of the Gene in Development and Evolution. Historical and Epistemological Perspectives*, edited by P. Beurton, R. Falk, and H.-J. Rheinberger. Cambridge: Cambridge University Press, pp. 91–113.

Dretske, F. 1977. Laws of Nature. *Philosophy of Science* 44: 248–68.

Dubinin, N. P. 1932. Stepallelomorphism in *Drosophila melanogaster*: The Allelomorphs achaete2-scute10, achaete1-scute11, and achaete3-scute13. *Journal of Genetics* 25: 163–81.

Dubochet, J., A. W. McDowall, B. Menge, E. N. Schmid, and K. G. Lickfeld. 1983. Electron Microscopy of Frozen-Hydrated Bacteria. *Journal of Bacteriology* 155: 381–90.

Duhem, P. 1954. *The Aim and Structure of Physical Theory*. Translated by P. Wiener. Princeton, N.J.: Princeton University Press.

Dupré, J. 1993. *The Disorder of Things: Metaphysical Foundations of the Disunity of Science*. Cambridge, Mass.: Harvard University Press.

Earman, J. 1992. *Bayes or Bust? A Critical Examination of Bayesian Confirmation Theory*. Cambridge, Mass.: MIT Press.

Earman, J., and J. Roberts. 1999. Ceteris Paribus, There Is No Problem of Provisos. *Synthese* 118: 439–78.

East, E. M. 1912. The Mendelian Notation as a Description of Physiological Facts. *American Naturalist* 46: 633–95.

Ellis, B. 2001. *Scientific Essentialism*. Cambridge: Cambridge University Press.

Ellis, B. 2002. *The Philosophy of Nature: A Guide to the New Essentialism*. Chesham: Acumen.

Epp, C. D. 1997. Definition of a Gene. *Nature* 389: 537.

Ereshefsky, M. (ed.) 1992. *The Units of Evolution: Essays on the Nature of Species*. Cambridge, Mass.: MIT Press.

Ernster, L., and G. Schatz. 1981. Mitochondria: A Historical Review. *Journal of Cell Biology* 91: 227s–255s.

Bibliography

Eytan, G. D., M. J. Matheson, and E. Racker. 1976. Incorporation of Mitochondrial Membrane Proteins into Liposomes Containing Acidic Phospholipids. *Journal of Biological Chemistry* 251: 6831–7.

Falk, R. 1986. What Is a Gene? *Studies in History and Philosophy of Science* 17: 133–73.

Farquhar, M. G., and G. E. Palade. 1981. The Golgi Apparatus Complex (1954–1981): From Artifact to Center Stage. *Journal of Cell Biology* 91: 77s–103s.

Feyerabend, P. K. 1962. Explanation, Reduction and Empiricism. In *Scientific Explanation, Space, and Time*. Minnesota Studies in the Philosophy of Science, Vol. III, edited by H. Feigl and G. Maxwell. Minneapolis: University of Minnesota Press, pp. 28–97.

Feyerabend, P. K. 1988. *Against Method*. London: Verso.

Field, H. 1973. Theory Change and the Indeterminacy of Reference. *The Journal of Philosophy* 70: 462–81.

Fine, A. 1984. The Natural Ontological Attitude. In *Scientific Realism*, edited by J. Leplin. Berkeley: University of California Press, pp. 83–107.

Fitz-James, P. C. 1960. Participation of the Cytoplasmic Membrane in the Growth and Spore Formation of Bacilli. *Journal of Biophysical and Biochemical Cytology* 8: 507–29.

Fodor, J. A. 1974. Special Sciences or the Disunity of Science as a Working Hypothesis. *Synthese* 28: 97–115.

Fogle, T. 2000. The Dissolution of Protein Coding Genes in Molecular Biology. In *The Concept of the Gene in Development and Evolution. Historical and Epistemological Perspectives*, edited by P. Beurton, R. Falk, and H.-J. Rheinberger. Cambridge: Cambridge University Press, pp. 3–25.

Franklin, A. 1990. *Experiment, Right or Wrong*. Cambridge: Cambridge University Press.

Frege, G. 1892. Über Sinn und Bedeutung. *Zeitschrift für Philosophie und philosophische Kritik* NF 100: 25–50.

Freund, J. N., W. Zerges, P. Schedl, B. P. Jarry, and W. Vergis. 1986. Molecular Organization of the Rudimentary Gene of *Drosophila melanogaster*. *Journal of Molecular Biology* 189: 25–36.

Galison, P., and D. J. Stump (eds.) 1996. *The Disunity of Science: Boundaries, Contexts, and Power*. Stanford: Stanford University Press.

Gall, J. G., and M. L. Pardue. 1969. Formation and Detection of RNA–DNA Hybrid Molecules in Cytological Preparations. *Proceedings of the National Academy of Sciences of the United States of America* 63: 378–83.

Gannett, L. 1999. What's in a Cause? The Pragmatic Dimension of Genetic Explanations. *Biology and Philosophy* 14: 349–74.

Garber, R. L., A. Kuroiwa, and W. J. Gehring. 1983. Genomic and cDNA Clones of the Homeotic Locus Antennapedia in *Drosophila*. *The EMBO Journal* 2: 2027–36.

Garcia-Bellido, A. 1979. Genetic Analysis of the achaete-scute System of *Drosophila melanogaster*. *Genetics* 91: 491–520.

Gasper, P. 1992. Reduction and Instrumentalism in Genetics. *Philosophy of Science* 59: 499–513.

Gehring, W. J. 1987. Homeo Boxes in the Study of Development. *Science* 236: 1245.

Gehring, W. J. 1998. *Master Control Genes in Development and Evolution: The Homeobox Story*. New Haven: Yale University Press.

328

Bibliography

Gehring, W. J., and K. Ikeo. 1999. Pax 6: Mastering Eye Morphogenesis and Eye Evolution. *Trends in Genetics* 15: 371–7.

Geison, G. L., and A. N. H. Creager. 1999. Introduction: Research Materials and Model Organisms in the Biological and Biomedical Sciences. *Studies in History and Philosophy of Biological and Biomedical Sciences* 30C: 315–18.

Geison, G. L., and M. D. Laubichler. 2001. The Varied Lives of Organisms: Variation in the Historiography of the Biological Science. *Studies in History and Philosophy of Biological and Biomedical Sciences* 32: 1–29.

Ghysen, A., and C. Dambly-Chaudière. 1988. From DNA to Form: The *achaete-scute* Complex. *Genes & Development* 2: 495–501.

Giere, R. N. 1985. Constructive Realism. In *Images of Science. Essays on Realism and Empiricism, with a Reply from Bas C. Van Fraassen*, edited by P. M. Churchland and C. A. Hooker. Chicago: University of Chicago Press, pp. 75–98.

Giere, R. N. 1988. *Explaining Science: A Cognitive Approach*. Chicago: University of Chicago Press.

Giere, R. N. 1995. The Skeptical Perspective: Science without Laws of Nature. In *Laws of Nature. Essays on the Philosophical, Scientific and Historical Dimensions*, edited by F. Weinert. Berlin: de Gruyter, pp. 120–38.

Giere, R. N. 1997. Scientific Inference: Two Points of View. *Philosophy of Science* (Proceedings) 64: S180–S184.

Giere, R. N. 1999. *Science without Laws*. Chicago: University of Chicago Press.

Gilbert, G. N., and M. J. Mulkay. 1984. *Opening Pandora's Box: A Sociological Analysis of Scientists' Discourse*. Cambridge: Cambridge University Press.

Gillies, D. 1996. *Artificial Intelligence and Scientific Method*. Oxford: Oxford University Press.

Glennan, S. S. 1996. Mechanisms and the Nature of Causation. *Erkenntnis* 44: 49–71.

Glennan, S. S. 1997. Capacities, Universality, and Singularity. *Philosophy of Science* 64: 605–26.

Glennan, S. S. 2002. Rethinking Mechanistic Explanation. *Philosophy of Science* (Proceedings) 69: S342–S353.

Glymour, C. 1980. *Theory and Evidence*. Princeton, N.J.: Princeton University Press.

Godfrey-Smith, P. 2000a. Explanatory Symmetries, Preformation, and Developmental Systems Theory. *Philosophy of Science* (Proceedings) 67: S322–31.

Godfrey-Smith, P. 2000b. Information, Arbitrariness, and Selection: Comments on Maynard Smith. *Philosophy of Science* 67: 202–7.

Godfrey-Smith, P. 2000c. On the Theoretical Role of "Genetic Coding." *Philosophy of Science* 67: 26–44.

Godfrey-Smith, P. 2000d. The Replicator in Retrospect. *Biology and Philosophy* 15: 403–23.

Goldberg, M. L., R. Paro, and W. J. Gehring. 1982. Molecular Cloning of the white Locus Region of *Drosophila melanogaster*. *The EMBO Journal* 1: 93–8.

Goldschmidt, R. 1946. Position Effect and the Theory of the Corpuscular Gene. *Experientia* 2: 197–232.

Gooding, D. 1992. Putting Agency Back into Experiment. In *Science as Practice and Culture*, edited by A. Pickering. Chicago: University of Chicago Press, pp. 65–112.

Gooding, D., T. Pinch, and S. Schaffer (eds.) 1989. *The Use of Experiment. Studies in the Natural Sciences*. Cambridge: Cambridge University Press.

Bibliography

Goosens, W. K. 1978. Reduction by Molecular Genetics. *Philosophy of Science* 45: 73–95.

Gould, S. J., and R. C. Lewontin. 1979. The Spandrels of San Marco and the Panglossian Paradigm: A Critique of the Adaptionist Programme. *Proceedings of the Royal Society of London* B 205: 581–98.

Grasshoff, G. 1995. The Methodological Function of Surprises. *Foundations of Science* 1: 204–8.

Grasshoff, G., R. Casties, and K. Nickelsen. 2000. *Zur Theorie des Experiments. Untersuchungen am Beispiel der Entdeckung des Harnstoffzyklus.* Bern Studies in the History and Philosophy of Science. Norderstedt: BoD Gmbtt.

Grasshoff, G., and M. May. 1995. Methodische Analyse wissenschaftlichen Entdeckens. *Kognitionswissenschaft* 5: 51–67.

Grasshoff, G., and K. Nickelsen. 2001. *Dokumente zur Entdeckung des Harnstoffzyklus. Band 1: Laborbuch Hans Krebs und Erstpublikationen.* Bern Studies in the History and Philosophy of Science. Norderstedt: BoD Gmbtt.

Greenawalt, J. W., and T. L. Whiteside. 1975. Mesosomes: Membranous Bacterial Organelles. *Bacteriological Review* 39: 405–63.

Greville, G. D. 1969. A Scrutiny of Mitchell's Chemiosmotic Hypothesis of Respiratory Chain and Photosynthetic Phosphorylation. In *Current Topics in Bioenergetics,* Vol. 3, edited by R. Sanadi. New York: Academic Press, pp. 1–78.

Griesemer, J. 2000. Development, Culture, and the Units of Inheritance. *Philosophy of Science* (Proceedings) 67: S348–S368.

Griffiths, P. E. 2001. Genetic Information: A Metaphor in Search of a Theory. *Philosophy of Science* 68: 394–412.

Griffiths, P. E. Forthcoming. The Fearless Vampire Conservator: Philip Kitcher, Genetic Determinism and the Informational Gene. In *Genes in Development,* edited by E. M. Neumann-Held and C. Rehmann-Sutter. Durham: Duke University Press.

Griffiths, P. E., and R. D. Gray. 1994. Developmental Systems and Evolutionary Explanation. *The Journal of Philosophy* 91: 277–304.

Griffiths, P. E., and R. D. Gray. 1997. Replicator II – Judgment Day. *Biology and Philosophy* 12: 471–92.

Griffiths, P. E., and R. D. Knight. 1998. What Is the Developmentalist Challenge? *Philosophy of Science* 65: 253–8.

Grünbaum, A. 1964. The Bearing of Philosophy on the History of Science. *Science* 143: 1406–12.

Grunstein, M., and D. S. Hogness. 1975. Colony Hybridization: A Method for the Isolation of Cloned DNAs That Contain a Specific Gene. *Proceedings of the National Academy of Sciences of the United States of America* 72: 3961–5.

Grush, R. 2001. The Semantic Challenge to Computational Neuroscience. In *Theory and Method in the Neurosciences.* Pittsburgh–Konstanz Series in the Philosophy and History of Science, edited by P. K. Machamer, R. Grush, and P. McLaughlin. Pittsburgh: University of Pittsburgh Press, pp. 155–72.

Gurdon, J. B., and P.-Y. Bourillot. 2001. Morphogen Gradient Interpretation. *Nature* 413: 797–803.

Hacking, I. 1983. *Representing and Intervening. Introductory Topics in the Philosophy of Natural Science.* Cambridge: Cambridge University Press.

Bibliography

Hacking, I. 1985. Do We See through a Microscope? In *Images of Science: Essays on Realism and Empiricism, with a Reply from Bas C. Van Fraassen*, edited by P. A. Churchland and C. A. Hooker. Chicago: University of Chicago Press, pp. 132–52.

Hacking, I. 1992. The Self-Vindication of the Laboratory Sciences. In *Science as Practice and Culture*, edited by A. Pickering. Chicago: University of Chicago Press, pp. 29–64.

Hacking, I. 1999. *The Social Construction of What?* Cambridge, MA: Harvard University Press.

Hagner, M., and H.-J. Rheinberger. 1998. Experimental Systems, Object of Investigation, and Spaces of Representation. In *Experimental Essays – Versuche zum Experiment*, edited by M. Heidelberger and F. Steinle. Baden-Baden: Nomos, pp. 355–73.

Halder, G., P. Callaerts, and W. J. Gehring. 1995. Induction of Ectopic Eyes by Targeted Expression of the eyeless Gene in *Drosophila*. *Science* 267: 1788–92.

Hanson, N. R. 1958. *Patterns of Discovery*. Cambridge: Cambridge University Press.

Heidelberger, M., and F. Steinle (eds.) 1998. *Experimental Essays – Versuche zum Experiment*. Baden-Baden: Nomos.

Hempel, C. G. 1945. Studies in the Logic of Confirmation. *Mind* 54: 1–26, 97–121.

Hempel, C. G. 1965. *Aspects of Scientific Explanation and Other Essays in the Philosophy of Science*. New York: The Free Press.

Hempel, C. G., and P. Oppenheim. 1948. Studies in the Logic of Explanation. *Philosophy of Science* 15: 135–75.

Higgins, M. L., and L. Daneo-Moore. 1974. Factors Influencing the Frequency of Mesosomes Observed in Fixed and Unfixed Cells of *Streptococcus faecalis*. *Journal of Cell Biology* 61: 288–300.

Hogeboom, G. H., W. C. Schneider, and G. E. Pallade. 1948. Cytochemical Studies of Mammalian Tissues. 1. Isolation of Intact Mitochondria from Rat Liver; Some Biochemical Properties of Mitochondria and Submicroscopic Particulate Material. *Journal of Biological Chemistry* 172: 619–35.

Holland, A. 2001. Am Anfang war das Wort: Eine Kritik von Informationsmetaphern in der Genetik. In *Ethische Probleme in den Biowissenschaften*, edited by M. Weber and P. Hoyningen-Huene. Heidelberg: Synchron, pp. 93–105.

Holmes, F. L. 1991. *Hans Krebs. Volume I: The Formation of a Scientific Life, 1900–1933*. Oxford: Oxford University Press.

Holmes, F. L. 1992. Manometer, Tissue Slices, and Intermediary Metabolism. In *The Right Tools for the Job. At Work in Twentieth-Century Life Sciences*, edited by A. E. Clarke and J. H. Fujimura. Princeton, N.J.: Princeton University Press, pp. 151–71.

Holmes, F. L. 1993a. *Hans Krebs. Volume II: Architect of Intermediary Metabolism, 1933–1937*. Oxford: Oxford University Press.

Holmes, F. L. 1993b. The Old Martyr of Science: The Frog in Experimental Physiology. *Journal of the History of Biology* 26: 311–28.

Holmes, F. L. 2000a. The Logic of Discovery in the Experimental Life Sciences. In *Biology and Epistemology*, edited by R. Creath and J. Maienschein. Cambridge: Cambridge University Press, pp. 167–90.

Holmes, F. L. 2000b. Seymour Benzer and the Definition of the Gene. In *The Concept of the Gene in Development and Evolution. Historical and Epistemological Perspectives*, edited by P. Beurton, R. Falk, and H.-J. Rheinberger. Cambridge: Cambridge University Press, pp. 115–55.

Bibliography

Holmes, F. L. 2001. *Meselson, Stahl, and the Replication of DNA: A History of 'The Most Beautiful Experiment in Biology'*. New Haven: Yale University Press.

Hooker, C. A. 1981. Towards a General Theory of Reduction. Part I: Historical and Scientific Setting. *Dialogue* 20: 38–58.

Horan, B. L. 1994. The Statistical Character of Evolutionary Theory. *Philosophy of Science* 61: 76–95.

Howson, C., and P. Urbach. 1989. *Scientific Reasoning: The Bayesian Approach*. La Salle, Ill.: Open Court.

Hoyningen-Huene, P. 1987. Context of Discovery and Context of Justification. *Studies in History and Philosophy of Science* 18: 501–15.

Hoyningen-Huene, P. 1988. Diskussionsbemerkung zum Beitrag von Wolfgang Balzer. In *Wozu Wissenschaftsphilosophie? Positionen und Fragen zur gegenwärtigen Wissenschaftsphilosophie*, edited by P. Hoyningen-Huene and G. Hirsch. Berlin: De Gruyter, pp. 76–83.

Hoyningen-Huene, P. 1989. Epistemological Reductionism in Biology: Intuitions, Explications and Objections. In *Reductionism and Systems Theory in the Life Sciences*, edited by P. Hoyningen-Huene and F. M. Wuketits. Dordrecht: Kluwer Academic, pp. 29–44.

Hoyningen-Huene, P. 1992. On the Way to a Theory of Antireductionist Arguments. In *Emergence or Reduction? Essays on the Prospects of Nonreductive Physicalism*, edited by A. Beckermann, H. Flohr and J. Kim. Berlin: deGruyter, pp. 289–301.

Hoyningen-Huene, P. 1993. *Reconstructing Scientific Revolutions. The Philosophy of Science of Thomas S. Kuhn*. Chicago: University of Chicago Press.

Hoyningen-Huene, P. 1994. Zu Emergenz, Mikro- und Makrodetermination. In *Kausalität und Zurechnung. Über Verantwortung in komplexen kulturellen Prozessen*, edited by W. Lübbe. Berlin: De Gruyter, pp. 165–95.

Hoyningen-Huene, P. 1995. Two Letters of Paul Feyerabend to Thomas S. Kuhn on a Draft of *The Structure of Scientific Revolutions*. *Studies in History and Philosophy of Science* 26: 353–87.

Hoyningen-Huene, P. 1997. Comment on J. Kim's "Supervenience, Emergence, and Realization in the Philosophy of Mind." In *Mindscapes: Philosophy, Science, and the Mind*. Pittsburgh–Konstanz Series in the Philosophy and History of Science, edited by M. Carrier and P. K. Machamer. Pittsburgh: University of Pittsburgh Press, pp. 294–302.

Hoyningen-Huene, P., and H. Sankey (eds.) 2001. *Incommensurability and Related Matters*. Dordrecht: Kluwer.

Hudson, R. G. 1999. Mesosomes: A Study in the Nature of Experimental Reasoning. *Philosophy of Science* 66: 289–309.

Hull, D. L. 1972. Reduction in Genetics – Biology or Philosophy? *Philosophy of Science* 39: 491–9.

Hull, D. 1974. *Philosophy of Biological Science*. Englewood Cliffs: Prentice-Hall.

Hull, D. L. 1976. Informal Aspects of Theory Reduction. In *PSA 1974*, edited by R. S. Cohen, C. A. Hooker, A. C. Michalos and J. W. v. Evra. Dordrecht/Boston: Reidel, pp. 653–70.

Hull, D. 1981. Units of Evolution: A Metaphysical Essay. Reprinted in *Genes Organisms and Populations: Controversies over the Units of Selection*, edited by R. Brandon and R. M. Burian, Cambridge, Mass.: MIT Press, pp. 142–60.

International Human Genome Sequencing Consortium. 2001. Initial Sequencing and Analysis of the Human Genome. *Nature* 409: 860–921.

Jablonka, E. 2002. Information: Its Interpretation, Its Inheritance, and Its Sharing. *Philosophy of Science* 69: 578–605.

Jacob, F. 1988. *The Statue Within: An Autobiography*. New York: Basic Books.

Jacob, F., and J. Monod. 1961. Genetic Regulatory Mechanisms in the Synthesis of Proteins. *Journal of Molecular Biology* 3: 318–56.

Jagendorf, A. T., and E. Uribe. 1966. ATP Formation Caused by Acid–Base Transition of Spinach Chloroplasts. *Proceedings of the National Academy of Sciences of the United States of America* 55: 170–7.

Johannsen, W. 1913. *Elemente der exakten Erblichkeitslehre*, 2. Auflage. Jena: Gustav Fischer.

John, H. A., M. L. Birnstiel, and K. W. Jones. 1969. RNA–DNA Hybrids at the Cytological Level. *Nature* 223: 582–7.

Judson, H. F. 1979. *The Eighth Day of Creation. Makers of the Revolution in Biology*. New York: Simon and Schuster.

Kagawa, Y., and E. Racker. 1966a. Partial Resolution of the Enzymes Catalyzing Oxidative Phosphorylation. IX. Reconstruction of Oligomycin-Sensitive Adenosine Triphosphatase. *Journal of Biological Chemistry* 241: 2467–74.

Kagawa, Y., and E. Racker. 1966b. Partial Resolution of the Enzymes Catalyzing Oxidative Phosphorylation. VIII. Properties of a Factor Conferring Oligomycin Sensitivity on Mitochondrial Adenosine Triphosphatase. *Journal of Biological Chemistry* 241: 2461–6.

Kagawa, Y., and E. Racker. 1966c. Partial Resolution of the Enzymes Catalyzing Oxidative Phosphorylation. X. Correlation of Morphology and Function in Sumitochondrial Particles. *Journal of Biological Chemistry* 241: 2475–82.

Kagawa, Y., and E. Racker. 1971. Partial Resolution of the Enzymes Catalyzing Oxidative Phosphorylation. XXV. Reconstitution of Vesicles Catalyzing 32Pi–Adenosine Triphosphate Exchange. *Journal of Biological Chemistry* 246: 5477–87.

Kalckar, H. (ed.) 1969. *Biological Phosphorylations: Development of Concepts*. Englewood Cliffs: Prentice-Hall.

Kant, I. 1998. *Critique of Pure Reason*. Translated and edited by Paul Guyer and Allen E. Wood. Cambridge: Cambridge University Press.

Kay, L. E. 2000. *Who Wrote the Book of Life? A History of the Genetic Code*. Stanford: Stanford University Press.

Keilin, D. 1966. *The History of Cell Respiration and Cytochrome*. Cambridge: Cambridge University Press.

Keller, E. F. 1996. *Drosophila* Embryos as Transitional Objects: The Work of Donald Poulson and Christiane Nüsslein-Volhard. *Historical Studies in the Physical and Biological Sciences* 26: 313–46.

Keller, E. F. 2002. *Making Sense of Life: Explaining Biological Development with Models, Metaphors, and Machines*. Cambridge, Mass.: Harvard University Press.

Kim, J. 1996. *Philosophy of Mind*. Boulder: Westview.

Kimbrough, S. O. 1979. On the Reduction of Genetics to Molecular Biology. *Philosophy of Science* 46: 389–406.

Kitcher, P. 1978. Theories, Theorists and Theoretical Change. *Philosophical Review* 87: 519–47.

Bibliography

Kitcher, P. 1981. Explanatory Unification. *Philosophy of Science* 48: 507–31.

Kitcher, P. 1982. Genes. *British Journal for the Philosophy of Science* 33: 337–59.

Kitcher, P. 1984. 1953 and All That. A Tale of Two Sciences. *Philosophical Review* 93: 335–73.

Kitcher, P. 1989. Explanatory Unification and the Causal Structure of the World. In *Scientific Explanation*. Minnesota Studies in the Philosophy of Science, Vol. XIII, edited by P. Kitcher and W. C. Salmon. Minneapolis: University of Minnesota Press, pp. 410–505.

Kitcher, P. 1993. *The Advancement of Science. Science without Legend, Objectivity without Illusions*. Oxford: Oxford University Press.

Kitcher, P. 2001. Battling the Undead. How (and How Not) to Resist Genetic Determinism. In *Thinking About Evolution: Historical, Philosophical and Political Perspectives*, edited by R. Singh, C. Krimbas, D. B. Paul, and J. Beatty. Cambridge: Cambridge University Press, pp. 396–414.

Kitcher, P., and W. C. Salmon. 1987. Van Fraassen on Explanation. *The Journal of Philosophy* 84: 315–30.

Klemenz, R., U. Weber, and W. J. Gehring. 1987. The white Gene as a Marker in a New P-Element Vector for Gene Transfer in *Drosophila*. *Nucleic Acids Research* 15: 3947–59.

Knorr Cetina, K. 1981. *The Manufacture of Knowledge: An Essay on the Constructivist and Contextual Nature of Science*. Oxford: Pergamon.

Knorr Cetina, K. 1999. *Epistemic Cultures: How the Sciences Make Knowledge*. Cambridge, MA: Harvard University Press.

Kohler, R. E. 1991. Systems of Production: Drosophila, Neurospora and Biochemical Genetics. *Historical Studies in the Physical and Biological Sciences* 22: 87–129.

Kohler, R. E. 1993. Drosophila: A Life in the Laboratory. *Journal of the History of Biology* 26: 281–310.

Kohler, R. E. 1994. *Lords of the Fly. Drosophila Genetics and the Experimental Life*. Chicago: University of Chicago Press.

Kornberg, T. B., and M. A. Krasnow. 2000. The *Drosophila* Genome Sequence: Implications for Biology and Medicine. *Science* 287: 2218–20.

Krebs, H. A. 1981. *Reminiscences and Reflections*. Oxford: Clarendon.

Krebs, H. A., and K. Henseleit. 1932. Untersuchungen über die Harnstoffbildung im Tierkörper. *Klinische Wochenschrift* 11: 757–9.

Kripke, S. A. 1971. Identity and Necessity. In *Identity and Individuation*, edited by M. K. Munitz. New York: New York University Press, pp. 135–64.

Kripke, S. A. 1980. *Naming and Necessity*. Cambridge: Harvard University Press.

Kuhn, T. S. 1970. *The Structure of Scientific Revolutions*. 2d ed. Chicago: University of Chicago Press.

Kulkarni, D., and H. A. Simon. 1988. The Processes of Scientific Discovery: The Strategy of Experimentation. *Cognitive Science* 12: 139–75.

Kuroiwa, A., E. Hafen, and W. J. Gehring. 1984. Cloning and Transcriptional Analysis of the Segmentation Gene fushi tarazu of *Drosophila*. *Cell* 37: 825–31.

Langley, P., H. A. Simon, G. L. Bradshaw, and J. M. Zytlow. 1987. *Scientific Discovery: Computational Explorations of the Creative Process*. Cambridge, Mass.: MIT Press.

Latour, B. 1987. *Science in Action: How to Follow Scientists and Engineers through Society*. Cambridge, Mass.: Harvard University Press.

Latour, B. 1999. For David Bloor . . . and Beyond: A Reply to David Bloor's "Anti-Latour." *Studies in History and Philosophy of Science* 30A: 113–29.

Latour, B., and S. Woolgar. 1979. *Laboratory Life. The Social Construction of Scientific Facts.* London: Sage.

Laubichler, M. D., and G. P. Wagner. 2001. How Molecular Is Molecular Developmental Biology? A Reply to Alex Rosenberg's "Reductionism Redux: Computing the Embryo." *Biology and Philosophy* 16: 53–68.

Laudan, L. 1984. A Confutation of Convergent Realism. In *Scientific Realism,* edited by J. Leplin. Berkeley: University of California Press, pp. 218–49.

Laudan, L. 1990. Demystifying Underdetermination. In *Scientific Theories.* Minnesota Studies in the Philosophy of Science, Vol. XIV, edited by C. W. Savage. Minneapolis: University of Minnesota Press, pp. 267–97.

Laudan, L. 1996. *Beyond Positivism and Relativism: Theory, Method and Evidence.* Boulder: Westview.

Lawrence, P. A. 1992. *The Making of a Fly: The Genetics of Animal Design.* London: Blackwell Scientific.

Lederman, M., and R. M. Burian. 1993. Introduction: The Right Organism for the Job. *Journal of the History of Biology* 26: 235–7.

Lederman, M. and S. A. Tolin. 1993. OVATOOMB: Other Viruses and the Origins of Molecular Biology. *Journal of the History of Biology* 26: 239–54.

Lee, C.-P., and L. Ernster. 1966. The Energy-Linked Nicotinamide Nucleotide Transhydrogenase Reaction: Its Characteristics and Its Use as a Tool for the Study of Oxidative Phosphorylation. In *Regulation of Metabolic Processes in Mitochondria,* edited by J. M. Tager, S. Papa, W. Quagliariello, and E. C. Slater. Amsterdam: Elsevier, pp. 218–34.

Lehninger, A. L. 1964. *The Mitochondrion.* New York: W. A. Benjamin.

Leplin, J. 1997. *A Novel Defense of Scientific Realism.* Oxford: Oxford University Press.

Lewis, E. B. 1978. A Gene Complex Controlling Segmentation in *Drosophila. Nature* 276: 565–70.

Lewis, E. B. 1992. Clusters of Master Control Genes Regulate the Development of Higher Organisms. *Journal of the American Medical Association* 267: 1524–31.

Lipmann, F. 1941. Metabolic Generation and Utilization of Phosphate Bond Energy. *Advances in Enzymology* 18: 99–162.

Lipton, P. 1991. *Inference to the Best Explanation.* London: Routledge.

Loomis, W. F., and F. Lipmann. 1948. Reversible Inhibition of the Coupling Between Phosphorylation and Oxidation. *Journal of Biological Chemistry* 173: 807–8.

Luria, S., and M. Delbrück. 1943. Mutations of Bacteria from Virus Sensitivity to Virus Resistance. *Genetics* 28: 491–511.

Lyko, F., and R. Paro. 1999. Chromosomal Elements Conferring Epigenetic Inheritance. *Bioessays* 21: 824–32.

Lyons, T. D. 2003. Explaining the Success of a Scientific Theory. *Philosophy of Science* 70: 891–901.

Machamer, P., L. Darden, and C. Craver. 2000. Thinking About Mechanisms. *Philosophy of Science* 67: 1–25.

Mackenzie, S. M., M. R. Brooker, T. R. Gill, G. B. Cox, A. J. Howells, and G. D. Ewart. 1999. Mutations in the white Gene of *Drosophila melanogaster* Affecting ABC

Transporters that Determine Eye Colouration. *Biochimica et Biophysica Acta* 1419: 173–85.

Mackie, J. L. 1980. *The Cement of the Universe. A Study of Causation.* Oxford: Oxford University Press.

Mahner, M., and M. Bunge. 2001. Function and Functionalism: A Synthetic Perspective. *Philosophy of Science* 68: 75–94.

Mann, R. S., and G. Morata. 2000. The Developmental and Molecular Biology of Genes that Subdivide the Body of *Drosophila. Annual Review of Cell and Developmental Biology* 16: 243–71.

Maull, N. L. 1980. Comment on Schaffner. In *Scientific Discovery: Case Studies,* edited by T. Nickles. Dordrecht: Reidel, pp. 207–9.

Maxwell, G. 1970. Theories, Perception, and Structural Realism. In *The Nature and Function of Scientific Theories,* edited by R. Colodny. Pittsburgh: University of Pittsburgh Press, pp. 3–34.

Maynard Smith, J. 2000. The Concept of Information in Biology. *Philosophy of Science* 67: 177–94.

Maynard Smith, J., and E. Szathmáry. 1995. *The Major Transitions in Evolution.* Oxford: Freeman.

Mayo, D. G. 1996. *Error and the Growth of Experimental Knowledge.* Chicago: University of Chicago Press.

Mayo, D. G. 1997. Error Statistics and Learning from Error: Making a Virtue of Necessity. *Philosophy of Science* (Proceedings) 64: S195–S212.

Mayo, D. G. 2000. Experimental Practice and an Error Statistical Account of Evidence. *Philosophy of Science* (Proceedings) 67: S193–S207.

Mayr, E. 1961. Cause and Effect in Biology. *Science* 134: 1501–6.

Mayr, E. 1982. *The Growth of Biological Thought.* Cambridge, Mass.: Harvard University Press.

Mayr, E. 1997. *This Is Biology.* Cambridge, Mass.: Harvard University Press.

McGinnis, W., R. L. Garber, J. Wirz, A. Kuroiwa, and W. J. Gehring. 1984a. A Homologous Protein-Coding Sequence in *Drosophila* Homeotic Genes and its Conservation in Other Metazoans. *Cell* 37: 403–8.

McGinnis, W., C. P. Hart, W. J. Gehring, and F. H. Ruddle. 1984b. Molecular Cloning and Chromosome Mapping of a Mouse DNA Sequence Homologous to Homeotic Genes of *Drosophila. Cell* 38: 675–80.

McGinnis, W., M. S. Levine, E. Hafen, A. Kuroiwa, and W. J. Gehring. 1984. A Conserved DNA Sequence in Homoeotic Genes of the *Drosophila* Antennapedia and Bithorax Complexes. *Nature* 308: 428–33.

McLaughlin, P. 1993. Der neue Experimentalismus in der Wissenschaftstheorie. In *Die Experimentalisierung des Lebens: Experimentalsysteme in den biologischen Wissenschaften 1850/1950,* edited by H.-J. Rheinberger and M. Hagner. Berlin: Akademie-Verlag, pp. 207–18.

McLaughlin, P. 2001. *What Functions Explain: Functional Explanation and Self-Reproducing Systems.* Cambridge: Cambridge University Press.

Mill, J. S. 1996 [1843]. *Collected Works of John Stuart Mill:VII. System of Logic: Ratiocinative and Inductive.* London: Routledge.

Millikan, R. G. 1989. Biosemantics. *The Journal of Philosophy* 86: 281–97.

Bibliography

Mitchell, P. 1957. A General Theory of Membrane Transport from Studies of Bacteria. *Nature* 180: 134–6.

Mitchell, P. 1961. Coupling of Phosphorylation to Electron and Hydrogen Transfer by a Chemiosmotic Type of Mechanism. *Nature* 191: 144–8.

Mitchell, P. 1981. Bioenergetic Aspects of Unity in Biochemistry: Evolution of the Concept of Ligand Conduction in Chemical, Osmotic, and Chemiosmotic Reaction Mechanisms. In *Of Oxygen, Fuels, and Living Matter*, edited by G. Semenza. Chichester: Wiley, pp. 1–56.

Mitchell, P., and J. Moyle. 1958. Group-Translocation: A Consequence of Enzyme-Catalyzed Group-Transfer. *Nature* 372–3.

Mitchell, P., and J. Moyle. 1965a. Evidence Discriminating between the Chemical and the Chemiosmotic Mechanisms of Electron Transport Phosphorylation. *Nature* 208: 1205–6.

Mitchell, P., and J. Moyle. 1965b. Stoichiometry of Proton Translocation through the Respiratory Chain and Adenosine Triphosphatase Systems of Rat Liver Mitochondria. *Nature* 208: 147–51.

Mitchell, P., and J. Moyle. 1967. Proton-Transport Phosphorylation: Some Experimental Tests. In *Biochemistry of Mitochondria*, edited by E. C. Slater, Z. Kaniuga, and L. Wojtczak. New York: Academic Press, pp. 53–74.

Mitchell, P., J. Moyle, and L. Smith. 1968. Bromthymol Blue as a pH Indicator in Mitochondrial Suspensions. *European Journal of Biochemistry* 4: 9–19.

Mitchell, S. 2000. Dimensions of Scientific Law. *Philosophy of Science* 67: 242–65.

Mitman, G., and A. Fausto-Sterling. 1992. Whatever Happened to Planaria? C. M. Child and the Physiology of Inheritance. In *The Right Tools for the Job. At Work in Twentieth-Century Life Sciences*, edited by A. E. Clarke and J. H. Fujimura. Princeton, N.J.: Princeton University Press, pp. 172–97.

Monod, J. 1966. From Enzymatic Adaptation to Allosteric Transitions. *Science* 154: 475–83.

Morange, M. 1998. *A History of Molecular Biology*. Cambridge, Mass.: Harvard University Press.

Morange, M. 2000. The Developmental Gene Concept: History and Limits. In *The Concept of the Gene in Development and Evolution*, edited by P. Beurton, R. Falk, and H.-J. Rheinberger. Cambridge: Cambridge University Press, pp. 193–215.

Morgan, G. J. 2001. Bacteriophage Biology and Kenneth Schaffner's Rendition of Developmentalism. *Biology and Philosophy* 16: 85–92.

Morgan, T. H. 1910. Sex-Limited Inheritance in *Drosophila*. *Science* 32: 120–2.

Morgan, T. H. 1911. Random Segregation versus Coupling in Mendelian Inheritance. *Science* 34: 384.

Morgan, T. H., H. J. Muller, A. H. Sturtevant, and C. B. Bridges. 1915. *The Mechanism of Mendelian Heredity*. New York: Henry Holt & Co.

Muller, H. J. 1920. Are the Factors of Heredity Arranged in a Line? *American Naturalist* 54: 97–121.

Muller, H. J., and A. A. Prokofyeva. 1935. The Individual Gene in Relation to the Chromomere and the Chromosome. *Proceedings of the National Academy of Sciences of the United States of America* 21: 16–26.

Bibliography

Musgrave, A. 1988. The Ultimate Argument for Scientific Realism. In *Relativism and Realism in Science*, edited by R. Nola. Dordrecht: Kluwer Academic. 229–52.

Nagel, E. 1961. *The Structure of Science. Problems in the Logic of Scientific Explanation.* London: Routledge and Kegan Paul.

Nanninga, N. 1968. Structural Features of Mesosomes (Chondrioids) of *Bacillus subtilis* after Freeze-Etching. *Journal of Cell Biology* 39: 251–63.

Nanninga, N. 1971. The Mesosome of *Bacillus subtilis* as Affected by Chemical and Physical Fixation. *Journal of Cell Biology* 48: 219–24.

Nickles, T. 1980a. Scientific Discovery and the Future of Philosophy of Science. In *Scientific Discovery, Logic and Rationality*, edited by T. Nickles. Dordrecht: Reidel, pp. 1–60.

Nickles, T. (ed.) 1980b. *Scientific Discovery: Case Studies*. Dordrecht: Reidel.

Nickles, T. (ed.) 1980c. *Scientific Discovery, Logic and Rationality*. Dordrecht: Reidel.

Nola, R. 1980. Fixing the Reference of Theoretical Terms. *Philosophy of Science* 47: 505–31.

Nüsslein-Volhard, C., and E. Wieschaus. 1980. Mutations Affecting Segment Number and Polarity in *Drosophila*. *Nature* 287: 795–801.

Nye, M. J. 1972. *Molecular Reality. A Perspective on the Scientific Work of Jean Perrin.* London: MacDonald/New York, American Elsevier.

Oberheim, E., and P. Hoyningen-Huene. 1997. Incommensurability, Realism, and Meta-incommensurability. *Theoria* 12: 447–65.

O'Hare, K., C. Murphy, R. Levis, and G. M. Rubin. 1984. DNA Sequence of the white Locus of *Drosophila melanogaster*. *Journal of Molecular Biology* 180: 437–55.

Olby, R. C. 1979. Mendel No Mendelian? *History of Science* 17: 53–72.

Olby, R. 1985. *Origins of Mendelism*, 2d ed. Chicago: University of Chicago Press.

Olby, R. C. 1994. *The Path to the Double Helix*, 2d ed. New York: Dover.

Oppenheim, P., and H. Putnam. 1958. The Unity of Science as a Working Hypothesis. In *Concepts, Theories and the Mind–Body Problem*. Minnesota Studies in the Philosophy of Science Vol. II, edited by H. Feigl, M. Scriven, and G. Maxwell. Minneapolis: University of Minnesota Press, pp. 3–36.

Ownby, C. L. 1988. Foreword. In *Artifacts in Biological Electron Microscopy*, edited by R. F. E. Crang and K. L. Klomparens. New York: Plenum, pp. vii–x.

Oyama, S. 1985. *The Ontogeny of Information: Developmental Systems and Evolution.* Cambridge: Cambridge University Press.

Oyama, S. 2000. Causal Democracy and Causal Constributions in Developmental Systems Theory. *Philosophy of Science* (Proceedings) 67: S332–S347.

Oyama, S., P. E. Griffiths, and R. D. Gray (eds.) 2001. *Cycles of Contingency: Developmental Systems and Evolution*. Cambridge, Mass.: MIT Press.

Painter, T. S. 1933. A New Method for the Study of Chromosome Rearrangements and the Plotting of Chromosome Maps. *Science* 78: 585–6.

Painter, T. S. 1934. Salivary Chromosomes and the Attack on the Gene. *Journal of Heredity* 25: 464–76.

Pardee, A. B., F. Jacob, and J. Monod. 1959. The Genetic Control and Cytoplasmic Expression of "Inducibility" in the Synthesis of β-Galactosidase by *E. coli*. *Journal of Molecular Biology* 1: 165–78.

338

Pardue, M. L., and J. G. Gall. 1969. Molecular Hybridization of Radioactive DNA to the DNA of Cytological Preparations. *Proceedings of the National Academy of Sciences of the United States of America* 64: 600–4.

Penefsky, H. S., M. E. Pullman, A. Datta, and E. Racker. 1960. Partial Resolution of the Enzymes Catalyzing Oxidative Phosophorylation II. Participation of the Soluble Adenosine Triphosphatase in Oxidative Phosphorylation. *Journal of Biological Chemistry* 235: 3330–6.

Pickering, A. (ed.) 1992. *Science as Practice and Culture*. Chicago: University of Chicago Press.

Pickering, A. 1999. *Constructing Quarks: A Sociological History of Particle Physics*. Reprint ed. Chicago: University of Chicago Press.

Pontecorvo, G. 1952. Genetic Formulation of Gene Structure and Gene Action. *Advances in Enzymology* 13: 121–49.

Popper, K. R. 1959. *The Logic of Scientific Discovery*. London: Hutchinson Education.

Portin, P. 1993. The Concept of the Gene: Short History and Present Status. *Quarterly Review of Biology* 68: 173–223.

Prebble, J., and B. Weber. 2003. *Wandering in the Gardens of the Mind: Peter Mitchell and the Making of Glynn*. Oxford: Oxford University Press.

Prebble, J. 2001. The Philosophical Origins of Mitchell's Chemiosmotic Concepts: The Personal Factor in Scientific Theory Formulation. *Journal of the History of Biology* 34: 433–60.

Psillos, S. 1994. A Philosophical Study of the Transition from the Caloric Theory of Heat to Thermodynamics: Resisting the Pessimistic Meta-induction. *Studies in History and Philosophy of Science* 25: 159–90.

Psillos, S. 1996. Scientific Realism and the "Pessimistic Induction." *Philosophy of Science* 63 (Proceedings): S306–S314.

Psillos, S. 1999. *Scientific Realism: How Science Tracks Truth*. London: Routledge.

Pullman, M. E., H. S. Penefsky, A. Datta, and E. Racker. 1960. Partial Resolution of the Enzymes Catalyzing Oxidative Phosphorylation I. Purification and Properties of Soluble Dinitrophenol-stimulated Adenosine Triphosphatase. *Journal of Biological Chemistry* 235: 3322–9.

Punnett, R. C. (ed.) 1928. *Scientific Papers of William Bateson*. Cambridge: Cambridge University Press.

Purdue, P. E., and P. B. Lazarow. 2001. Peroxisome Biogenesis. *Annual Review of Cell and Developmental Biology* 17: 701–52.

Putnam, H. 1975a. Explanation and Reference. In *Mind, Language and Reality*, Philosophical Papers, Vol. 2. Cambridge: Cambridge University Press, pp. 196–214.

Putnam, H. 1975b. *Mathematics, Matter, and Method*. Philosophical Papers, Vol. 1. Cambridge: Cambridge University Press.

Putnam, H. 1975c. The Meaning of "Meaning." In *Language, Mind and Knowledge*. Minnesota Studies in the Philosophy of Science, Vol. VII, edited by K. Gunderson. Minneapolis: University of Minnesota Press, pp. 131–93.

Putnam, H. 1975d. The Nature of Mental States. In *Mind, Language and Reality*. Philosophical Papers, Vol. 2. Cambridge: Cambridge University Press 1975, pp. 139–52.

Quine, W. V. O. 1953. Two Dogmas of Empiricism. In *From A Logical Point of View*. Cambridge, Mass.: Harvard University Press, pp. 20–46.

Bibliography

Racker, E. 1970. Function and Structure of the Inner Membrane of Mitochondria and Chloroplasts. In *Membranes of Mitochondria and Chloroplasts*, edited by E. Racker. New York: Van Nostrand Reinhold, pp. 127–71.

Racker, E. 1976. *A New Look at Mechanisms in Bioenergetics*. New York: Academic Press.

Racker, E., and A. Kandrach. 1971. Reconstitution of the Third Site of Oxidative Phosphorylation. *Journal of Biological Chemistry* 246: 7069–71.

Racker, E., and A. Kandrach. 1973. Partial Resolution of the Enzymes Catalyzing Oxidative Phosphorylation. XXXIX. Reconstitution of the Third Segment of Oxidative Phosphorylation. *Journal of Biological Chemistry* 248: 5841–7.

Racker, E., and I. Krimsky. 1952. Mechanism of Oxidation of Aldehydes by Glyceraldehyde-3-Phosphate Dehydrogenase. *Journal of Biological Chemistry* 198: 731–43.

Racker, E., M. E. Pullman, H. S. Penefsky, and M. Silverman. 1963. A Reconstructed System of Oxidative Phosphorylation. In *Proceedings of the Fifth International Congress of Biochemistry*, Vol. V, edited by E. C. Slater. Oxford: Pergamon, pp. 303–12.

Racker, E., and F. W. Racker. 1981. Resolution and Reconstitution. A Dual Autobiographical Sketch. In *Of Oxygen, Fuels, and Living Matter*, edited by G. Semenza. Chichester: Wiley, pp. 265–87.

Racker, E., and W. Stoeckenius. 1974. Reconstitution of Purple Membrane Vesicles Catalyzing Light-Driven Proton Uptake and Adenosine Triphosphate Formation. *Journal of Biological Chemistry* 249: 662–3.

Rader, K. A. 1999. Of Mice, Medicine, and Genetics: C. C. Little's Creation of the Inbred Laboratory Mouse. *Studies in History and Philosophy of Biological and Biomedical Science* 30C: 319–44.

Raffel, D., and H. J. Muller. 1940. Position Effect and Gene Divisibility Considered in Connection with Three Strikingly Similar scute Mutations. *Genetics* 25: 541–83.

Ragan, C. I., and E. Racker. 1973. Partial Resolution of the Enzymes Catalyzing Oxidative Phosphorylation. 28. The Reconstitution of the First Site of Energy Conservation. *Journal of Biological Chemistry* 248: 2563–9.

Rasmussen, N. 1993. Facts, Artifacts, and Mesosomes: Practicing Epistemology with the Electron Microscope. *Studies in History and Philosophy of Science* 24: 227–65.

Rasmussen, N. 2001. Evolving Scientific Epistemologies and the Artifacts of Empirical Philosophy of Science: A Reply Concerning Mesosomes. *Biology and Philosophy* 16: 629–54.

Rea, S., Eigenhober, F., O'Carroll, D., Strahl, B. D., Sun, Z.-W., Schmid, M., Opravil, S., Mechtler, K., Ponting, C. S., Allis, C. D., and T. Jenuwein. 2000. Regulation of Chromatin Structure by Site-Specific Histone H3 Methyltransferases. *Nature* 406: 593–9.

Reik, W., and M. Constancia. 1999. Genomic Imprinting: Making Sense or Antisense? *Nature* 389: 669–71.

Reik, W., and W. Dean. 2003. Silent Clones Speak Up. *Nature* 423: 390–1.

Reiner, R., and R. Pierson. 1995. Hacking's Experimental Realism: An Untenable Middle Ground. *Philosophy of Science* 62: 60–9.

Rescher, N. 1987. *Scientific Realism. A Critical Reappraisal*. Dordrecht: Reidel.

Bibliography

Rheinberger, H.-J. 1993. Experiment and Orientation: Early Systems of in Vitro Protein Synthesis. *Journal of the History of Biology* 26: 443–71.

Rheinberger, H.-J. 1996. Comparing Experimental Systems: Protein Synthesis in Microbes and in Animal Tissue at Cambridge (Ernest F. Gale) and at the Massachusetts General Hospital (Paul C. Zamecnik), 1945–1960. *Journal of the History of Biology* 29: 387–416.

Rheinberger, H.-J. 1997. *Toward a History of Epistemic Things: Synthesizing Proteins in the Test Tube.* Stanford: Stanford University Press.

Rheinberger, H.-J. 2000a. Ephestia: The Experimental Design of Alfred Kühn's Physiological Developmental Genetics. *Journal of the History of Biology* 33: 535–76.

Rheinberger, H.-J. 2000b. Gene Concepts: Fragments from the Perspective of Molecular Biology. In *The Concept of the Gene in Development and Evolution,* edited by P. Beurton, R. Falk, and H.-J. Rheinberger. Cambridge: Cambridge University Press, pp. 219–39.

Robinson, J. D. 1986. Appreciating Key Experiments. *British Journal for History of Science* 19: 51–6.

Robinson, J. D. 1997. *Moving Questions: A History of Membrane Transport and Bioenergetics.* New York: Published for the American Physiological Society by Oxford University Press.

Roll-Hansen, N. 1979. Reductionism in Biological Research: Reflections on Some Historical Case Studies in Experimental Biology. In *Perspectives in Metascience,* edited by J. Bärmark. Göteborg: Kungl. Veteskaps- och Vitterhets- Samhället, pp. 157–72.

Rosenberg, A. 1978. The Supervenience of Biological Concepts. *Philosophy of Science* 45: 368–86.

Rosenberg, A. 1985. *The Structure of Biological Science.* Cambridge: Cambridge University Press.

Rosenberg, A. 1994. *Instrumental Biology or the Disunity of Science.* Chicago: University of Chicago Press.

Rosenberg, A. 1997a. Reductionism Redux: Computing the Embryo. *Biology and Philosophy* 12: 445–70.

Rosenberg, A. 1997b. Can Physicalist Antireductionism Compute the Embryo? *Philosophy of Science* (Proceedings) 64: S359–S371.

Rowen, L. 1986. Normative Epistemology and Scientific Research: Reflections on the "Ox-Phos" Controversy, a Case History in Biochemistry. Ph.D. Thesis, Vanderbilt University.

Rubin, G. M., et al. 2000. Comparative Genomics of the Eukaryotes. *Science* 287: 2204–15.

Ruse, M. 1973. *The Philosophy of Biology.* London: Hutchinson.

Ruse, M. 1976. Reduction in Genetics. In *PSA 1974,* edited by R. S. Cohen, C. A. Hooker, A. C. Michalos, and J. W. v. Evra. Dordrecht/Boston: Reidel, pp. 633–51.

Salmon, W. C. 1984. *Scientific Explanation and the Causal Structure of the World.* Princeton, N.J.: Princeton University Press.

Salmon, W. C. 1989. Four Decades of Scientific Explanation. In *Scientific Explanation.* Minnesota Studies in the Philosophy of Science Vol. XIII, edited by P. Kitcher and W. C. Salmon. Minneapolis: University of Minnesota Press, pp. 43–94.

Salmon, W. C. 1991. The Appraisal of Theories: Kuhn Meets Bayes. In *PSA 1990*, edited by A. Fine, M. Forbes, and L. Wessels. East Lansing, MI: Philosophy of Science Association, pp. 325–32.

Sánchez-Herrero, E., I. Vernós, R. Marco, and G. Morata. 1985. Genetic Organization of *Drosophila* Bithorax Complex. *Nature* 313: 108–13.

Sankey, H. 1994. *The Incommensurability Thesis*. Aldershot: Ashgate.

Sankey, H. 2001. Scientific Realism: An Elaboration and a Defence. *Theoria* 98: 35–54.

Sankey, H. 2002. Realism, Method, and Truth. In *The Problem of Realism*, edited by M. Marsonet. Ashgate: Aldershot, pp. 64–81.

Sarkar, S. 1996. Biological Information: A Sceptical Look at Some Central Dogmas of Molecular Biology. In *The Philosophy and History of Molecular Biology: New Perspectives*, edited by S. Sarkar. Dordrecht: Kluwer, pp. 187–231.

Sarkar, S. 1998. *Genetics and Reductionism*. Cambridge: Cambridge University Press.

Sarkar, S. 2000. Information in Genetics and Developmental Biology: Comments on Maynard Smith. *Philosophy of Science* 67: 208–13.

Scalenghe, F., E. Turco, J. E. Edstrom, V. Pirrotta, and M. Melli. 1981. Microdissection and Cloning of DNA from a Specific Region of *Drosophila melanogaster* Polytene Chromosomes. *Chromosoma* 82: 205–16.

Schaffner, K. F. 1967. Approaches to Reduction. *Philosophy of Science* 34: 137–47.

Schaffner, K. F. 1969. The Watson–Crick Model and Reductionism. *British Journal for the Philosophy of Science* 20: 325–48.

Schaffner, K. F. 1974a. Logic of Discovery and Justification in Regulatory Genetics. *Studies in History and Philosophy of Science* 4: 349–85.

Schaffner, K. F. 1974b. The Peripherality of Reductionism in the Development of Molecular Biology. *Journal for the History of Biology* 7: 111–39.

Schaffner, K. F. 1976. Reduction in Biology: Prospects and Problems. In *PSA 1974*, edited by R. S. Cohen, C. A. Hooker, A. C. Michalos, and J. W. v. Evra. Dordrecht/Boston: Reidel, pp. 613–32.

Schaffner, K. F. 1993. *Discovery and Explanation in Biology and Medicine*. Chicago: University of Chicago Press.

Schaffner, K. F. 1998. Genes, Behavior, and Developmental Emergentism: One Process, Indivisible? *Philosophy of Science* 65: 209–52.

Schaffner, K. F. 2000. Behavior at the Organismal and Molecular Levels: The Case of *C. elegans*. *Philosophy of Science* (Proceedings) 67: S273–S288.

Schaffner, K. F. 2001. Extrapolation from Animal Models. Social Life, Sex, and Super Models. In *Theory and Method in the Neurosciences*. Pittsburgh–Konstanz Series in the Philosophy and History of Science, edited by P. K. Machamer, R. Grush, and P. McLaughlin. Pittsburgh: University of Pittsburgh Press, pp. 200–30.

Schatz, G. 1997. Efraim Racker: 28 June 1913 to 9 September 1991. In *Selected Topics in the History of Biochemistry: Personal Recollections*, Vol. V (Comprehensive Biochemistry, Vol. 40), edited by G. Semenza and R. Jaenicke. Amsterdam: Elsevier, pp. 253–76.

Scheffler, I. 1967. *Science and Subjectivity*. Indianapolis: Bobbs-Merrill.

Schier, A. F., and W. J. Gehring. 1992. Direct Homeodomain-DNA Interaction in the Autoregulation of the fushi tarazu Gene. *Nature* 356: 804–7.

Schröder, R. 2003. The Genes orthodenticle and hunchback Substitute for bicoid in the Beetle *Tribolium*. *Nature* 422: 621–5.

Bibliography

Schwartz, S. 2000. The Differential Concept of the Gene: Past and Present. In *The Concept of the Gene in Development and Evolution. Historical and Epistemological Perspectives*, edited by P. Beurton, R. Falk, and H.-J. Rheinberger. Cambridge: Cambridge University Press, pp. 26–39.

Scott, M. P. 1987. Complex Loci of Drosophila. *Annual Reviews of Biochemistry* 56: 195–227.

Scott, M. P., Welner, A. J., Hazelrigg, T. I., Polisky, B. A., Pirrotta, V., Scalenghe, F. and T. C. Kaufman. 1983. The Molecular Organization of the Antennapedia Locus of *Drosophila. Cell* 35: 763–76.

Segraves, W. A., C. Louis, S. Tsubota, P. Schedl, J. M. Rawls, and B. P. Jarry. 1984. The rudimentary Locus of *Drosophila melanogaster. Journal of Molecular Biology* 175: 1–17.

Shapin, S., and S. Schaffer. 1985. *Leviathan and the Air-Pump: Hobbes, Boyle and the Experimental Life*. Princeton:

Silva, M. T., J. C. Sousa, J. J. Polonia, M. A. Macedo, and A. M. Parente. 1976. Bacterial Mesosomes. Real Structures or Artifacts? *Biochimica et Biophysica Acta* 443: 92–105.

Simon, H. A. 1977. *Models of Discovery*. Dordrecht: Reidel.

Simon, H. A., P. Langley, and G. Bradshaw. 1981. Scientific Discovery as Problem-Solving. *Synthese* 47: 1–27.

Sklar, L. 1993. *Physics and Chance. Philosophical Issues in the Foundations of Statistical Mechanics*. Cambridge: Cambridge University Press.

Slater, E. C. 1953. Mechanism of Phosphorylation in the Respiratory Chain. *Nature* 172: 975–8.

Slater, E. C. (ed.) 1963. *Proceedings of the Fifth International Congress of Biochemistry*, Vol. V. Oxford: Pergamon.

Slater, E. C. 1967. An Evaluation of the Mitchell Hypothesis of Chemiosmotic Coupling in Oxidative and Photosynthetic Phosphorylation. *European Journal of Biochemistry* 1: 317–26.

Slater, E. C. 1971. The Coupling Between Energy-Yielding and Energy-Utilizing Reactions in Mitochondria. *Quarterly Reviews in Biophysics* 4: 35–71.

Slater, E. C. 1997. An Australian Biochemist in Four Different Countries. In *Selected Topics in the History of Biochemistry: Personal Recollections* Vol. V (Comprehensive Biochemistry, Vol. 40), edited by G. Semenza and R. Jaenicke. Amsterdam: Elsevier, pp. 69–203.

Smart, J. J. C. 1963. *Philosophy and Scientific Realism*. London: Routledge and Kegan Paul.

Smith, K. C. 1992. The New Problem of Genetics: A Response to Gifford. *Biology and Philosophy* 7: 331–48.

Sober, E. 1984. *The Nature of Selection. Evolutionary Theory in Philosophical Focus*. Cambridge, Mass.: MIT Press.

Sober, E. 1988. *Reconstructing the Past: Parsimony, Evolution, and Inference*. Cambridge, Mass.: MIT Press.

Sober, E., and D. S. Wilson. 1994. A Critical Review of Philosophical Work on the Units of Selection Problem. *Philosophy of Science* 61: 534–55.

Sperry, R. W. 1986. Discussion: Macro- versus Micro-Determinism. *Philosophy of Science* 53: 265–70.

Stadler, L. J. 1954. The Gene. *Science* 120: 811–19.

Stanford, P. K. 2000. An Antirealist Explanation of the Success of Science. *Philosophy of Science* 67: 266–84.

Stanford, P. K. 2003. No Refuge for Realism: Selective Confirmation and the History of Science. *Philosophy of Science* 70: 913–25.

Stegmann, U. Forthcoming. The Arbitrariness of the Genetic Code. *Biology and Philosophy.*

Steinle, F. 1997. Entering New Fields: Exploratory Uses of Experimentation. *Philosophy of Science* (Proceedings) 64: S65–S74.

Steinle, F. 1998. Exploratives vs. theoriebestimmtes Experimentieren:Ampères erste Arbeiten zum Elektromagnetismus. In *Experimental Essays – Versuche zum Experiment*, edited by M. Heidelberger and F. Steinle. Baden-Baden: Nomos, pp. 272–97.

Stent, G. S. 1980. The Genetic Approach to Developmental Neurobiology. *Trends in Neurosciences* 3: 51.

Stent, G. S. 1985. Hermeneutics and the Analysis of Complex Biological Systems. In *Evolution at a Crossroad: The New Biology and the New Philosophy of Science*, edited by D. J. Depew and B. Weber. Cambridge, Mass.: MIT Press, pp. 209–25.

Sterelny, K. 2000a. Development, Evolution, and Adaptation. *Philosophy of Science* (Proceedings) 67: S369–S387.

Sterelny, K. 2000b. The "Genetic Program" Program: A Commentary on Maynard Smith on Information in Biology. *Philosophy of Science* 67: 195–201.

Sterelny, K. Forthcoming. Symbiosis, Evolvability and Modularity. In *Modularity in Development and Evolution*, edited by G. Schlosser and G. P. Wagner. Chicago: University of Chicago Press.

Sterelny, K., and P. E. Griffiths. 1999. *Sex and Death: An Introduction to Philosophy of Biology.* Chicago: University of Chicago Press.

Sterelny, K., K. Smith, and M. Dickison. 1996. The Extended Replicator. *Biology and Philosophy* 11: 377–403.

Strahl, B. D., and D. Allis. 2000. The Language of Covalent Histone Modifications. *Nature* 403: 41–5.

Stryer, L. 1981. *Biochemistry.* 2d ed. New York: Freeman.

Sturtevant, A. H. 1913. The Linear Arrangement of Six Sex-Linked Factors in *Drosophila,* as Shown by Their Mode of Association. *Journal of Experimental Zoology* 14: 43–59.

Sturtevant, A. H. 1965. *A History of Genetics.* New York: Harper & Row.

Summers, W. C. 1993. How Bacteriophage Came to Be Used by the Phage Group. *Journal of the History of Biology* 26: 255–67.

Surani, M. A. 2001. Reprogramming of Genome Function through Epigenetic Inheritance. *Nature* 414: 121–8.

Szilard, L. 1960. The Control and the Formation of Specific Proteins in Bacteria and in Animal Cells. *Proceedings of the National Academy of Sciences of the United States of America* 46: 277–92.

Tager, J. M., R. D. Veldsema-Currie, and E. C. Slater. 1966. Chemi-osmotic Theory of Oxidative Phosphorylation. *Nature* 212: 376–9.

Thagard, P. 1988. *Computational Philosophy of Science.* Cambridge, Mass.: MIT Press.

Tooley, M. 1987. *Causation: A Realist Approach.* Oxford: Clarendon.

Turing, A. M. 1952. The Chemical Basis of Morphogenesis. *Philosophical Transactions of the Royal Society of London. Series B, Biological Sciences* 237: 37–72.

Bibliography

van Fraassen, B. C. 1980. *The Scientific Image*. Oxford: Clarendon.

van Fraassen, B. C. 1989. *Laws and Symmetry*. Oxford: Clarendon.

Vance, R. E. 1996. Heroic Antireductionism and Genetics: A Tale of One Science. *Philosophy of Science* 63 (Proceedings): S36–S45.

Verner, K., and M. Weber. 1989. Protein Import into Mitochondria in a Homologous Yeast in vitro System. *Journal of Biological Chemistry* 264: 3877–9.

Vicedo, M. 1995. How Scientific Ideas Develop and How to Develop Scientific Ideas. *Biology and Philosophy* 10: 489–99.

Waters, C. K. 1990. Why the Anti-reductionist Consensus Won't Survive the Case of Classical Mendelian Genetics. In *PSA 1990*. East Lansing: Philosophy of Science Association, pp. 125–39.

Waters, C. K. 1991. Tempered Realism about the Force of Selection. *Philosophy of Science* 58: 553–73.

Waters, C. K. 1994. Genes Made Molecular. *Philosophy of Science* 61: 163–85.

Waters, C. K. 1998. Causal Regularities in the Biological World of Contingent Distributions. *Biology and Philosophy* 13: 5–36.

Waters, C. K. 2000. Molecules Made Biological. *Revue Internationale de Philosophie* 214: 9–34.

Waters, C. K. Forthcoming. A Pluralist Interpretation of Gene-Centered Biology. In *Scientific Pluralism*. Minnesota Studies in Philosophy of Science Vol. XIX, edited by S. Kellert, H. E. Longino, and C. K. Waters. Minneapolis: University of Minnesota Press.

Watson, J. D., and F. H. C. Crick. 1953. Molecular Structure of Nucleic Acids. A Structure for Deoxyribose Nucleic Acid. *Nature* 171: 737–8.

Weber, B. 1991. Glynn and the Conceptual Development of the Chemiosmotic Theory: A Retrospective and Prospective View. *Bioscience Reports* 11: 577–617.

Weber, M. 1996. Fitness Made Physical: The Supervenience of Biological Concepts Revisited. *Philosophy of Science* 63: 411–31.

Weber, M. 1998a. *Die Architektur der Synthese. Entstehung und Philosophie der modernen Evolutionstheorie*. Berlin: De Gruyter.

Weber, M. 1998b. Representing Genes: Classical Mapping Techniques and the Growth of Genetical Knowledge. *Studies in History and Philosophy of Biological and Biomedical Sciences* 29: 295–315.

Weber, M. 1999a. The Aim and Structure of Ecological Theory. *Philosophy of Science* 66: 71–93.

Weber, M. 1999b. Hans Drieschs Argumente für den Vitalismus. *Philosophia Naturalis* 36: 265–95.

Weber, M. 2001a. Determinism, Realism, and Probability in Evolutionary Theory. *Philosophy of Science* (Proceedings) 68: S213–S224.

Weber, M. 2001b. Kommentar zur Alan Holland. In *Ethische Probleme in den Biowissenschaften*, edited by M. Weber and P. Hoyningen-Huene. Heidelberg: Synchron, pp. 106–12.

Weber, M. 2001c. Under the Lamppost: Commentary on Schaffner. In *Theory and Method in the Neurosciences*. Pittsburgh–Konstanz Series in the Philosophy and History of Science, edited by P. K. Machamer, R. Grush, and P. McLaughlin. Pittsburgh: University of Pittsburgh Press, pp. 231–49.

Bibliography

Weber, M. 2002a. Incommensurability and Theory Comparison in Experimental Biology. *Biology and Philosophy* 17: 155–69.

Weber, M. 2002b. Theory Testing in Experimental Biology: The Chemiosmotic Mechanism of ATP Synthesis. *Studies in History and Philosophy of Biological and Biomedical Sciences* 33C: 29–52.

Weber, M. and M. Esfeld. 2003. Holism in the Sciences. In *Encyclopedia of Life Support Systems. Unity of Knowledge in Transdisciplinary Research for Sustainability,* edited by G. Hirsch Hadorn. Oxford: EOLSS Publishers.

Weiner, A. J., M. P. Scott, and T. C. Kaufman. 1984. A Molecular Analysis of fushi tarazu, a Gene in *Drosophila melanogaster* that Encodes a Product Affecting Embryonic Segment Number and Cell Fate. *Cell* 37: 843–51.

Weiner, J. 1999. *Time, Love, Memory. A Great Biologist and His Journey from Genes to Behavior.* London: Faber and Faber.

Williams, G. C. 1966. *Adaptation and Natural Selection.* Princeton, N.J.: Princeton University Press.

Wimsatt, W. C. 1974. Complexity and Organization. In *PSA 1972,* edited by K. F. Schaffner and R. S. Cohen. Dordrecht: Reidel, pp. 67–86.

Wimsatt, W. C. 1976a. Reductionism, Levels of Organization, and the Mind–Body Problem. In *Consciousness and the Brain: A Scientific and Philosophical Inquiry,* edited by G. G. Globus, G. Maxwell, and I. Savodnik. New York: Plenum, pp. 199–267.

Wimsatt, W. C. 1976b. Reductive Explanation: A Functional Account. In *PSA 1974,* edited by R. S. Cohen, C. A. Hooker, A. C. Michalos, and J. W. v. Evra. Dordrecht/Boston: Reidel, pp. 671–710.

Wimsatt, W. C. 1981. Robustness, Reliability, and Overdetermination. In *Scientific Inquiry and the Social Sciences,* edited by M. B. Brewer and B. E. Collins. San Francisco: Jossey-Bass, pp. 124–63.

Wimsatt, W. C. 1987. False Models as Means to Truer Theories. In *Neutral Models in Biology,* edited by M. H. Nitecki and A. Hoffman. Oxford: Oxford University Press, pp. 23–55.

Winnie, J. 2000. Information and Structure in Molecular Biology: Comments on Maynard Smith. *Philosophy of Science* 67: 517–26.

Wolpert, L. 1969. Positional Information and the Spatial Pattern of Cellular Differentiation. *Journal of Theoretical Biology* 25: 1–47.

Wolpert, L. 1989. Positional Information Revisited. *Development* 107 (Suppl.): 3–12.

Wolpert, L., R. Beddington, T. Jessell, P. Lawrence, E. Meyerowith, and J. Smith. 2002. *Principles of Development.* 2d ed. Oxford: Oxford University Press.

Woodward, J. 1997. Explanation, Invariance, and Intervention. *Philosophy of Science* (Proceedings) 64: S26–S41.

Woodward, J. 2000. Explanation and Invariance in the Special Sciences. *British Journal for the Philosophy of Science* 51: 197–254.

Woodward, J. 2001. Law and Explanation in Biology: Invariance is the Kind of Stability that Matters. *Philosophy of Science* 68: 1–20.

Woodward, J. 2002. What Is a Mechanism? A Counterfactual Account. *Philosophy of Science* (Proceedings) 69: S366–S377.

Worrall, J. 1989. Fresnel, Poisson, and the White Spot: The Role of Successful Predictions in the Acceptance of Scientific Theories. In *The Use of Experiment. Studies in the Natural Sciences*, edited by D. Gooding, T. Pinch, and S. Schaffer. Cambridge: Cambridge University Press, pp. 135–57.

Wright, L. 1973. Functions. *Philosophical Review* 82: 139–68.

Yanofsky, C., B. C. Carlton, J. R. Guest, D. R. Helinski, and U. Henning. 1964. On the Colinearity of Gene Structure and Protein Structure. *Proceedings of the National Academy of Sciences of the United States of America* 51: 266–72.

Zahar, E. 1973. Why Did Einstein's Programme Supersede Lorentz's? *British Journal for the Philosophy of Science* 24: 95–123.

Zallen, D. T. 1993. The "Light" Organism for the Job: Green Algae and Photosynthesis Research. *Journal of the History of Biology* 26: 269–79.

Zubay, G. 1983. *Biochemistry*. Reading, MA: Addison–Wesley.

Index

Index

Printed in the United States
By Bookmasters